Lecture Notes in Control and Information Sciences

Edited by A. V. Balakrishnan and M. Thoma

For further listing of published volumes please turn over to inside of back cover.

Lecture Notes in Control and Information Sciences

Edited by A.V. Balakrishnan and M. Thoma

24

Methods and Applications in Adaptive Control

Proceedings of an International Symposium
Bochum, 1980

Edited by H. Unbehauen

Springer-Verlag Berlin Heidelberg GmbH 1980

Editor
Prof. Dr.-Ing. H. Unbehauen
Lehrstuhl für Elektrische Steuerung und Regelung
Ruhr-Universität Bochum
Postfach 10 21 48, 4630 Bochum 1

ISBN 978-3-540-10226-7   ISBN 978-3-540-38285-0 (eBook)
DOI 10.1007/978-3-540-38285-0

Adaptive control was first proposed more than 25 years ago. For the most part, however, it has been so far a province of experience and art. Only in the last five years have sound theories of adaptive control been developed. For the realization of adaptive systems we now have several theoretical approaches at our disposal. There are the self-tuning regulator and the model reference approaches. Both have already been used successfully in several applications. The major developments during the last two years include the resolution of the long standing stability problem and the realization that the two approaches mentioned above are essentially equivalent. In addition, with the advent of cheap realization possibilities using the recent major advances in microprocessor technology it seems that today adaptive control is about ready for industrial application.

In view of this development an International Symposium on Adaptive Systems was held at the Department of Electrical Engineering of the Ruhr-University Bochum during March 20th and 21st 1980 together with GMR of the VDI/VDE. The aim of this symposium was the discussion of the actual situation and the future development of adaptive systems. About 180 specialists from 13 countries came together. All papers presented at this conference are published in this volume.

The papers of the symposium during the first day were concerned with methods in adaptive control and during the second day with applications. The primary intention of the conference was therefore to bring together researchers and practising engineers. Theorists on the one side could learn from the relevant features of practical applications, whereas on the other side practising engineers could learn what possibilities are offered by adaptive control theories, especially how to apply specific theoretical methods.

Three of the 26 presented papers represent survey papers. The survey paper of K.J. Åström gave a unified description of many types of self-tuning regulators and their design principles. K.S. Narendra and B.B. Peterson pointed out in their survey the recent developments in adaptive control, especially the solution of stability problems in adaptive systems. The survey paper of P.C. Parks et al. reviewed the application of adaptive control in three areas: aircraft control systems, process control and electrical drives.

Besides the presentation of the papers a round table panel discussion on the future of adaptive control was held at the end of the symposium. The main result of this discussion was that adaptive control is now ready for practical application. However, during the next few years most of the work probably has to be done in a close collaboration between universities and industry, because there is still much to be learned before adaptive control can be considered a routine industrial technique.

There are many people whom I have to thank for their assistance in arranging this symposium. First I say a big "thank you" to the university authorities and those industrial companies which gave us the necessary financial support. Next I would like to thank my colleagues Professor Parks and Professor Schaufelberger who served on the steering committee and who selected with me very carefully the papers submitted for this symposium. I also would like to thank the editor of these lecture notes, Professor Thoma, for his willingness to publish this volume. Finally it is a pleasure to acknowledge the contribution of my assistants and secretaries, who prepared most of the administrative details of this successful symposium.

Bochum, May 1980   H. Unbehauen

C O N T E N T S

METHODS IN ADAPTIVE CONTROL

DESIGN PRINCIPLES FOR SELF-TUNING REGULATORS

K.J. ÅSTRÖM
Department of Automatic Control
Lund Institute of Technology
Lund, Sweden

ABSTRACT

A unified description of many types of self-tuners is given. Relations to design of controllers for systems with known parameters and recursive estimation methods are emphasized. The distinction between self-tuners based on identification of explicit and implicit process models are discussed as well as the relations between Self-Tuning Regulators (STR), and Model Reference Adaptive Systems (MRAS). An overview of practical problems and operational issues is given. The particular problems of integral action and estimator windup are covered in more detail.

1. INTRODUCTION

Adaptive control has been a challenge to control engineers for a long time. Many adaptive control schemes have been proposed. In spite of this progress in the field has been comparatively slow. One reason is that it is difficult to understand how adaptive systems work because they are inherently nonlinear. Another reason is that it has been costly and fairly complicated to implement adaptive controllers. The situation has changed drastically with the advent of microprocessors which makes implementation of adaptive controllers feasible. Recently there has also been progress in theory of adaptive control. See Ljung (1977), Egardt (1979), Goodwin et al (1978), Morse (1979) and Narendra et al (1979).

Self-tuning regulators (STR) and model reference adaptive systems (MRAS) are two popular approaches. An overview of STR is given in Section 2. It is shown that self-tuning regulators can be derived in a simple way which has a strong intuitive appeal. It is then shown by examples, how many different types of self-tuners can be generated. Relations between STR and MRAS are also discussed in Section 2. Practical aspects on self-tuners are discussed in Section 3. This includes different ways to use STR as well as abuses of self-tuners. Two particular practical problems namely how to introduce integral action and how to avoid estimator windup are discussed in Sections 4 and 5. The parametrization problem is discussed in Section 6.

2. SELF-TUNING REGULATORS

This section gives a brief description of self-tuning regulators. The discussion is limited to control of single-input single-output systems described by

$$A(q^{-1}) \; y(t) = B(q^{-1}) \; u(t) \qquad\qquad\qquad (2.1)$$

where u is the input, y the output and $A(q^{-1})$ and $B(q^{-1})$ polynomials in the back-ward shift-operator. For further details we refer to the original papers Peterka (1970), and Åström and Wittenmark (1973) and the recent review Åström (1979a), where many references are given. The principles are first discussed. A self-tuner based on classical control design is then presented as an example. The notion of explicit and implicit algorithms is also discussed.

Principles

A block diagram of self-tuning regulator is shown in Fig.1.

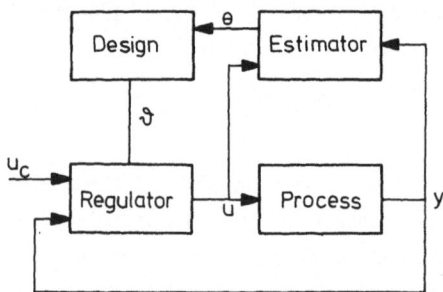

Figure 1. Schematic diagram of a self-tuning regulator

The self-tuner can be thought of as being composed of three parts, a parameter esti-mator, a design calculation and a regulator with adjustable parameters. The design calculation computes the parameters of the regulator from the parameters which de-scribe the process. The parameter estimator determines the parameters which charac-terize the process and its environment from measurements of the process input and output.

The regulator structure shown in Fig.1 is very flexible because it allows many different combinations of design and estimation methods. So far, only a small number of the possible combinations have been explored. Intuitively it seems reasonable to choose a design method, which gives desired performance when the parameters of the process are known, and an estimation method which will work well for the particular disturbances. It turns out, however, that the structure shown in Fig.1 also has un-expected properties. The regulator shown in Fig.1 is a *certainty equivalence* con-trol in the terminology of stochastic control theory because the fact that the para-meter estimates are not exact is disregarded. It is possible to introduce modifica-tions which also take the uncertainties of the parameter estimates into account *(cautious control)* and modifications which introduce extra *probing signals* when the parameter estimates are uncertain. The principles will be illustrated by a few simple examples.

A self-tuning servo

Consider a servoproblem. A classical formulation of the design problem is to find a regulator which gives the desired transfer function from the command signal to the output. Let the desired transfer function be

$$G_M = \frac{Q}{P} \tag{2.2}$$

A self-tuning servo which gives this transfer function is given by

ALGORITHM E1   *(Basic explicit algorithm)*

Data: The polynomials P, T_1, and Q_1 are given.

Step 1: Estimate the parameters of the model

$$Ay(t) = Bu(t) \tag{2.1}$$

by least squares.

Step 2: Factor the estimated polynomial $\hat{B}$ into $\hat{B}^+$ and $\hat{B}^-$ where all zeros of $\hat{B}^+$ are well damped and all zeros of $\hat{B}^-$ are unstable or poorly damped.

Step 3: Solve the linear equation.

$$\hat{A}R_1 + \hat{B}^-S = PT_1. \tag{2.3}$$

(Notice that there are many solutions and that a choice has to be made).

Step 4: Calculate the control variable u from

$$Ru = Tu_c - Sy \tag{2.4}$$

where $R = R_1\hat{B}^+$, and $T = T_1Q_1$.

The steps 1, 2, 3, and 4 are repeated at each sampling period.   □

The algorithm is discussed in detail in Aström and Wittenmark (1979). Similar algorithms for regulation are discussed in Wellstead et al (1979). If the parameter estimates converge the closed loop transfer function will be

$$\frac{Q_1B^-}{P}.$$

Notice that this is the best that can be obtained because it is not possible to cancel unstable or poorly damped process zeros.

The algorithm E1 is called an algorithm based on *estimation of process* parameters or an algorithm with *explicit identification,* because the parameters of the process model (3.1) in the standard form are estimated. Using the terminology of model reference adaptive systems the algorithm is also called *indirect*, because the parameters of the regulator are updated indirectly via estimation of the process parameters (Step 1) and the design calculations (Steps 2, 3, and 4). See Narendra, Lin and Valavani (1979).

The algorithm E1 can be simplified little in two special cases. If it is known that the process has no unstable zeros apart from a known number of time-delays it follows that $B^-(q^{-1}) = q^{-k}$. Step 2 is then not necessary. The second step in the algorithm is also avoided if all process zeros are considered as unstable or poorly damped. In that case $\hat{B}^- = \hat{B}$.

Implicit algorithms

It is possible to construct algorithms where the design calculations are avoided and the parameters of the regulator are updated directly. The basic self-tuning regulator in Åström and Wittenmark (1973) is a prototype for algorithms of this type. The idea is to rewrite the process model in such a way that the design step is trivial. By a proper choice of model structure the regulator parameters are updated directly and the design calculations are thus eliminated. Algorithms of this type are called algorithms based on *implicit identification* of a process model. In the terminology of model reference adaptive systems the corresponding algorithms are also called *direct methods* because the parameters of the regulator are updated directly.

An example of an explicit algorithm will now be given. Consider a process described by (3.1) with $B^- = q^{-k}$. Assume that it is desired to find a feedback such that the transfer function from the reference value to the output is

$$\frac{z^{-k}}{P(z^{-1})}$$

This means that all process zeros have to be cancelled. Assuming that the process model is known the design equation (2.3) becomes

$$PT_1 = AR_1 + q^{-k}S$$

Hence

$$PT_1y = AR_1y + q^{-k}Sy = q^{-k}R_1Bu + q^{-k}Sy = q^{-k}(Ru + Sy) \qquad (2.5)$$

where (2.1) is used to obtain the second equality. The process can thus be represented either by (2.1) or by (2.5). The representation (2.5) has the advantage that the polynomials R and S, which appear in the feedback law, occur explicitly. The following self-tuning control algorithm is then obtained

ALGORITHM 12 (*Implicit algorithm with all·process zeros cancelled*)

Data: Given the polynomials P and T, where P is normalized such that P(1) = 1.

Step 1: Estimate the parameters of the polynomials R and S in the model

$$PT \ y = q^{-k}(Ru + Sy) \tag{2.5}$$

by least squares.

Step 2: Calculate the control signal using

$$\hat{R}u = T u_c - \hat{S}y, \tag{2.6}$$

where $\hat{R}$ and $\hat{S}$ are the polynomials estimated in Step 1.

The Steps 1 and 2 are repeated at each sampling period. □

This algorithm was originally proposed in Clarke and Gawthrop (1975). Since the specifications require that all process zeros are cancelled, they must be sufficiently well damped for the algorithm to function. The algorithm will thus not work for non-minimum-phase systems. It also requires that k is known apriori. Notice that T can be interpreted as the observer polynomial.

Implicit STR and MRAS

It will now be shown that the implicit self-tuning pole-placement algorithm 2 is equivalent to a model reference adaptive system (MRAS). For this purpose it is necessary to consider some details of the algorithm. Introduce

$$\varphi(t) = [y(t-k) \ \dots \ y(t-k-n_S) \ u(t-k) \ \dots \ u(t-k-n_R)]^T \tag{2.7}$$

where

$$n_S = \deg S \text{ and } n_R = \deg R.$$

In the implicit algorithm the estimated parameters are equal to the regulator parameters. Hence

$$\theta = [s_0 \dots s_{n_S} \ r_0 \dots r_{n_R}]. \tag{2.8}$$

The residual ε can then be written as

$$\varepsilon(t) = PT\ y(t) - \hat{R}u(t-k) - \hat{S}y(t-k) = PT\ y(t) - \varphi^T(t)\theta \qquad (2.9)$$

The least squares formula for updating the parameter estimates can be written as

$$\theta(t+1) = \theta(t) + P(t+1)\ \varphi(t+1)\ \varepsilon(t+1) \qquad (2.10)$$

Equation (2.10) can clearly be interpreted as an adjustment rule for the regulator parameters θ. Notice that it follows from (2.9) that

$$\varphi(t) = -\ grad_\theta\ \varepsilon(t) \qquad (2.11)$$

The vector φ can thus be interpreted as a sensitivity derivative, and the least squares updating formula can be written as

$$\theta(t+1) = \theta(t) - P(t+1)\ \varepsilon(t+1)\ grad_\theta\ \varepsilon(t+1) \qquad (2.12)$$

This is identical to the 'MIT rule' used to design MRAS, provided that the model error is replaced by the least squares residual.

LQG self-tuners

Optimal control methods are popular design techniques. Such methods can of course also be used to generate self-tuning regulators. The idea is illustrated using a simple example. Consider a system described by

$$A(q^{-1})\ y(t) = B(q^{-1})\ u(t) + C(q^{-1})\ e(t) \qquad (2.13)$$

where e is white noise. Assume that it is desired to find a control law such that the criterion

$$J = \lim_{N\to\infty} \frac{1}{N} \sum_{n=0}^{\infty} [y^2(t) + \rho u^2(t)] \qquad (2.14)$$

is minimal. A self-tuning regulator for this problem is given below.

ALGORITHM (Explicit LQG)

Data: Given ρ and the sampling period h.

Step 1: Estimate the parameters of the model (2.12) by extended least squares or by recursive maximum likelihood.

Step 2: Determine a stable polynomial P such that

$$PP* = \rho\hat{A}\hat{A}* + \hat{B}\hat{B}* \tag{2.15}$$

where $\hat{A}$ and $\hat{B}$ are the estimates obtained in Step 1 and A* denotes the reciprocal of the polynomial A. Find a solution to the diophantine equation

$$\hat{A}R + \hat{B}S = \hat{C}P \tag{2.16}$$

such that deg S = deg $\hat{B}$ + deg C – deg P

Step 3: Use the control law

$$Ru = -Sy \tag{2.17}$$

The steps 1, 2 and 3 are repeated at each sampling period □

Notice that there are many variants. Instead of performing the spectral factorization (2.15) and solving the linear equation (2.16) the feedback law (2.17) can be obtained from a Riccati equation. See Åström (1974).

3. PRACTICAL ASPECTS

Some practical aspects on simple regulators are first reviewed briefly. The corresponding problems for self-tuners are then discussed. Operational issues and abuses of self-tuners are also covered.

Simple Regulators

The basic algorithm for a PID regulator is very simple:

$$u = K[e + \frac{1}{T_I} \int^t e(s)\,ds + T_D \frac{de}{dt}]. \tag{3.1}$$

An implementation of this algorithm in analog or digital hardware does, however, not necessarily give a good controller. In practice it is also necessary to consider operator interface, filtering of the signals, automatic/manual transfer, bumpless parameter changes, reset windup, nonlinear output, (gap, saturation etc). How well a PID regulator works in an industrial environment depends very much upon these considerations.

Self-tuners

All things that apply to the simple regulators also apply to the self-tuners. For self-tuners there are, however, more things to be considered because the basic algorithm is more complicated than the PID algorithm. For example windup occurs in a PID regulator because the integrator in the algorithm could achieve large values if the control value saturates or if it is driven manually. In a self-tuner with a forgetting factor windup can also occur in the estimator. Some of these problems are discussed in more detail in the following sections. The self-tuning regulator can operate in many different modes like estimation only, tuning etc. The problem of operator interface is particularly important. A key problem is how the specifications are entered and how an operator should interact with the controller. There are many different possibilities ranging from the case where there are no buttons at all on the panel to fairly complicated operator interfaces. Instead of just having manual and automatic modes it maybe useful to have several automatic modes e.g. fixed gain, estimate process parameters but do not update controller parameters, estimate and update controller parameters. Certainly there are many interesting possibilities as is illustrated on self-tuning regulators which are already on the market or which are in the process of coming out.

Operational issues

Self-tuning regulators can be used in many different ways. Since the regulator becomes an ordinary constant gain feedback regulator if the parameter estimates are kept constant, the self-tuner can be used as a *tuner* to adjust the parameters of a control loop. In such an application the self-tuner is connected to the process and run until satisfactory performance is obtained. The self-tuner is then disconnected and the system is left with the constant parameter regulator obtained. This mode of using the self-tuner is convenient to implement in a package for direct digital control (DDC-package). The DDC-package is simply provided with a tuning routine which can be connected to an arbitrary loop in the package.

The self-tuner can also be used *to build up a gain schedule.* In such a case the system is run at different operating points and the controller parameters obtained are stored. When the process has been run at a sufficient number of operating points a table for scheduling the controller parameters can be generated by interpolating and smoothing   the parameter values obtained.

The self-tuner can also be used as a truly *adaptive controller* for systems with varying parameters. In cases where rapid adaptation over widely varying operating conditions are required combinations between gain-scheduling and self-tuning can also be considered.

Abuses of self-tuners

Compared with a three-term controller the self-tuner is a sophisticated controller.

Such a controller can of course be misused. The self-tuner should of course not be used if a simpler controller will do the job. Before considering a self-tuning regulator it is therefore useful to check if a simpler regulator will work. The following list may help to decide.

PI or PID
Linear MISO (What order?)
Nonlinear
Fixed Gain
Gain Schedule
Self-tuning or Adaptive

Notice that it is not always easy to decide if a constant gain regulator will work based on the open loop characteristics of the process. Two examples illustrate the point.

Example 1

Fig. 2 shows the step responses of systems with the transfer function

$$G(s) = \frac{1}{(s+1)(s+a)}$$

(3.2)

for a = 0, 0.01 and 0.02.

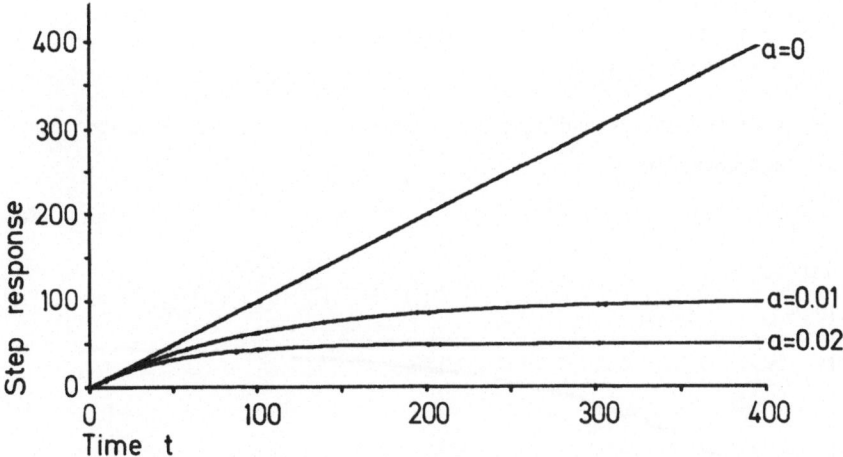

Figure 2. Step responses of open loop systems with transfer function (3.2).

The step responses of the corresponding closed loop systems obtained with the constant parameter feed-back

$$u(t) = y_r - y(t)$$

are shown in Fig.3.

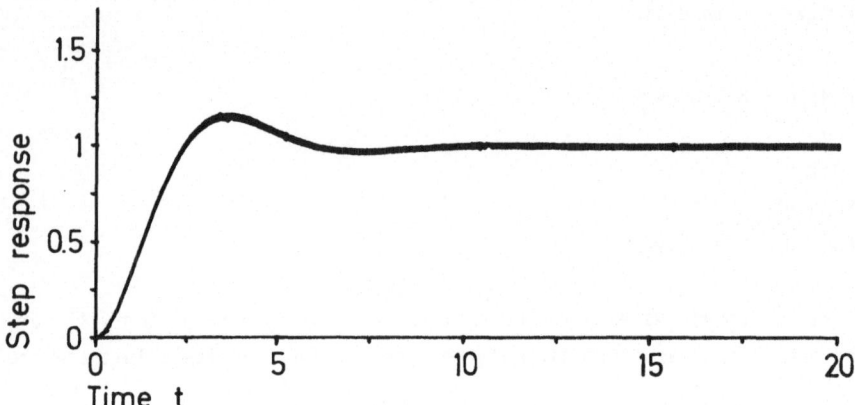

Figure 3. Unit step responses for closed loop systems

□

Example 2

Fig.4 shows the step responses of systems with the transfer function

$$G(s) = \frac{20(1-sT)}{(s+1)(s+20)(1+sT)} \tag{3.3}$$

for T = 0, 0.01, 0.02

The step responses of the corresponding closed loop systems obtained with the constant parameter feed-back law

$$u(t) = 20(y_r - y(t))$$

are shown in Fig.5.

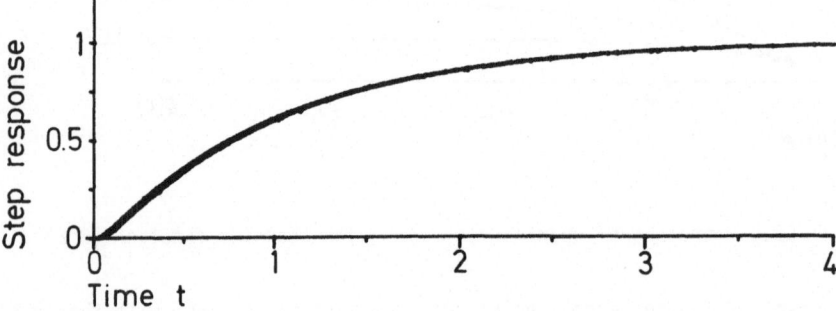

Figure 4. Unit step responses for systems with transfer function (3.3).

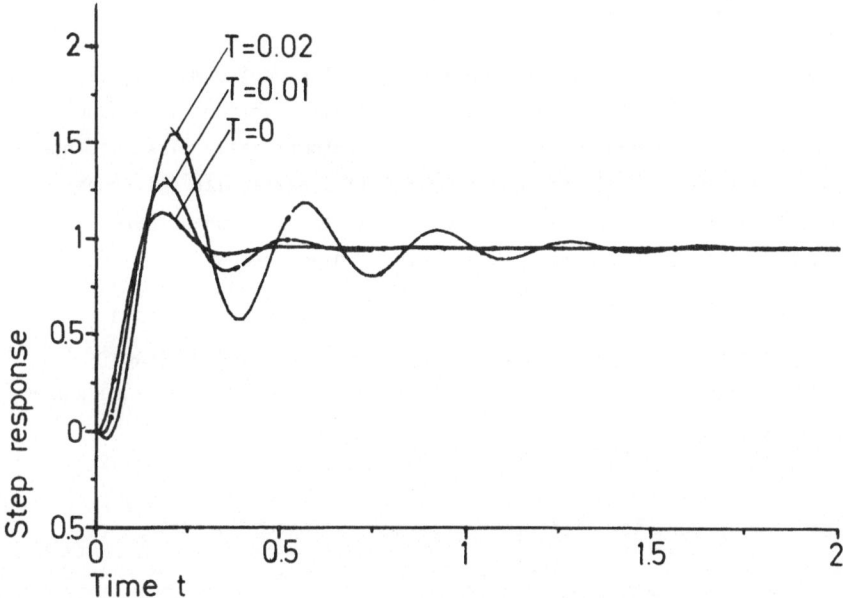

Figure 5. Unit step responses for closed loop systems with transfer function (3.3)

□

When designing a self-tuning regulator it is useful to consider the particular application carefully and decide upon a design method which is suitable for the particular
problem if a model for the process and its environment are known. A parameter estimation scheme which works well for the particular problem should also be chosen before
the details of the design are considered.

4. INTEGRAL ACTION

The reason for introducing reset and integral action is to eliminate steady state
erros in the closed loop system. Steady state errors can be generated by many different mechanisms, calibration errors, nonlinearities, load disturbances etc. Irre-
spective of the origin of the disturbances it has been found empirically that the
errors can be eliminated simply by letting the feedback signal have a term which is
proportional to the integral of the error. It is also well known that integral feedback can lead to difficulties. It destabilizes the system and may lead to oscillations
with large amplitudes. Since the integral is an unstable system it may happen that
the integral can assume very large values if the control signal saturates (due to
nonlinearities or manual control) when there is an error. This is called *reset windup*. Special precautions have to be taken in ordinary regulators to avoid windup of
the integrator. There are several ways to provide reset in self-tuning regulators.
Since there is no method which is uniformly best a few different schemes will be
discussed.

Automatic reset provided by the STR

Since many self-tuning regulators estimate models of the environment it can be expected that the self-tuner will attempt to model slowly drifting disturbances and
compensate for them by introducing integral action automatically. This is indeed the
case for many configurations. It is easy to check if a particular self-tuner has this
ability simply by investigating possible stationary solutions when there is an offset or a drifting disturbance. A typical example is given below.

EXAMPLE 3

Consider the simple implicit self-tuner discussed in Åström and Wittenmark (1973),
which is based on least squares parameter estimation and minimum variance control.
The self-tuner is based on the model

$$y(t+k) = Ru(t) + Sy(t)$$

The conditions for an equilibrium of the parameter estimates is that

$$\frac{1}{N} \sum_{t=1}^{N} y(t+\tau) \, y(t) = 0 \qquad \tau = k,\ldots,k + \deg S$$

$$\frac{1}{N} \sum_{t=1}^{N} y(t+\tau) \, u(t) = 0 \qquad \tau = k,\ldots, k + \deg R$$

These conditions can clearly not be satisfied unless the mean value of the output y
is zero. When there is an off-set or a disturbance the parameter estimates will
assume values such that $\hat{R}(1) = 0$.

□

Another example which shows that reset can also be provided automatically in explicit
algorithms is given in Åström (1979b).

In many cases it is thus not necessary to make any special provisions to obtain reset action. The self-tuner will automatically introduce reset when needed. The main
drawback of such a scheme is that the response of the system to sudden variations
in the load level may be slow. The problem is particularly severe if the nature of
the disturbances change drastically with time. The method is also inconvenient when
the STR is used as a tuner. It could easily happen that the disturbances encountered
during the tuning have a small low frequency component. The regulator obtained will
then not necessarily have sufficient gain at low frequencies. When integration is provided automatically it is necessary to introduce facilities to avoid reset windup i.e.
to ensure that the regulator state which correspond to the integral will not grow
without bounds when the output saturates. One possibility is to replace the control
law (2.4) by

$$u(t) = sat[Tu_c(t) - Sy(t) - (R-r_0)u(t)]/r_0 \qquad (4.1)$$

where sat is a saturation function which saturates before the actuator. Another way to avoid reset windup is discussed in Andersson and Åström (1978).

Estimation of a Bias

A simple way to model the off-set errors is to replace the model (2.1) by

$$A \ y(t) = Bu(t) + b \qquad (4.2)$$

where the bias term b represent the errors. With a model like (4.2) it is natural to estimate the bias b and to compensate for it. Such a scheme was proposed by Clarke and Gawthrop (1979). An advantage is that the estimation of b is simple. The drawbacks are that an extra parameter has to be estimated. The estimate $\hat{b}$ will converge slowly unless special precautions are taken. If forgetting factors are used it is useful to have separate forgetting factors for $\hat{b}$ and the other parameters. See Åström (1979b). If bias is eliminated in this way it is not possible to use the STR simply as a tuner because there will be no reset when estimation is switched off.

Forced Integral Action by Use of a Special Model Structure

One possibility to obtain reset is to choose a model structure so that the regulator designed from the model will always contain an integrator. For explicit self-tuners based on pole-placement design this can be done by using the lack of uniqueness in the equation (2.3) to impose the condition that $1 - q^{-1}$ should be a factor of R. This can always be done. For implicit self-tuners integral action can be imposed by replacing the model (2.5) by

$$PT_1 \ y(t) = R\nabla u(t-k) + Sy(t-k) + b \qquad (4.3)$$

where $\nabla = 1 - q^{-1}$.

The control law (2.6) is then replaced by

$$R\nabla u(t) = Tu_c(t) - Sy(t-k) \qquad (4.4)$$

Notice that it follows from the design procedure that $T(1) = S(1)$. Notice also that it is useful to include estimation of the bias b although the estimate is not used when calculating the control signal.

The main advantage of this scheme is that the controller will always have integral action. If the STR is used as a tuner the regulator obtained when the tuning is

switched off will always have integral action. A drawback is that there will be one
additional mode in the controller. In the pole-placement design it is then an additional pole to position. This pole is not entirely trivial to choose. If it is
placed at the origin the controller will have an unnecessarily high gain. The
scheme also requires special tricks to avoid reset windup. Another drawback with the
scheme is that the self-tuner may try to eliminate the integral action when it is
not needed. The estimated polynomial S then has the factor $\nabla = (1 - q^{-1})$. This means
that the regulator transfer function has an unstable mode which is cancelled, and
the system will be unstable. An example where this happens is discussed in Åström
and Gustavsson (1978).

Integration in Inner Loop

Steady state errors can be avoided by the scheme shown in Fig.6. The process is provided with a fixed gain feedback loop with integrating action.

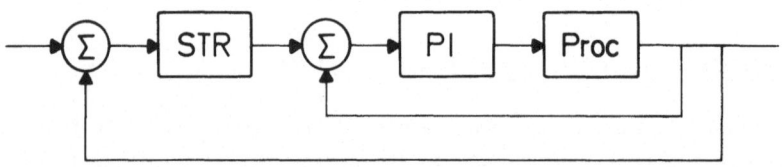

Figure 6. Block-diagram of a regulator with an integrating regulator in an
inner-loop and an outer-loop with self-tuning.

The use of an inner loop was originally proposed by Wittenmark (1973). The arrangement shown in Fig.6 was applied by Dumont and Belanger (1978). One drawback of the
scheme is that it may be difficult to tune the regulator in the inner loop. Another
drawback is that it is not good practice to have integration in an inner loop even
for systems with fixed parameters.

Integration in an Outer Loop

Another possibility to avoid steady state errors is shown in Fig.7. A self-tuner
is first connected to the process. An outer loop with integral action is then introduced. Since the self-tuner makes the inner loop invariant to changes in process
dynamics it is possible to have fixed gain in the outer loop.

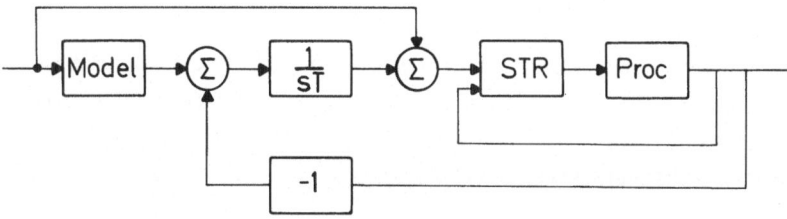

Figure 7. Block-diagram of a regulator structure with a self-tuner in an inner loop and an outer loop with integration.

The approach is particularly attractive for self-tuners whose specifications are directly related to properties of the closed loop transfer function because the outer loop gain can then be set automatically. Another advantage is that standard methods can be used to avoid reset windup. The major disadvantage is that is is not convenient to use the scheme for self-tuners whose performance are not directly related to the bandwidth of the closed loop system. In such a case the integrator gain cannot be set automatically.

5. ESTIMATOR WINDUP

The problem of windup can occur whenever there is an unstable mode in a regulator. In a self-tuning regulator there may be unstable modes associated with the parameter estimator. The problem is closely connected with the design of the estimator, and the way in which control signals are limited.

Input Saturation

There are several mechanisms which can cause instability. Consider for example the case when the actuator saturates. If no precautions are taken it could easily happen that the control signal calculated by the regulator is outside the saturation limits. The estimated process model will then have too low gain. The calculated controller gain will be too large. Saturation effects will be even more pronounced etc. This simple intuitive argument has been supported by simulations. In simple cases it can also be verified analytically. There is a simple remedy. Introduce a saturation in the controller where the limits are set tighter than the actuator saturation. e.g. as in (4.1). The parameter estimator will then have a faithful representation of the actual process variable.

Covariance Windup

Another mechanism which can cause instability will now be discussed. For this purpose the equations describing the parameter estimator are needed. They are

$$\theta(t+1) = \theta(t) + P(t+1) \; \varphi(t+1) \; \varepsilon(t+1) \tag{5.1}$$

$$P(t+1) = [P(t) - P(t) \; \varphi(t) \; R(t) \; \varphi^T(t) \; P(t)]/\lambda \tag{5.2}$$

$$R(t) \;\; = [\lambda + \varphi^T(t) \; P(t) \; \varphi(t)]^{-1} \tag{5.3}$$

Consider the equation (5.2). The negative term in the right hand side represents the
reduction in uncertainty due to the last measurements. If the control signal and the
output are zero the vector $P(t)\varphi(t)$ will then be zero. There will not be any changes
in the parameter estimate and the negative term in the right hand side of (5.2) will
be zero. The equation (5.2) then reduces to

$$P(t+1) = \frac{1}{\lambda} \; P(t)$$

and the matrix P will thus grow exponentially if $\lambda < 1$. If there are no changes for
a long time the matrix P may thus become very large. Since P represents the gain in
the parameter estimator (5.1) a change in the command signal may then lead to large
changes in the parameter estimates and in the process output. The large values of the
matrix P may also lead to numerical problems. The problem will occur whenever the
vector $P(t)\varphi(t)$ is zero or sufficiently small over a period of time. The problem is
closely associated with identifiability conditions and the selection of the forgetting
factor λ.

Excitation of the Process

Identifiability depends on the input signal u and the number of estimated parameters.
In typical regulation problems where the system is continuously excited by the disturbances the problem will not occur provided that the number of estimated parameters
is not too high. The problem will be much severe in a typical servo problem where
the major excitation comes from the command signal which may be constant for long
periods of time. The situation is similar for regulation problems where the major
disturbances are constant over long time periods. One possibility to ensure that
the process is properly excited is to introduce perturbation signals or to use a
dual control law.

The Forgetting Factor

Covariance windup is closely related to the choice of the forgetting factor λ. If
$\lambda = 1$ the problem will not occur. For $\lambda = 1$ the estimator gain will, however, de-

crease and the estimator will be very sluggish. There are, however, several other possibilities to obtain estimators with non-decreasing gain. The matrix P could simply be chosen as a fixed matrix. This is commonly done in model reference systems. Another possibility is to replace equation (5.3)

$$P(t+1) = P(t) + P(t) \; \varphi(t) \; R(t) \; \varphi^T(t)P(t) + R_1 \qquad\qquad (5.4)$$

In this case the matrix P will grow linearly instead of exponentially when $P\varphi$ is zero. A third possibility is to replace the equation (5.3) by

$$P^{-1}(t) = [\alpha I + \sum_{k=1}^{t} \lambda^{t-k} \; \varphi(k)\varphi^T(k)]^{-1}$$

where α is a small number. This ensures that P stays bounded. The size of P is determined by α.

A fourth possibility is to simply put a bound on P e.g. by restricting it so that the trace of the matrix P is constant in each iteration. This has been proposed by Irving (1979).

A fifth possibility is to adjust the forgetting factor automatically. It can e.g. be chosen as

$$\lambda = 1 - \alpha \; \varepsilon^2/\overline{\varepsilon^2}$$

where $\overline{\varepsilon^2}$ is the mean value of ε^2 over a certain period. More complicated formula for adjusting λ have also been proposed. See Fortescue et al (1978). An automatic adjustment of λ does not guarantee that the matrix P stays bounded. The period where P has a reasonable size may, however, increase substantially.

It has also been proposed to eliminate covariance windup by stopping the updating of θ and P when $P\varphi$ or ε is sufficiently small. See Egardt (1979).

In Goodwin et al (1978) it is proposed to analyse the conditioning number of the matrix P and to switch to a stochastic approximation algorithm when the matrix P becomes poorly conditioned.

6. THE PARAMETERIZATION PROBLEM

A mathematical model can be parametrized in many different ways. The choice of parameters is important for the design of self-tuners. For example when discussing implicit and explicit algorithms for self-tuning servos in section 2 it was found that the algorithm could be simplified substantially if the model was parametrized in the regulator parameters.

Although the parametrization problem is important it has been given little attention in literature. The general tendency, both as far as MRAC and STR are concerned, is to parametrize in such a way that the estimation problem becomes simple e.g. linear in the parameters. In Aström (1979c) an example is given which shows that it may be advantageous to use other parametrizations.

The parametrization of the minimum variance self-tuner (Åström and Wittenmark (1973)) or its model reference equivalent has been given some attention. For minimum variance regulation the variable y is often chosen as the control error. Since $PT_1 = 1$ for minimum variance control the estimation model (2.5) then reduces to

$$y(t+k) = S(q^{-1}) y(t) + R(q^{-1}) u(t) \qquad (6.1)$$

and the control law becomes

$$\hat{R}(q^{-1}) u(t) = -\hat{S}(q^{-1}) y(t) \qquad (6.2)$$

This control law has one redundant parameter because the polynomials $\hat{R}$ and $\hat{S}$ can be multiplied by a constant without changing the control law. The redundant parameter can be eliminated by reparametrizing the estimation model (6.1) as

$$y(t+k) = r_0[u(t) + r_1' u(t-1) + \ldots + r_{n_R}' u(t-n_R)]$$
$$\qquad (6.3)$$
$$+s_0 y(t) + s_1 y(t-1) + \ldots + s_{n_S} y(t-n_R)$$

The control law (6.2) then becomes

$$u(t) = -\frac{1}{\hat{r}_0}[\hat{s}_0 y(t) + \ldots + \hat{s}_{n_S} y(t-n_S)]$$
$$\qquad (6.4)$$
$$- \hat{r}_1' u(t-1) - \ldots - \hat{r}_{n_R}' u(t-n_R)$$

It is shown in Åström and Wittenmark (1973) that the estimate $\hat{r}_0$ can be fixed apriori if

$$0.5 \leqslant \hat{r}_0/r_0 < \infty$$

without influencing the equilibrium condition. In Ljung (1977) it is shown that if the algorithm converges for $\hat{r}_0 = r_0$ it will still converge if (6.5) holds. The convergence rate is, however, influenced by $\hat{r}_0$. It is fastest for $\hat{r}_0 = r_0$

For minimum variance self-tuners either of the models can be used. The algorithm based on (6.4) with fixed $\hat{r}_0$ is most robust provided that apriori knowledge to choose r_0 subject to (6.5) is available. If this is not possible the parameters in (6.1) can be estimated. Identifiability is poor because of the feedback. The estimates of the parameter combinations r_i/r_0 and s_i/r_0 converge as 1/t. The estimate of r_0 converges, however, at a slower rate. Algorithms which treat r_0 in a special way are therefore also used.

7. CONCLUSIONS

The word self-tuning regulator may lead to the false conclusion that such regulators can be switched on and used blindly without any apriori considerations. This is not true. The self-tuning regulator is a fairly complex control law. A proper design involves the choices of gross features like underlying design and estimation methods and decisions on details like initialization, selection of parameters, and safeguard methods. Proper choices require insight and knowledge. There are known cases where bad choices have been disastrous.

There has recently been considerable progress in the theory of adaptive control. Stability results have been proven for simple self-tuners (implicit minimum variance and pole-placement) connected to linear systems. The theory requires assumptions which are hard to verify in a practical situation. The theory is also limited to simple self-tuners. The theory required to use self-tuners confidently is thus not available A cautious person would then perhpas be inclined not to try a self-tuner. To get some perspective it may be useful to reflect on the role of theory in similar situations. The properties of the closed loop system obtained when a PID regulator is connected to a linear system are fully understood theoretically provided that the regulator operates in the linear region. As soon as nonlinearities associated with gap, saturation and anti-windup are introduced there is, however, little theory which tell theoretically what happens. In spite of this, large systems with many interconnected PID regulators are designed, sold, commissioned and used routinely.

Based on attempts to develop suitable theory and experiences from a few applications I believe, however, that self-tuning regulators can and will be used profitably, even if all their properties are not fully understood theoretically. I hope that this paper may inspire some of you to acquire the appropriate knowledge and try some schemes. I also hope that some of you will tackle the important theoretical problems that remain.

8. ACKNOWLEDGEMENTS

This work was supported by the Swedish Board of Technical Development (STU) under contract STU 78-3763.

9. REFERENCES

Andersson L and Åström K J (1978): An interactive MISO regulator. Dept of Automatic Control, Lund Institute of Technology, Lund, Sweden, CODEN:LUTFD2/(TFRT-7154)/ 1-034/(1978).

Åström K J (1974): A self-tuning regulator for non-minimum phase systems. Dept of Automatic Control, Lund Institute of Technology, Lund, Sweden, Report TFRT-7411.

Åström K J (1979a): Self-tuning regulators - design principles and applications. Proc Yale Workshop on Applications of Adaptive Control, Yale University.

Åström K J (1979b): Simple self-tuners I. Dept of Automatic Control, Lund Institute of Technology, Lund, Sweden, CODEN:LUTFD2/(TFRT-7184))1-052/(1979).

Åström K J (1979c): New implicit adaptive pole-zero-placement algorithms for non-minimum phase systems. Dept of Automatic Control, Lund Institute of Technology, Lund, Sweden, CODEN:LUTFD2/(TFRT-7172)/1-121/(1979).

Åström K J and Gustavsson I (1978): Analysis of a self-tuning regulator in a servo-loop. Dept of Automatic Control, Lund Institute of Technology, Lund, Sweden, CODEN:LUTFD2/(TFRT-3150)/1-058/(1978).

Åström K J and Wittenmark B (1973): On self-tuning regulators. Automatica 9, 185-199.

Åström K J and Wittenmark B (1979): Self-tuning controllers based on pole-zero placement. Submitted to IEE Proceedings.

Clarke D W and Gawthrop B A (1975): Self-tuning controller. Proc IEE 122, 929-934.

Clarke D W and Gawthrop P J (1979): Implementation and application of microprocessor based self-tuners. Preprints 5th IFAC Symposium on Identification and System Parameter Estimation, Darmstadt, September 1979, p 197-208.

Dumont G A and Bélanger P R (1978): Self-tuning control of a titanium dioxide kiln. IEEE Trans AC-23,532-537.

Egardt B (1979). Stability of adaptive controllers. Lecture Notes in Control and Information Sciences, Vol 20, Springer-Verlag, Berlin.

Fortescue T R, Kershenbaum L S and Ydstie B E (1978): Implementation of self-tuning regulators with variable forgetting factors. Report, Dept of Chemical Eng and Chemical Technology, Imperial College, London.

Goodwin G C, Ramadge P J and Caines P E (1978): Discrete time stochastic adaptive control. Report Div of Applied Science, Harward University Cambridge Mass., USA, December, 1978.

Irving E (1979): Improving power network stability with adaptive generator control - new developments. Proc Workshop on Applications of Adaptive Control, Yale University, New Haven.

Ljung L (1977): Analysis of recursive stochastic algorithms. IEEE AC-22, 551-575.

Morse A A (1979): Global stability of parameter adaptive control systems. S&IS Report 7902R, Yale University, March, 1979.

Narendra K S, Lin Y-H, and Valavani L S (1979): Stable adaptive controller design - Part II. Proof of Stability. S&IS Report 7904, Yale University, April, 1979.

Peterka V (1979: Adaptive digital regulation of noisy systems. Proc 2nd IFAC Symp on Identification and Process Parameter Estimation, Prague.

Wellstead P E, Prager D and Zanker P (1979): Pole assignment self-tuning regulator. Proc IEE, 126, 781-787.

Wittenmark B (1973b): Self-tuning regulators. Report TFRT-7321 (thesis). Dept of Automatic Control, Lund Institute of Technology, Lund, Sweden.

SIMPLE SELF-TUNING CONTROLLERS

B. Wittenmark   K.J. Åström

Department of Automatic Control
Lund Institute of Technology
Lund, Sweden

ABSTRACT

The problem of design of simple self-tuning controllers is discussed. The basic idea
is to estimate a low order model and to use pole-placement in order to obtain a desired closed loop performance. The controller has a three mode action and can be regarded as a generalized PID-controller. It is shown that it is possible to obtain a
controller with only one tuning knob. This knob can be calibrated in the desired
bandwidth of the closed loop system. Simulated examples as well as an experiment on
a laboratory process illustrates the properties of the controller.

1. INTRODUCTION

One of the advantages of the well-known PID-controller is that it is a sufficiently
flexible controller for many applications. The three parameters of the controller
are generally tuned with the process in closed loop. The tuning is often easy. It
may, however, be cases when tuning is difficult and time-consuming. Automatic tuning
of the controllers is therefore of interest. The idea of self-tuning regulators was
introduced in order to simplify the tuning of industrial controllers. The self-tuning
regulators have, however, also tuning parameters. It can thus be said that one set
of tuning parameters has been replaced by an other set. Hopefully the new parameters
are easier to choose. In the early applications (Åström et al 1977), the self-tuners
were applied to special problems. Good rules for choosing the parameters could then
be found, Wittenmark (1973). Some parameters are, however, critical. For instance
for self-tuners based on minimum variance control and least squares parameter estimation, it is crucial to have an upper bound on the time delay of the process.

The suitable parameterization of a self-tuner has been discussed widely. It has been
suggested that there should be no adjustable parameters at all. A moment of reflex-
ion shows that it is at least necessary to provide the controller with information
about the desired specifications. The main idea is that the parameters selected by
the operator should be related to the desired performance of the closed loop system.
Such parameters are easier to choose than to choose parameters in the control law.

This paper describes a simple self-tuner intended for simple servo applications. It
is assumed that the process can be described by a second order model. The regulator
is based on recursive least squares estimation and pole-placement design, see Åström
and Wittenmark (1979). The tuning parameters are the bandwidth of the closed loop

system and póssibly also the desired relative damping. In the paper it is only possible to give a brief description of the algorithm and its properties. Further details about simple self-tuners can be found in Wittenmark (1979) and Åström (1979c).

2. ALGORITHM DESIGN

The simple self-tuner is intended to solve simple servo problems for system which can be described by low order models. It is natural to characterize the performance of the servo by the bandwidth and the relative damping of the closed loop system. A servo problem is conveniently formulated as a pole-placement problem. It is then natural to use the formulation of self-tuning servos discussed in Åström and Wittenmark (1979). Since the low frequency properties of a system often can be approximated by a low order model it can be expected that a self-tuner based on a low order model will behave satisfactorily provided that the chosen bandwidth is sufficiently small, see Åström (1979a).

Problem formulation

Assume that the process can be described by the model

$$y(t) + a_1 y(t-h) + a_2 y(t-2h) = b_1 u(t-h) + b_2 u(t-2h) + b_3 \qquad (2.1)$$

where h is the sampling time and b_3 is a bias. Introduce the polynomials $A(q^{-1})$ = $= 1 + a_1 q^{-1} + a_2 q^{-2}$ and $B(q^{-1}) = b_1 + b_2 q^{-1}$ where q^{-1} is the backward shift operator.

The problem can be formulated as to find a feedback such that the closed loop system has poles that corresponds to the poles of a continuous time system with the characteristic polynomial $s^2 + 2\zeta\omega s + \omega^2$. For a sampled data system this means that the characteristic polynomial should be

$$P(q^{-1}) = 1 + p_1 q^{-1} + p_2 q^{-2} \qquad (2.2)$$

where

$$p_1 = -2e^{-\zeta\omega h} \cos\omega h\sqrt{1-\zeta^2}$$

$$p_2 = e^{-2\zeta\omega h}$$

The process model (2.1) has a zero at $z = -b_2/b_1$. If this corresponds to a well damped mode the factor $b_1 + b_2 q^{-1}$ can be cancelled by the regulator. This will be the case if

$$z_1 \leqslant -b_2/b_1 \leqslant z_2 \qquad (2.3)$$

where the choice of z_1 and z_2 is discussed in Section 3. The desired closed loop response is then characterized by the pulse transfer function

$$G_d = \frac{q^{-1}(1 + p_1 + p_2)}{1 + p_1 q^{-1} + p_2 q^{-2}} \tag{2.4}$$

If the process zero corresponds to an unstable or poorly damped mode the zero cannot be cancelled and the desired pulse transfer function is instead

$$G_d = \frac{1 + p_1 + p_2}{b_1 + b_2} \cdot \frac{b_1 q^{-1} + b_2 q^{-2}}{1 + p_1 q^{-1} + p_2 q^{-2}} \tag{2.5}$$

Control design for known parameters

The calculation of the control law when the process model is known is straight forward, see e.g. Åström (1979b). The control law is given by

$$Ru(t) = Ty_r(t) - Sy(t) \tag{2.6}$$

where y_r is the reference signal and R, S and T are polynomials in the backward shift operator, q^{-1}. In order to eliminate the bias term we assume that $R = R_1(1-q^{-1})$, i.e. there is an integrator in the controller. Other ways to eliminate the bias are discussed in Section 3. In order to treat the two cases above simultaneously we introduce

$$P'(q^{-1}) = \begin{cases} P(q^{-1})(1 + b_2/b_1 q^{-1}) & \text{if } z_1 \leq -b_2/b_1 \leq z_2 \\ P(q^{-1}) & \text{otherwise} \end{cases}$$

The control law is obtained by solving the polynomial equation

$$AR_1(1 - q^{-1}) + q^{-1}BS = P' \tag{2.7}$$

where R_1 and S are of order 1 and 2 respectively. The identity (2.7) has a unique solution provided $A(1-q^{-1})$ and B do not have a common factor. The correct steady state gain is obtained if we choose

$$T(q^{-1}) = S(1) = s_0 + s_1 + s_2 \tag{2.8}$$

The controller has four parameters, the coefficients of the polynomials $R_1 = 1 + r_1 q^{-1}$ and $S = s_0 + s_1 q^{-1} + s_2 q^{-2}$. The closed loop system obtained when (2.6) is used will be

$$y(t) = \frac{q^{-1}TB}{AR+q^{-1}BS} y_r(t) + \frac{R\, b_3}{AR+q^{-1}BS} = \frac{q^{-1}TB}{P'} y_r(t) + \frac{R\, b_3}{P'} \tag{2.9}$$

The system will have the desired transfer function (2.4) or (2.5). Further if y_r is constant $y(t) \to y_r$ as $t \to \infty$.

Common factors in the process model

The polynomials A and B have a common facor if

$$T_{cf} = b_2^2 - a_1 b_1 b_2 + a_2 b_1^2 = 0$$

and B will contain the factor $1 - q^{-1}$ if $B(1) = 0$. If there is a almost common factor the solution of (2.7) will be poorly conditioned and that may result in very large control signals. To get dimension free test quantities the following test is used

$$T_{cf} \text{ or } (b_1 + b_2)^2 \le \varepsilon \max(b_1^2, b_2^2) \tag{2.10}$$

to test for common or nearly common factor. The number ε is related to the maximum size of the feedback gain. When cancelling a common factor the transfer function of the process will be reduced to $b/(1+aq^{-1})$ where $b=b_1$ and $a=a_2 b_1/b_2 = a_1 - b_2/b_1$. The identity (2.7) can now be solved if R_1 and S both are of first order or if $R_1=1$ and S is of second order.

The sampling time

The desired performance of the closed loop system is determined by the damping, ζ, and the bandwidth or equivalently the natural frequency ω. It is then natural to have the sampling time inversely proportional to the bandwidth. A reasonable choice is

$$h = \frac{2\pi}{N\omega\sqrt{1-\zeta^2}} \tag{2.11}$$

where N is the number of samples per period. The choice of N is discussed in Åström (1979c). It is found that a reasonable choice is N=10-20. Further if we assume that the damping is $\zeta=1/\sqrt{2}$ then the sampling time should be choosen as

$$\omega h \approx 0.45 - 0.9.$$

If the parameter ω is changed during an experiment then the sampling time also changes. This will then influence the values of the parameters in the model (2.1).

The estimator in the self-tuner will of course adjust to these changes. It is, in principle, easy to compute how the model is changed. This can be done by transforming the model to a continuous time system and then sample this system with the new sampling time. It is, however, possible that the estimated model does not have a continuous time counterpart. A simplified method is to approximate z=exp(sh) as z=1+sh. Simple calculations will lead to a transformation which relates the parameters of the model for different sampling times, see Åström (1979c).

Estimation procedure

A self-tuning controller contains a parameter estimator. In this case a recursive least squares estimation with exponential forgetting of old data is used. The controller discussed here contains an integrator. This implies that the bias term b_3 in (2.1) does not need to be estimated. The other parameters are estimated from the differences of the inputs and outputs, i.e. using u(t)−u(t−h) and y(t)−y(t−h) respectively.

A simplified self-tuning controller

The discussion above can now be summarized into the following algorithm, where Steps 1–4 are repeated at each sampling time.

Data: The operator selects ω and ζ which determines the closed loop characteristic polynomial. The sampling time is choosen according to (2.11).

Step 1:   Estimation. The parameters a_1, a_2, b_1 and b_2 in the process model are estimated. The previous estimates are transformed if the sampling time has been changed.

Step 2:   Test of the model. Common or nearly common pole and zero are removed using the test (2.10). The desired characteristic polynomial P' is determined based on the test (2.3).

Step 3:   Controller parameter determination. The parameters of the controller are determined by solving the polynomial equation (2.7) and using (2.8).

Step 4:   Control. The control signal is determined from
$$u(t)=sat[t_0 y_r(t)-s_0 y(t)-s_1 y(t-h)-s_2 y(t-2h)+(1-r_1)u(t-h)+r_1 u(t-2h)]$$
to avoid saturation and reset windup.

3. DISCUSSION OF THE ALGORITHM

The algorithm presented in the previous section contains some parameters that have to be determined. This together with a discussion of the properties of the algorithm are given in this section.

Choice of parameters

The choice of the parameters in the algorithm is discussed and examplified in Åström (1979c). Some nominal values that can be used are given below. The initial values in the estimator can be $\hat{a}_1(0)=-1.5$, $\hat{a}_2(0)=0.7$, $\hat{b}_1(0)=0.1$ and $\hat{b}_2(0)=0$. The inital covariance matrix in the estimator can be 100 times a unit matrix and the exponential forgetting factor approximately 0.95-0.99. If it is desirable to have only one tuning parameter the damping could be fixed to $\zeta=0.7$. A reasonable value of ε in (2.10) is 0.01. The zero of the process may be removed if $(z_1,z_2)=(-0.1,\ 0.99)$. If a smaller value of z_1 is used the control signal usually starts to oscillate. The values given above are reasonable rules of thumb values. It has been found in simulations that none of the values are very critical. The inital estimates of $\hat{b}_1(0)$ and $\hat{b}_2(0)$ will, however, have crucial influence on the initial transient.

Reset action

There are several ways to eliminate steady state errors due to bias or load disturbances. The way used here is to postulate that the controller has an integrator. In Åström (1979c) it is shown that the parameter estimator can take care of the bias automatically. This will in general give unsymmetrical responses for positive and negative steps. A third way is to estimate the bias b_3 in the model (2.1) and compensate for it. It has been found advantageous to have a smaller forgetting factor for the bias than for the dynamic parameters. Finally, the bias can be eliminated by having a self-tuning controller in an inner loop and a fixed integral controller in an outer loop. All methods have been investigated through simulations and there are no drastic differences in the performances.

Interpretation of the controller

The controller (2.6) with the number of parameters used here can be interpreted as a PID-controller with a special structure. Consider the PID-controller given by

$$u(t) = \frac{\alpha_1}{(1-q^{-1})R_1}\ (y_r(t) - y(t)) - \alpha_0 y(t) - \frac{\beta(1-q^{-1})}{R_1}\ y(t)$$

The three terms on the right hand side are the integral, proportional and derivative parts. The factor R_1 can be interpreted as the filter that should be used to obtain the derivative. Notice that the proportional and derivative parts only works on the output and not on the error. The controller can be written as

$$(1-q^{-1})R_1\ u(t) = \alpha_1\ y_r(t) - [\alpha_1 + (1-q^{-1})(\alpha_0 R_1 + \beta(1-q^{-1}))]y(t)$$

This controller has exactly the same structure as (2.6) and the parameters α_0, α_1 and β can be obtained from the parameters s_0, s_1 and s_2 if $r_1 \neq -1$. This and other structures for self-tuning PID-controllers are discussed in Wittenmark (1979).

Higher order processes

The discussed self-tuning controller will work well if the process can be well approximated by the second order model (2.1) and if the desired bandwidth is not too large.
In Åström (1979a) results are given which show that the closed loop system designed
on the basis of an approximative model will be stable if the desired bandwidth of the
closed loop system is sufficiently small.

The tuning rule for the regulator is thus very simple. Start with a small bandwidth.
Establish the possible range of bandwidths for which the regulator will work by increasing the specified bandwidth until the performance deterioates. If the desired
bandwidth is outside the range found it is necessary to use a more complex regulator or to change the specifications. Tuning is simple because it involves only one
parameter.

4. EXAMPLES

Three examples will be given which will illustrate some of the properties of the
simple self-tuner. The first two examples are simulations while the the third is
level control of a laboratory process

Example 4.1   Second order system

The system

$$G(s) = \frac{1}{(s+1)^2}$$

is controlled with the self-tuner described in the previous sections. The specifications are $\omega = 1.5$ and $\zeta = 1/\sqrt{2}$. Fig.1 shows the output, the reference value, and
the control signal. Already at the second step there is a good agreement between the
desired output and the process output. The first transient will of course depend on
the chosen initial values in the estimator. For $t \geq 15$ a load disturbance $v = 1$ is
added to the input of the process. The controller eliminates the effect of the disturbance.

Example 4.2 Fourth order system

The system has the transfer function

$$G(s) = \frac{1}{(s+1)^4}$$

In this case it is more difficult to find a good second order approximation of the
process. The desired bandwidth has to be chosen quite small. Fig. 2 shows the output and the control signal at a step in the reference signal when the estimator has

converged. For ω = 0.3 the control is good. The behaviour starts to deteriorate when ω is increased to 0.4 and further to 0.45. In all three cases the desired damping has been 0.7.

Example 4.3 Level control

One variant of the simple self-tuner has been implemented on a LSI-11 computer. The communication with the operator is done through commands. The different parameters in the controller can be easily changed on-line. The controller and operator communication is written in Pascal. Further details about the implementation is given in Wittenmark, Hagander and Gustavsson (1980).

As an example the controller has been used to control a laboratory process consisting of a pneumatic valve and a small water tank. The position of the valve is the control signal and the output is the level in the tank. Fig. 3 shows the level and the control signal when the reference level is changed in steps about each 45 second. Each step is about 10 % of the maximum level which is 0.5 m. The specifications where ω = 0.45 and ζ = 1 and the sampling time was 1 s. From the figure it can be seen that the controller gives the same response over the whole range of levels. This is not possible with a fixed controller. The parameters in the controller changed about 20 – 40 % going from the minimum to the maximum level.

5. CONCLUSIONS

The report presents a simple self-tuner for typical servo problems. The self-tuner has one major adjustable parameter which is proportional to the desired bandwidth of the closed loop system. All other parameters are fixed or related to the bandwidth. It is shown by simulations that the algorithm works well in many circumstances. However, the simple self-tuning controller which can be interpreted as a PID-controller cannot control all processes. It can only behave as a well tuned PID-controller. It is thus possible to use the self-tuner on the same type of processes as the conventional PID-controller can be used on. Many common processes in practice belong to this class and manual tuning can thus often be eliminated.

6. ACKNOWLEDGEMENTS

This work has been stimulated through discussions with many collegues at the Department of Automatic Control. Especially we want to thank Leif Andersson, Per Hangader, Ivar Gustavsson and Carl Fredrik Mannerfelt. This work was partially supported by the Swedish Board for Technical Development (STU) under contract No 78-3763.

7. REFERENCES

Åström, K J (1979a): Robustness of a design method based on assignment of poles
and zeroes. IEEE Transaction on Automatic Control. To appear.

Åström, K J (1979b): Algebraic system theory as a tool for regulator design.
In Acta Polytechnica Scandinavia, Ma31, Helsinki: Topics in System Theory.
Publication in honour of Professor Hans Blomberg on the occasion of his
sixtieth birthday on December 18th, 1979.

Åström, K J (1979c): Simple self-tuners I, Department of Automatic Control, Lund
Institute of Technology, Sweden, CODEN: LUTFD2/(TFRT-7184)/(1-052)/(1979).

Åström, K J, Borisson, U, Ljung, L and Wittenmark, B (1977): Theory and applications
of self-tuning regulators, Automatica, 13, 457-476.

Åström, K J, Wittenmark, B (1979): Self-tuning controllers based on pole-zero
placement. Submitted to IEE Proceedings Part D, Control Theory and Applications.

Wittenmark, B (1973): A self-tuning regulator, TFRT-1003, Department of Automatic
Control, Lund Institute of Technology, Sweden.

Wittenmark, B (1979): Self-tuning PID-controllers based on pole placement. Department
of Automatic Control, Lund Institute of Technology, Sweden. CODEN:LUTFD2/
(TFRT-7179)/1-037/(1979).

Wittenmark, B, Hagander, P and Gustavsson, I (1980): STUPID - Implementation of a
self-tuning PID-controller, Department of Automatic Control, Lund Institute
of Technology, Sweden. To appear

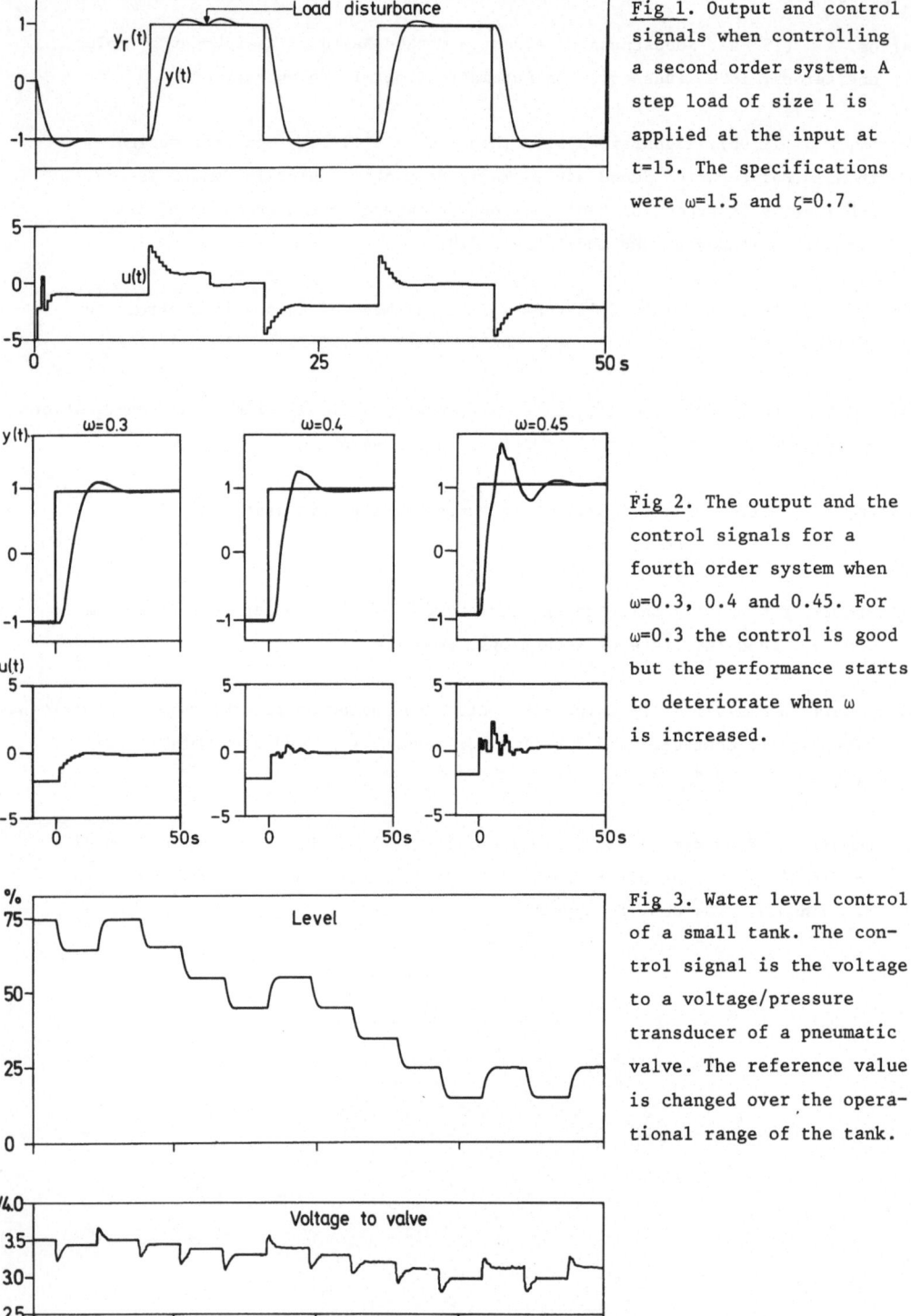

Fig 1. Output and control signals when controlling a second order system. A step load of size 1 is applied at the input at t=15. The specifications were ω=1.5 and ζ=0.7.

Fig 2. The output and the control signals for a fourth order system when ω=0.3, 0.4 and 0.45. For ω=0.3 the control is good but the performance starts to deteriorate when ω is increased.

Fig 3. Water level control of a small tank. The control signal is the voltage to a voltage/pressure transducer of a pneumatic valve. The reference value is changed over the operational range of the tank.

SOME RELATIONS IN DISCRETE ADAPTIVE CONTROL

D. Matko

Faculty of Electrical Engineering, Ljubljana
61000 Ljubljana, Tržaška c.25, Yugoslavia

INTRODUCTION

In the past 20 years a large amount of work has been done on the design
of adaptive control systems. The adaptive control can be quite different
with the respect to the procedure for the detection of process changes
as well as with the respect to their compensation. Among many kinds of
adaptive systems, the parameter adaptive (called also self tunning regulators) and model reference systems are most frequently used in practice. Parameter adaptive systems include two computer procedures, namely identification and optimization. Model reference systems are simpler
in their structure, they include only one computer procedure, namely the
minimization of the error between the output of the process and the reference model. Many authors /1,2,3,4/ describe the connections between
the identification with an adaptive model and model reference adaptive
procedures. They treat the model reference system as the identification
of the reference model by the adjustable plant. It has been shown for
continuous case /5/ that in the cases where adaptive systems are synthe-
sized using the same criterion (linear optimal control with quadratic
criterion function) and procedures (Lyapunov functions) both approaches
are not only inverse but lead to the identical regulator structure.

The purpose of this work is to show the connections between discrete
optimal model reference and parameter adaptive systems. It is shown that
in the case of optimal adaptive systems using a quadratic criterion
function

$$W = \underline{x}_p^{\ T}(T)\underline{F}\underline{x}_p(T) + \sum_{k=0}^{N-1} \left[\underline{x}_p^T(k)\underline{Q}\underline{x}_p(k) + Ru^2(k)\right] \tag{1}$$

both approaches lead to the same regulator structure. In eqn.(1) $\underline{x}_p$ is
the vector of the process state, u the control signal $\underline{F}$ a constant positive semidefinite matrix, $\underline{Q}$ a positive semidefinite matrix and R a positive constant. It is shown, that the state of the controlled plant is
needed only for its compensation but not for the identification of its
parameters. In the second part of the paper the possibility of elimina-

ting the request for the measurability of the plant state is discussed. It is shown, that this request can be overcome by increasing the order of the compensated plant.

OPTIMAL PARAMETER ADAPTIVE SYSTEMS·

The process, which has to be optimally controlled can be described in the following form:

$$x_p(k+1) = \underline{A}_p \underline{x}_p(k) + \underline{b}_p u(k) \tag{2}$$
$$y_p(k) = \underline{c}_p^T \underline{x}_p(k)$$

where $\underline{A}_p$ and $\underline{b}_p$ are unknown, slowly changing matrix of order nxn and vector of order n respectively. The process can be identified using a model described by

$$x_e(k+1) = \underline{D}\underline{x}_e(k) + (\underline{A}_e - \underline{D}) \underline{x}_p(k) + \underline{b}_e u(k) \tag{3}$$
$$y_e(k) = \underline{c}_e^T \underline{x}_e(k)$$

where $\underline{A}_e$, $\underline{b}_e$ and $\underline{c}_e$ are estimates of $\underline{A}_p, \underline{b}_p$ and $\underline{c}_p$ respectively and x_e is the state vector of the model. $\underline{D}$ is a constant matrix with all eigenvalues inside the unit circle. Defining the error

$$e(k) = y_p(k) - y_e(k) , \tag{4}$$

writing process and model in the observability cannonical form and using z transform on eq (3), the following equation is obtained

$$y_e = \frac{D(z^{-1}) - A_e(z^{-1})}{1+D(z^{-1})} y_p + \frac{B_e(z^{-1})}{1+D(z^{-1})} u, \tag{5}$$

where

$$A_e(z^{-1}) = a_{e1}z^{-1} + a_{e2}z^{-2} + \ldots + a_{en}z^{-n} \tag{6}$$
$$B_e(z^{-1}) = b_{e1}z^{-1} + b_{12}z^{-2} + \ldots + b_{en}z^{-n} \tag{7}$$
$$D(z^{-1}) = d_1 z^{-1} + d_2 z^{-2} + \ldots + d_n z^{-n} \tag{8}$$

Eq. (5) can be rewritten in the form

$$y_e = -A_e(z^{-1})y_p + B_e(z^{-1}) u + D(z^{-1})e \tag{9}$$

or in the time domain

$$y_e(k) = \underline{\Psi}^T(k)\underline{\hat{\theta}}(k-1) + \underline{e}^T\underline{d}, \tag{10}$$

where

$$\underline{e}^T(k) = \left[e(k-1), e(k-2), \ldots, e(k-n)\right] \tag{11}$$

$$\underline{\mathbf{Y}}^T(k) = \left[-y_p(k-1), \ldots, -y_p(k-n, u(k-1), \ldots, u(k-n)\right] \qquad (12)$$

$$\hat{\underline{\theta}}^T = \left[a_{e1}, \ldots, a_{en}, b_{e1}, \ldots, b_{en}\right] \qquad (13)$$

The considered method of identification is recursive extended least squares method /6/ with constant matrix $\underline{D}$ and well known results

$$\hat{\underline{\theta}}(k+1) = \hat{\underline{\theta}}(k) + \underline{\gamma}(k)\underline{P}(k)\underline{\mathbf{Y}}(k+1)e(k+1) \qquad (14)$$

$$\underline{P}(k+1) = \underline{P}(k) - \underline{\gamma}(k)\underline{P}(k)\underline{\mathbf{Y}}(k+1)\underline{\mathbf{Y}}^T(k+1)\underline{P}(k) \qquad (15)$$

$$\underline{\gamma}(k) = \left[\underline{\mathbf{Y}}^T(k+1)\underline{P}(k)\underline{\mathbf{Y}}(k+1) + 1\right]^{-1} \qquad (16)$$

If $\underline{\underline{A}}_p$ and $\underline{b}_p$ are slowly changing, $\underline{\underline{A}}_e$ and $\underline{b}_e$ track them. So we can estimate the optimal control vector

$$u^*_e(k) = -(R+\underline{b}_e^T\underline{\underline{K}}_e(k+1)\underline{b}_e)^{-1}\underline{b}_e^T\underline{\underline{K}}_e(k+1)\underline{\underline{A}}_e\underline{x}_p(k) \qquad (17)$$

where $\underline{\underline{K}}_e$ is the solution of the Riccati equation

$$\underline{\underline{K}}_e(k) = \underline{Q}+\underline{\underline{A}}_e^T\underline{\underline{K}}_e(k+1)\left[\underline{\underline{I}}-\underline{b}_e(R+\underline{b}_e^T\underline{\underline{K}}_e(k+1)\underline{b}_e)^{-1}\underline{b}_e^T\underline{\underline{K}}_e(k+1)\right]\underline{\underline{A}}_e \qquad (18)$$

with the final condition

$$\underline{\underline{K}}_e(N) = \underline{Q} \qquad (19)$$

Due to the time changes of $\underline{\underline{A}}_e$ and $\underline{b}_e$ the difference equation (18) has to be solved repetitively. Figure 1 represents the block diagram of the described optimal parameter adaptive system. The state of the controlled plant is needed only for the optimal control, but not for the identification of parameters.

OPTIMAL MODEL REFERENCE SYSTEM

In the case of model reference systems the best possible process output tracking of the reference model output is desired. The model is described by the following difference equations

$$\underline{x}_m(k+1) = \underline{\underline{A}}_m\underline{x}_m(k) + \underline{b}_m v(k) \qquad (20)$$

$$y_m(k) = \underline{c}_m^T \underline{x}_m(k)$$

where $\underline{x}_m$, v, $\underline{\underline{A}}_m$ and $\underline{b}_m$ represent the model state vector, model input, feedback matrix and input vector respectively. The process parameters are compensated by the input signal

$$u = b_k(\underline{a}_k^T\underline{x}_p + v) \qquad (21)$$

Using this equation the difference equation of the compensated plant is obtained

$$\underline{x}_p(k+1) = (\underline{A}_p + \underline{b}_p b_k \underline{a}_k^T) \underline{x}_p + (\underline{b}_p b_k) v \tag{22}$$

The existence of the such b^*_k and $\underline{a}^*_k$, such that the equations

$$\underline{b}_p b^*_k = \underline{b}_m \tag{23}$$

and

$$\underline{A}_p + \underline{b}_m \underline{a}^{*T}_k = \underline{A}_m \tag{24}$$

are satisfied at each instant, represents the condition for the model output to be perfectly followed by the process output. This condition is fulfilled in the case where the columns of error matrices, which are represented by the differences between process and model feedback matrices and the process input vector linearly depend on the vector $\underline{b}_m$. For the process and model in the observability cannonical form this condition is fulfilled if vectors $\begin{bmatrix} a_{mn} - a_{pn}, & \ldots, & a_{m1} - a_{p1} \end{bmatrix}$ and $\underline{b}_p$ linearly depend on vector $\underline{b}_m$. In this case the compensating vector $\underline{a}_k$ has only one

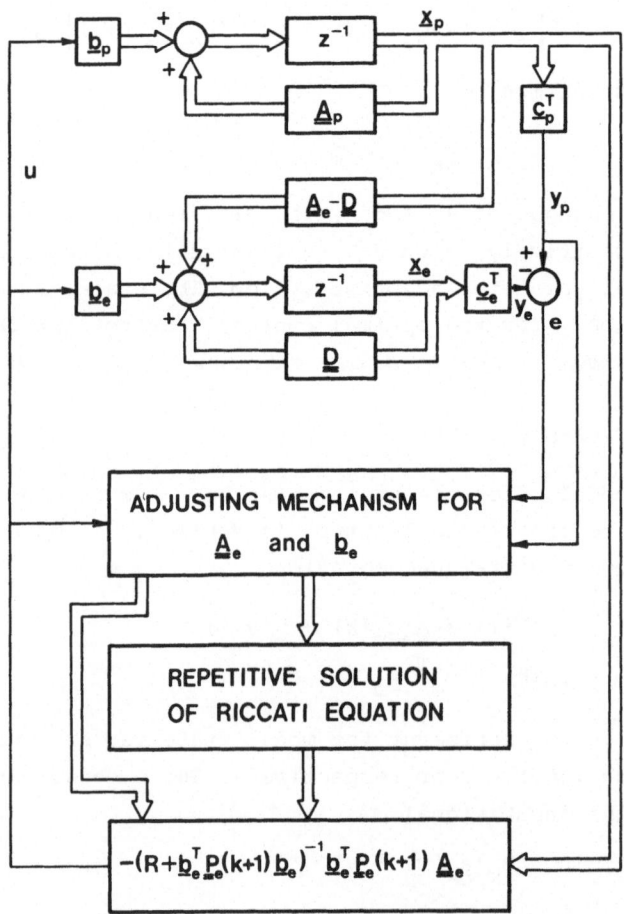

Fig. 1.   The block diagram of the optimal parameter adaptive system

element and for the compensation only the output of the process is needed. In the praxis this case is very seldom. For the process and model in the controllability cannonical form all states of the process are needed for its compensation. From eq. (21) and (20) we obtain relation

$$\underline{x}_m(k+1) = \underline{\underline{A}}_m \underline{x}_m - \underline{b}_m \underline{a}_k^T \underline{x}_p + \underline{b}_m b_k^{-1} u, \tag{25}$$

which can be rewritten using equations

$$e(k) = y_p(k) - y_m(k) \tag{26}$$

$$\underline{\underline{A}}_m - \underline{b}_m \underline{a}_k^T = \underline{\underline{A}}_k' \tag{27}$$

$$\underline{b}_m b_k^{-1} = \underline{b}_k' \tag{28}$$

in the following form

$$y_m(k) = \underline{\psi}^T(k)\, \underline{\hat{\theta}}(k-1) + \underline{e}^T \underline{a}_m, \tag{29}$$

where $\underline{e}$ and $\underline{\psi}^T$ are defined by eq. ((11) and (12) and

$$\underline{\hat{\theta}}^T = \begin{bmatrix} a'_{k1}, & \cdots, & a'_{kn}, & b'_{k1}, & \cdots, & b'_{kn} \end{bmatrix} \tag{30}$$

Using least squares method we obtain following adjusting mechanism:

$$\underline{\hat{\theta}}(k+1) = \underline{\hat{\theta}}(k) + \underline{\gamma}(k)\underline{p}(k)\underline{\psi}(k+1)\, e(k+1), \tag{31}$$

where

$\underline{\gamma}(k)$ and $\underline{p}(k)$ are defined in eq. (15) and (16). Solving eqns. (31,27, and 28) with respect to the compensating terms, the adjusting mechanism of the model reference adaptive system is obtained. If $\underline{\underline{A}}_p$ and $\underline{b}_p$ are slowly changing, $\underline{a}_k$ and b_k change in such a manner, that the output of the compensated process tracks the output of the reference model. So we can write the following equation

$$\underline{x}_p(k+1) = \underline{\underline{A}}_m \underline{x}_p(k) + \underline{b}_m v \tag{32}$$

From this equation follows the optimal control signal using eq. (21)

$$v^* = \left\{ -b_k^{-1} \left[R + (\underline{b}_m b_k^{-1})^T \underline{\underline{K}}(k+1)(\underline{b}_m b_k^{-1}) \right]^{-1} (\underline{b}_m b_k^{-1})\underline{\underline{K}}(k+1)(\underline{\underline{A}}_M - \underline{b}_m \underline{a}_k^T) - \right.$$
$$\left. - \underline{a}_k^T \right\} \underline{x}_p \tag{33}$$

where $\underline{\underline{K}}$ is the solution of Riccati type difference equation

$$\underline{\underline{K}}(k) = \underline{\underline{Q}} + (\underline{\underline{A}}_m - \underline{b}_m \underline{a}_k^T)^T \underline{\underline{K}}(k+1) \left[\underline{\underline{I}} - \underline{b}_m b_k^{-1}(R + (\underline{b}_m b_k^{-1})^T \underline{\underline{K}}(k+1)(\underline{b}_m b_k^{-1}))^{-1} \cdot \right.$$
$$\left. \cdot (\underline{b}_m b_k^{-1})^T K(k+1) \right] (\underline{\underline{A}}_m - \underline{b}_m \underline{a}_k^T) \tag{34}$$

The corresponding block diagram of the optimal model reference system is shown on Fig.2.

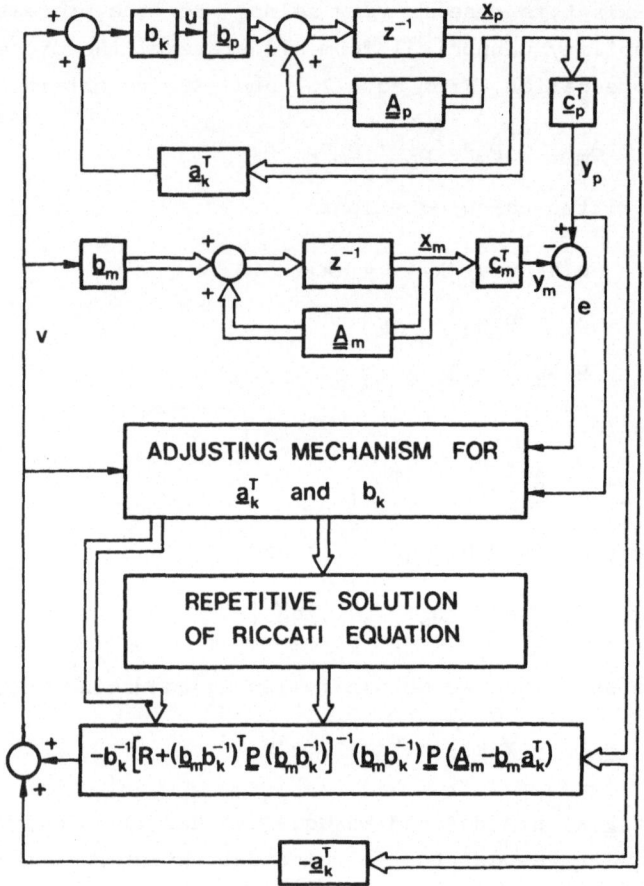

Fig.2.   The block diagram of the optimal model reference system

THE TRANSFORMATION OF OPTIMAL MODEL REFERENCE SYSTEMS TO THE PARAMETER
ADAPTIVE SYSTEMS

Figure 2 can be represented in a slightly different form shown in Fig.3.
By cancellation of $\underline{a}_k$ and b_k by $-\underline{a}_k$ and b_k^{-1} respectively, the system in
Fig. 4 is obtained. Defining equivalences

$$\underline{\underline{A}}_m - \underline{b}_m \underline{a}_k^T = \underline{\underline{A}}_e \qquad (35)$$

$$\underline{\underline{D}} = \underline{\underline{A}}_m \qquad (36)$$

and

$$\underline{b}_m b_k^{-1} = \underline{b}_e \qquad (37)$$

Figures 1 and 4 are identical. The Riccati difference equations (18) and
(34) and the equations for adjusting $\underline{a}_e$, $\underline{b}_e$ and $\underline{a}_k$, b_k respectively are

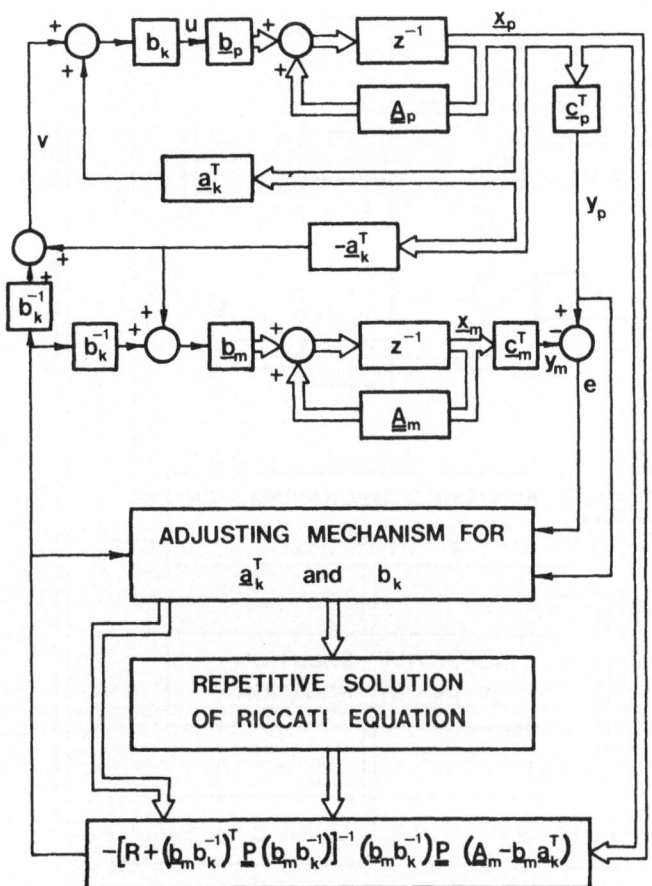

Fig.3. The corresponding modification of Fig.2

identical too. Every model reference system can be transformed into a parameter adaptive system, while the reverse transformation is possible only if b^*_k and $\underline{a}^*_k$ exist.

When the compensating terms $\underline{a}_k$ and b_k of the model reference systems reach their optimal values $\underline{a}^*_k$ and b^*_k, eqns. (35) and (37) can be rewritten in the form

$$\underline{\underline{A}}_m - \underline{b}_m \, \underline{a}^{T*}_k = \underline{\underline{A}}^*_e \tag{38}$$

and

$$\underline{b}_m \, b^{*-1}_k = \underline{b}^*_e \tag{39}$$

Comparing these expressions with the eqn. (23) and (24) we obtain

$$\underline{\underline{A}}^*_e = \underline{\underline{A}}_p \tag{40}$$

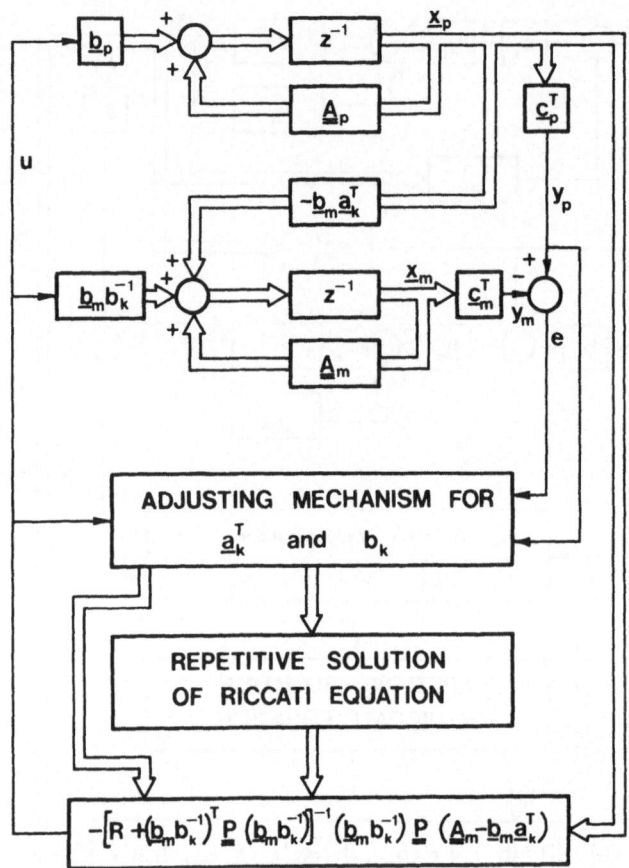

Fig. 4: The transformed block diagram of the optimal model reference system

and

$$\underline{b}^*_e = \underline{b}_p \qquad (41)$$

which means, that the parameter adaptive estimate matrix and vector reach the values of the process matrix and vector.

ADAPTIVE CONTROL OF PROCESSES WITH SOME UNACCESSIBLE STATES

The model reference approach described above represents a form of alge-braic (static) compensation controller. The estimation scheme represen-ted by eqn. (5) is shown in Fig. 5, where

$$G_{M1} = \frac{B_e(z^{-1})}{1+D(z^{-1})} \qquad (42)$$

and

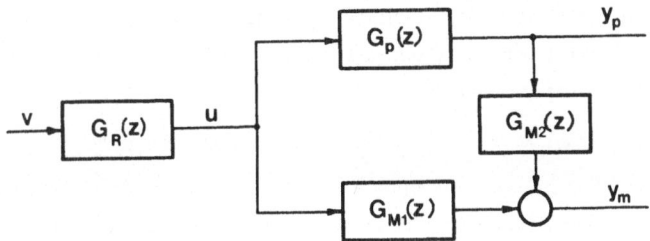

Fig. 5:   The estimation scheme

$$G_{M2} = \frac{D(z^{-1})-A_e(z^{-1})}{1+D(z^{-1})} \qquad (43)$$

If we generate the control signal in the folloving form

$$u = \frac{1+D(z^{-1})}{B_e(z^{-1})} \; G_M v \; - \; \frac{D(z^{-1})-A_e(z^{-1})}{B_e(z^{-1})} \; y_P \qquad (44)$$

the output of the estimation model

$$y_e = G_M v \qquad (45)$$

becomes the output of the reference model G_M. Eqn. (44) is realizable if
- the reference model has the same time delay as the process (d) and if
- the output of the process and its first d-1 predicted values are
  known.

In this case the measurability of all plant states is not required. Similar results in the case of continuous model reference systems are obtained using Kalman - Yacubovich lemma. The resulting controller is of compensating type and this fact claryfies the request for process zeros
to lie inside the unit circle. For the some reason the applicability of
such adaptive systems is limited to asymptotically stable and minimal
phase systems.

CONCLUSION

We can conclude, that the model reference systems do not include the identification procedure explicitly because the identification of the unknown
plant parameters is included implicitly in the minimization of the differences between the plant and model outputs. For optimal parameter adaptive systems the optimisation procedure is not necessary, because the lat-

ter is realized in the form of linear optimal regulator. This facts enables the transformation of optimal model reference systems into the optimal parameter adaptive systems.

REFERENCES

/1/ Landau, I.D., Unbiased recursive identification using model refe-
    rence adaptive techniques, IEEE Trans. Automat. Contr. vol. AC-21,
    pp. 194-202, Apr. 1976

/2/ Landau I.D., An Addendum to "Unbiased Recursive Identification
    Using Model Reference Adaptive Techniques", IEEE Trans. Automat
    Contr. vol. AC-23, pp. 97-99 February 1978

/3/ Larminat P., On overall stability of certain adaptive control sys-
    tems, Identification and system parameter identification, Fifth
    IFAC Symposium, Darmstadt, 1979

/4/ Dugard L., Landau I.D., Convergence analysis of M.R.A.S. schemes
    used for recursive identification, Identification and system pa-
    rameter identification, Fifth IFAC Symposium, Darmstadt, 1979

/5/ Matko D., Bremšak F., On the equivalence of parameter adaptive
    and model reference systems, Int. J. Control, vol. 30, pp. 203-211,
    No. 2

/6/ Isermann R., Digitale Regelsysteme, Springer Verlag, Berlin, 1977

MULTIVARIABLE SELF TUNING AUGMENTED REGULATOR

by

M.A. El-Bagoury and M.M. Bayoumi

Abstract

The self tuning regulator is applied to multivariable systems with unknown constant parameters. The disturbances acting on the system and the uncertainty about the parameter values prevent application of deterministic control strategies. Based upon the certainty equivalence principle, the parameters of the process model are estimated using the least squares method and then the estimated parameters are used in the control algorithm of the regulator. The control signal is then augmented with a random signal. The transient properties of this combination are considerably improved over the basic multivariable self tuning regulator (MVSTR). For systems with white noise disturbances, random signals with different characteristics are tested.

The results of these tests are compared against the basic MVSTR. It is evident from the investigation that an improvement of the response could be achieved. This is also reflected in the reduction of the performance criterion

1. Introduction

The self tuning regulator is a powerful, yet simple, approach to control processes with unknown constant parameters (Astrom et al 1977). The strategy is built upon the certainty equivalence principle. According to this principle, the control signal is derived by substituting the parameters estimates for the true parameters in the controller algorithm (Astrom and Wittenmark 1971, 1973). This strategy belongs to the passive learning class of adaptive control methods (Wittenmark 1975). On the other hand, an active learning (dual) control would try to minimize the parameters uncertainties. Computation of dual control signals is not practically feasible even for simple systems (Feldbaum 1960, 1961).

Previous attempts to approach the problem of designing a multi-variable self tuning regulator (MVSTRO)can be traced back to the work of Peterka and Astrom (1973). They converted the problem to the linear quadratic Gaussian (LQG) formulation with iterative solution of the discrete Riccatti equation in between the sampling instants. They noted that although their algorithm was not a dual control strategy in Feldbaum's sense, the computations were significantly smaller than that required to determine the optimal dual control strategy. However, it should be noted that the required computations are still prohibitive.

Borison (1975, 1979) and Keviczky et al (1977) adopted an approach to estimate the controller parameters instead of the process parameters using the method of least squares. This approach required a larger number of parameters than needed. Keviczky et al (1978) reported the application of the method to the control of raw material blending in a cement factory. Bayoumi and El-Bagoury (1979a,b) pointed out the disadvantages of identifying the controller parameters instead of the process parameters. Moreover, they considered a controller based on minimizing a quadratic function with penalties on the input magnitude and the expected output deviation from the reference vector.

In this study, the basic MVSTR will be reviewed in section 2, while the augmeted control strategy is presented and discussed in section 3. Simulation examples will then follow in section 4. Finally, a summary of the results is given in section 5.

2.   Problem Statement

Consider the general representation of a multivariable process:

$$A(z^{-1}) \ y_t \ = \ B(z^{-1}) \ u_{t-d} \ + \ C(z^{-1}) \ w_t \tag{1}$$

where

t	is the time given as an integer representing the number of sampling periods
d	is the process time delay
u_t	is the q-vector input at time t
y_t	is the q-vector output at time t
w_t	is a q-vector of white Gaussian noise with zero mean and a diagonal covariance matrix Λ.

$A(z^{-1})$, $B(z^{-1})$ and $C(z^{-1})$ are (q × q) polynomial matrices in the backward shift operator (z^{-1}), which can be written in the general form as:

$$A(z^{-1}) \ = \ I_q \ + \ \sum_{j=1}^{n} \ A_j \ z^{-j}$$

$$B(z^{-1}) \ = \ B_o \ + \ \sum_{j=1}^{m} \ B_j \ z^{-j}$$

$$C(z^{-1}) \ = \ I_q \ + \ \sum_{j=1}^{k} \ C_j \ z^{-j}$$

It is assumed that the degrees m and n of the process model are known from an independent investigation (Bayoumi et al 1980). The case treated here is when $C(z^{-1}) = I_q$: i.e. the case of additive white noise.

It is required to estimate the unknown process parameters of $A(z^{-1})$ and $B(z^{-1})$. These estimates are then used in the controller algorithm that minimizes a quadratic cost function of the inputs and outputs. The main objective for this scheme is output regulation as close as possible in the mean square sense to the reference vector y_r representing the set point vector.

The system equation can be rewritten in the form:

$$y_t \ = \ B_o \ u_{t-d} \ + \ \dots \ + \ B_m \ u_{t-d-m} \ - \ A_1 \ y_{t-1} \ - \ \dots \ - \ A_n \ y_{t-n} \ + \ w_t$$

$$= \ P \ x_{t-1} \ + \ w_t \tag{2}$$

where

$$P = [B_o, B_1, \ldots, B_m, -A_1, -A_2, \ldots, -A_n]$$

$$x_{t-1} = [u_{t-d}^T, u_{t-d-1}^T, \ldots, u_{t-d-m}^T, y_{t-1}^T, y_{t-2}^T, \ldots, y_{t-n}^T]^T$$

Consider the model

$$\hat{y}_j = \hat{P}_t \, x_{j-1} \qquad\qquad j = 1, 2, \ldots, t \qquad\qquad (3)$$

where the matrix $\hat{P}_t$ is the estimate of the parameter matrix P based on information obtained from the input-output record up to time t. $\hat{P}_t$ is sought to minimize the sum of squares of the residuals. The least squares estimate $\hat{P}_t$ may be written in the useful recursive form:

$$\hat{P}_t = \hat{P}_{t-1} + (y_t - \hat{P}_{t-1} \, x_{t-1}) \, x_{t-1}^T \, R_t \qquad\qquad (4)$$

$$R_t = R_{t-1} - \frac{R_{t-1} \, x_{t-1} \, x_{t-1}^T \, R_{t-1}}{1 + x_{t-1}^T \, R_{t-1} \, x_{t-1}} \qquad\qquad (5)$$

This procedure was explained in detail in several location, e.g., Young (1974) and Bayoumi and El-Bagoury (1976 b).

Equations (4) and (5) represent the least squares estimation procedure performed at each sampling step t to improve the estimates of the parameter matrix. The estimates converge asymptoticaly to the true parameters due to the assumption of white noise disturbances.

In the controller stage, the objective of the controller is to minimize the quadratic cost function

$$L_t = \| E(y_{t+d}|t) - y_r \|_{Q_1}^2 + \| u_t \|_{Q_2}^2 \qquad\qquad (6)$$

where Q_1 is a positive semidefinite symmetric matrix and Q_2 is positive definite and symmetric. The control effort and the estimated output deviation from the reference vector are penalized.

The expected output vector at t+d is expressed as:

$$E(y_{t+d}|t) = \hat{y}_{t+d|t} = \hat{P}_t \, \hat{x}_{t+d-1} \qquad\qquad (7)$$

$$= \hat{B}_o \, u_t + \hat{B}_1 \, u_{t-1} + \ldots + \hat{B}_m \, u_{t-m} - \hat{A}_1 \, \hat{y}_{t-1+d|t} - \ldots$$

$$- \hat{A}_{d-1} \, \hat{y}_{t+1|t} - \hat{A}_d \, y_t - \ldots - \hat{A}_n \, y_{t-n+d} \qquad\qquad (8)$$

Thus the estimated output deviation from the reference vector y_r is given by:

$$\hat{y}_{t+d|t} - y_r = \hat{B}_o u_t + \hat{C}_t \qquad (9)$$

where

$$\hat{C}_t = \sum_{j=1}^{m} \hat{B}_i u_{t-i} - \sum_{j=1}^{d-1} \hat{A}_j \hat{y}_{t-j+d|t} - \sum_{k=d}^{n} \hat{A}_k y_{t-k+d} - y_r \qquad (10)$$

Substituting (9) into (6) we get

$$L_t = \| \hat{B}_o u_t + \hat{C}_t \|^2_{Q_1} + \| u_t \|^2_{Q_2} \qquad (11)$$

Therefore, the minimum of L_t is reached for

$$u_t^o = -(\hat{B}_o^T Q_1 \hat{B}_o + Q_2)^{-1} \hat{B}_o^T Q_1 \hat{C}_t \qquad (12)$$

The optimal control law given by equation (12) can be easily obtained at each step as follows: estimate the next outputs at t+1, t+2, ..., t+d using the input-output record up to time t by applying the model equation (3). The estimated output at t+1 is used to estimate the output at t+2, and so on. Finally $\hat{C}_t$ is obtained by the relationship (10) which can also be interpreted as

$$\hat{C}_t = \{ \hat{y}_{t+d|t} - y_r \}_{u_t = 0} \qquad (13)$$

$\hat{C}_t$ is then used in (12) to obtain u_t^o in a straightforward manner. This simple algorithm uses less computer storage than other algorithms (Borison 1975), (Keviczky et al 1977) which require larger dimensions controller parameter matrices.

The matrix Q_2, being positive definite, ensures the existence of a solution to (12) for all estimates $\hat{B}_o$. However, a large Q_2 might lead to instability and hence it should be carefully dealt with.

3.   The Augmented Control Signal

The control signal applied at time t will affect the output d-steps ahead at time t+d. The objective of the control signal calculated by equation (12) is to minimize the weighted sum of squares of the expected output deviation and the control effort. During the time delay period, the noise acting on the system at t+1, t+2, ..., t+d will disturb that aim. On the other hand, the parameter estimates that are used in the derivation of the optimal control signal would have to be close to the true values in order that such a control may be effective.

Consider the one-step prediction relationship:

$$\hat{y}_{t+1|t} = \hat{P}_t \, x_t$$

whereas the actual output at t+1 would take the form

$$y_{t+1} = P \, x_t + w_{t+1}$$

The prediction error at t+1 is then defined as

$$\tilde{y}_{t+1} = y_{t+1} - \hat{y}_{t+1|t}$$
$$= (P - \hat{P}_t) \, x_t + w_{t+1} \tag{14}$$

Define the model error $\S_{t+1}$ as

$$\S_{t+1} = (P - \hat{P}_t) \, x_t \tag{15}$$

Examining equation (14), the d-step ahead prediction error can be expressed as:

$$\tilde{y}_{t+d} = y_{t+d} - \hat{y}_{t+d|t}$$
$$= P \, x_{t+d-1} + w_{t+d} - \hat{P}_t \, \hat{x}_{t+d-1}$$
$$= (P-\hat{P}_t) \, x_{t+d-1} + w_{t+d} + \hat{P}_t (x_{t+d-1} - \hat{x}_{t+d-1})$$
$$= \S_{t+d} + w_{t+d} + \sum_{j=1}^{d-1} (-A_j) \, \tilde{y}_{t+d-j}$$

Therefore,

$$(I + A_1 z^{-1} + \ldots + A_{d-1} z^{1-d}) \, \tilde{y}_{t+d} = \S_{t+d} + w_{t+d} \tag{16}$$

We assume that the estimate $\hat{P}_t$ has almost converged to the true value P such that the prediction error $\S_{t+d}$ is negligible in comparison with the noise w_{t+d}. Under this assumption, the prediction output error is related to the disturbance by:

$$(I + A_1 z^{-1} + \ldots + A_{d-1} z^{1-d}) \tilde{y}_{t+d} = w_{t+d} \tag{17}$$

It can be easily shown that (17) can be written in the form:

$$\tilde{y}_{t+d} = (F_o + F_1 z^{-1} + \ldots + F_{d-1} z^{1-d}) w_{t+d} = F(z^{-1}) w_{t+d} \tag{18}$$

where $F(z^{-1})$ is a polynomial matrix of degree (d-1) satisfying the identity.

$$I = A(z^{-1}) F(z^{-1}) + z^{-d} G(z^{-1}) \tag{19}$$

and $G(z^{-1})$ is a polynomial matrix of degree (n-1). Thus we have established the equivalence of this simple step-by-step approach to the one-time d-steps ahead controller predictor of Borison (1979) for the case of white noise disturbance. The case of correlated noise disturbance is currently under investigation to overcome the bias in the least squares estimate of the parameter matrix.

Consider the alternative cost function given by:

$$L'_t = \| y_{t+d} - y_r \|^2_{Q_1} + \| u_t \|^2_{Q_2} \tag{20}$$

$$= \| \hat{y}_{t+d|t} - y_r + \tilde{y}_{t+d} \|^2_{Q_1} + \| u_t \|^2_{Q_2}$$

$$= \| \hat{B}_o u_t + \hat{C}_t + \tilde{y}_{t+d} \|^2_{Q_1} + \| u_t \|^2_{Q_2} \tag{21}$$

Minimizing the new cost function L'_t means that we want to minimize the difference between the future output y_{t+d} (rather than its expected value $\tilde{y}_{t+d|t}$) and the reference vector y_r. In this case, the optimal control policy would be:

$$u_t^{op} = -(\hat{B}_o^T Q_1 \hat{B}_o + Q_2)^{-1} \hat{B}_o^T Q_1 (\hat{C}_t + \tilde{y}_{t+d}) \tag{22}$$

It should be noted however, that the quantity $\tilde{y}_{t+d}$ is totally unknown. Therefore, a suboptimal control policy is proposed. This new policy assigns

$$u_t = u_t^* + v_t^* \tag{23}$$

where

$$u_t^* = -(\hat{B}_o^T Q_1 \hat{B}_o + Q_2)^{-1} \hat{B}_o^T Q_1 \hat{C}_t$$

and v_t^* is a random vector to be characterized.

The idea of augmenting the control u_t^* with the random signal v_t^* is to counteract, at least partially, the effects of the prediction and estimation errors. On using the method of least squares, the estimation error is large in the start, but it goes asymptotically to zero. On the other hand, the properties of the prediction error is closely related to the properties of the disturbance noise. Different random signals v_t^* are investigated in the following examples in an attempt to characterize the desirable properties of v_t^*.

4.   Examples

These examples were simulation runs on a digital computer for 500
steps.   Each case is explained then the results are compared.   Consider
the system given by:

$$A(z^{-1}) \ y_t = B(z^{-1}) \ u_{t-d} + w_t$$

where

$$A(z^{-1}) = \begin{bmatrix} 1 & 0 \\ 0 & 1 \end{bmatrix} + \begin{bmatrix} -1.5 & 0.3 \\ 0.2 & -1.5 \end{bmatrix} z^{-1} + \begin{bmatrix} 0.54 & -0.1 \\ 0.1 & 0.56 \end{bmatrix} z^{-2}$$

$$B(z^{-1}) = \begin{bmatrix} 2 & -0.3 \\ 0.1 & 1 \end{bmatrix} + \begin{bmatrix} -1.8 & 0.2 \\ -0.2 & -0.2 \end{bmatrix} z^{-1}$$

the time delay d=2, and w_t is a random disturbance assumed to be white
Gaussian with zero mean and covariance $(0.05)^2 I_2$ where I_2 is the identity
matrix.   It is required to regulate the output y_t as close as possible
to the reference output y_r by applying the methods of sections 2 and
3 (namely MVSTR and MVSTAR).   The weighting matrices Q_1 and Q_2 were
chosen as:

$$Q_1 = I_2, \ Q_2 = 0.1 \ I_2.$$

The reference vector is $y_r = [1 \quad 1]^T$.
Eight cases were considered.   In each case, recursive least squares
estimation is used to identify the elements of the parameter matrix
$\hat{P}_t$ starting arbitrarily at the zero point.   The same noise sequence w_t
was used in all cases.   The online control signal was computed differently
in each example.

Case 1:   This is the reference case, where it is assumed that the process parameters matrix P is known to the controller.   The control signal provides the minimum variance strategy.   Identification task was
performed as a by-product of the program.   In this case the control signal was independent of the identified parameters.

Case 2:   This is the basic MVSTR algorithm of section 2.   The parameter
estimation matrix is used by the controller to derive the control signal
according to equation (12).   In this case and the following ones, identification and control operate in a closed loop.

Case 3: The control is derived from the parameter estimates and the auxiliary component v* is added. The signal v* is taken from a uniform distribution in the range $[-\frac{a}{2}, \frac{a}{2}]$. The magnitude a is chosen to make the variance of the components of $\tilde{v}*$ the same as the noise variance, $\sigma_{v*} = 0.05$.

Case 4: This case is similar to case 3 with the further condition that the magnitude of v* is reduced as the output vector approaches the reference. The scale factor multiplying $v*_t$ is proportional to $\varepsilon^2_y = \|y_t - y_r\|^2$ if ε^2_y is less than a certain value. Thus, the control signal u* is augmented by v* whose magnitude decreases with better regulation.

Cases 5 and 6: The idea is similar to that of cases 3 and 4 above except that v* is taken from a white Gaussian distribution with zero mean and variance $\sigma^2 = (0.05)^2$. Case 6 is characterized by scaling down v* with $\varepsilon^2_y = \|y_t - y_r\|^2$ if it becomes less than σ_{v*}.

Cases 7 and 8: These are similar to cases 5 and 6 except that components of v* are correlated. The signal v* is taken from a random sequence that has the same mean and covariance as the quantity $(B_o^T Q_1 B_o + Q_2)^{-1} B_o^T Q_1 \tilde{Y}_{t+2}$ appearing on the rhs of equation (22). In this case:

$$E[v*] = 0$$

$$E[v* \ v*^T] = \begin{bmatrix} 1.834 \text{ E-3} & -1.311 \text{ E-5} \\ -1.311 \text{ E-5} & 6.799 \text{ E-3} \end{bmatrix}$$

Case 8 is again characterized by scaling v* down as the quantity $\|y_t - y_r\|^2$ becomes less than σ_{v*}.

5. Summary of the Results

Table 1 shows the results of the simulation runs. The row labelled "Transients" lists the extremes of the input and output components during the transient stage starting at the zero point. Of course the least transients occured with case 1 when the parameters were known. We also note that case 1 gives the minimum variance in the output.

The rows labelled "Cost" denote the accumulated cost over the 500 steps (Final) and the last 100 steps. The closest cost of the last 100 steps to case 1 occurs with case 4.

The mean and standard deviation of the outputs are shown for steps 41-500 which excludes the initial transient effects. These rows show that case 4 is next to case 2.

Upon examining the results of these experiments, it is evident that augmenting the control signal by the additional signal v* improved the transient considerably. The peak excursions of inputs and outputs have been reduced. This was also reflected in the final cost. On the other hand, reducing the additional signal v* as was performed in cases 4,6 and 8 have resulted in a lower steady state variance of the output vector and helped keep closer control on the system. It is thus concluded that adding the signal v* to the control signal of the self-tuning regulator tends to improve the response of the system in terms of transient and steady. The optimal character of the signal v* still needs further investigation.

Acknowledgement

This work was partly supported by the National Science and Engineering Research Council of Canada under Grant 7509.

References

1. Astrom, K.J. and Wittenmark, B. (1971). "Problems of identification and control". J. Math. Anal. & Appl. Vol. 34, pp. 90-113.

2. Astrom, K.J. and Wittenmark, B. (1973). "On self-tuning regulators". Automatica, Vol. 9, pp. 185-199.

3. Astrom, K.J., Borisson, U., Ljung, L., and Wittenmark, B. (1977). Theory and applications of self tuning regulators". Automatica Vol. 13, pp. 457-476.

4. Bayoumi, M. and El-Bagoury, M. (1979a)." Comments on 'Self tuning adaptive control of cement raw material blending". Automatica Vol. 15, pp. 693-694.

5. Bayoumi, M. and El-Bagoury,M.(1979b)"A self tuning regulator for multivariable systems". Queen's University, research report.

6. Bayoumi, M., El-Bagoury, M., and Wong, K.Y. (1980). "A Self Tuning Regulator for Multivariable Systems", Submitted to Automatica.

7. Borisson, U. (1975). "Self tuning regulators: industrial applications and multivariable systems". Lund report-7513.

8. Borisson, U. (1979). "Self tuning regulators for a class of multivariable systems". Automatica, Vol. 15, pp. 209-215.

9. Feldbaum, A.A. (1960, 1961). "Dual control theory - Parts 1-4". Avt i Tel. (Translated), Vol. 21, pp. 1240-1249, 1453-1464 and Vol. 22, pp. 3-16, 129-142. Also in Optimal and self optimizing control, Editor, R. Oldenburger The MIT press 1966, pp. 458-495.

10. Keviczky, L. and J. Hetthessy (1977). "Self Tuning minimum variance control of MIMO discrete time systems". Aut. Control Theory and Applications, Vol. 5 (1), pp. 11-17.

11. Keviczky, L., J. Hetthessy, M. Hilger, J. Kolostori (1978). "Self tuning adaptive control of cement raw material blending". Automatica, Vol. 14, pp. 525-532.

12. Peterka, V., K.J. Astrom (1973). "Control of multivariable systems with unknown but constant parameters." Preprints of the IFAC Congress on Systems Identification and Parameter Estimation, The Hague, Netherlands.

13. Wittenmark, B. (1975). "Stochastic Adaptive Control Methods: A survey",Int. J. Control Vol. 21 (5) pp. 705-730.

14. Young, P. (1974). "Recursive approaches to time-series analysis. Bulletin of Inst. of Math. and its Applications ,Vol. 10, pp. 209-224.

Case		1	2	3	4	5	6	7	8
v*	Type			UNIFORM		WHITE		COLOURED	
	Variance	–	–	con-stant	vari-able	con-stant	vari-able	con-stant	vari-able
TRANSIENTS	Min	-0.53	-13.77	-0.94	-0.94	-5.72	-5.72	-11.5	-11.5
	Max	1.51	16.93	3.75	3.75	5.83	5.83	3.52	3.52
COST	Final	141.4	1103.2	185.8	179.6	238.7	234.3	404.8	399.7
	last 100	25.998	25.948	27.524	25.955	27.769	26.354	27.875	26.419
MEAN	y_1	.935	.924	.929	.929	.934	.935	.923	.925
	y_2	.921	.917	.920	.917	.920	.919	.919	.917
S.D.	y_1	.0838	.0852	.1290	.0867	.1290	.0990	.1200	.0973
	y_2	.0854	.0862	.1010	.0868	.1000	.0912	.1190	.1000

Table 1. Results of Simulations

UNCONDITIONAL STABILIZERS FOR NONMINIMUM PHASE SYSTEMS

Ph. de Larminat

Laboratoire d'Automatique de l'E.N.S.M.

(Equipe de Recherche associée au C.N.R.S.)

1, rue de la Noë - 44072 Nantes Cedex - FRANCE

Summary :

The overall stability of certain self-adaptive systems has not been demonstrated
until now, except in a few particular cases, generally limited to monovariable and/or
minimum-phase systems. It is shown here that a large class of identification methods
can be associated with a very large class of control methods in order to perform the
unconditional stabilization of deterministic linear systems.

1 - INTRODUCTION

This paper concerns unconditional stabilization of determinist discrete linear systems by self-adaptive control techniques, which theoretically require no prior knowledge of the system to be controlled.

The principle itself of real-time identification, in association with a control
method, is a basic notion which has been implemented many times with more or less
success. Certain difficulties relate to the principle itself of controlling a system
about which there is no prior information. Other difficulties relate to shortcomings
of a theoretical sort which lead to incorrect algorithms.

The first theoretical studies concerning stability were carried out on systems
associating relatively simple identification methods, of the least squares or stochastic gradient type, with equally simple control methods, of the perfect model-tracking
type. With such systems the transfer from the model to the control is relatively
elementary, at most a simple retranscription of the coefficients. It even sometimes
happens that the identification function is not explicitly apparent, what is referred
to as "direct" adaptive control, although that may well be an artificial distinction.

It should be noted that perfect tracking techniques impose the very severe constraint
of minimum phase on the systems to be controlled. Even within the limited scope
referred to above, theory has been at a standstill for years. In fact, the hyperstability, or Lyapunov theories applied by various authors do not go beyond the following
reasoning : if a certain state is bounded, it converges toward zero, then it is bounded !

The first study indicating the feasibility of a real demonstration of overall stability was presented by de Larminat in 1975 [1], but went completely unnoticed at the
time. The results of this study were taken up again at the EDF-IRIA summer school in
July 1978 [2], then extended and systematized in July and October 1979 [3] [4]. In
the meantime, Goodwin [5], followed by many other authors [6, 7, 8], obtained similar
results.

It should be noted that these demonstrations of overall stability concern algorithms
(sometimes unnecessarily complicated) associating particular identification methods
with equally particular control methods, limited, with some exceptions[9], to monovariable and minimum-phase systems.

The present paper represents a considerable advance, making most previous studies
obsolete. In effect, it defines a few sufficient conditions in which the closed-loop
system is asymptotically stable. These conditions are satisfied by most ordinary
recursive identification methods, for instance standard least squares or exponential
forgetting least squares. They are also satisfied by most classical control methods,
such as model following, pole placement, quadratic optimization, etc..., with no
particular restriction to the minimum-phase case. Moreover, the demonstration made
here applying to the monovariable case may be easily extended to the multivariable
case.

Finally, subject to satisfying a few elementary conditions, it may be said that any
identification method associated with any control method constitutes what can be
called an "unconditional stabilizer", that is, a feedback loop capable of stabilizing
any linear system, with no conditions as to the value of the parameters of that system.

2 - HYPOTHESES ABOUT THE PROCESSES TO BE CONTROLLED

Given a process with input u_k, output y_k ($k = 1,2,...$), introduce the linearity hypothesis :

HL : There is a parameter vector $\theta^T = [a_1...a_n \; b_1...b_n]$
bounded, with, $\forall k$:

$$y_k = a_1 y_{k-1} + ... + a_n y_{k-n} + b_1 u_{k-1} + ... + b_n u_{k-n} \qquad (1)$$

Equivalent notation will be used :

$$A(z) \; y_k = B(z) \; u_k$$

where : $A(z) = 1 - a_1 z^{-1} - ... - a_n z^{-n}$

$B(z) = b_1 z^{-1} + ... + b_n z^{-n}$

with z^{-1} = backward shift operator.

Defines $\qquad x_k^T = [y_{k-1} \cdots y_{k-n} \; u_{k-1} \cdots u_{k-n}]$

which then gives

$$y_k = \theta^T x_k$$

(1) can also be written in the state form :

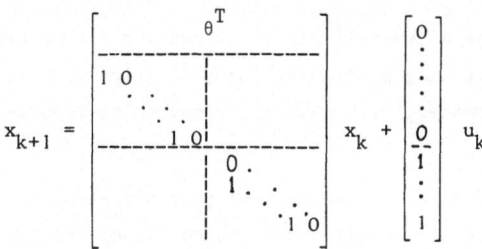

that can be written as :

$$x_{k+1} = F \, x_k + g \, u_k$$

The process defined above may be stable or not, minimum phase or not, unit time lag or not ($b_1, b_2 \ldots$ can be nil). Moreover, the formulation may be extended immediately to the multivariable case by replacing $A(z)$ and $B(z)$ with matrix polynomials.

The stabilizability hypothesis only is needed :

HS : The process is stabilizable, which may be expressed in the same way by saying that :
- The noncontrollable part of (F, g) is asymptotically stable.
- The common zeros of $A(z)$ and $B(z)$ are strictly inside the unit circle.

3 - HYPOTHESIS ABOUT THE IDENTIFICATION METHOD

Consider an identification algorithm I :

$$(u_1 \; y_1 \; \cdots \; u_{k-1} \; y_{k-1}) \xmapsto{\quad I \quad} \hat{\theta}_k$$

$\hat{\theta}_k$ is an estimation of θ on the basis of data until the time $k-1$.

The quantity e_k defined below is thus a *prediction* error :

$$e_k = y_k - \hat{\theta}_k^T x_k$$

The following hypotheses about I are introduced :

HI1 : $\hat{\theta}_k$ remains bounded $\forall k$

HI2 : $\lim_{k \to \infty} (\hat{\theta}_{k+1} - \hat{\theta}_k) = 0$

HI3 : There is a vector $\tilde{\theta}_k$ and a scalar v_k, bounded $\forall k$, with : $e_k = \tilde{\theta}_k^T x_k + v_k$

and with : $\lim_{k \to \infty} (\tilde{\theta}_k, v_k) = (0,0)$

Comments :

- $\tilde{\theta}_k$ is not defined by $\theta - \hat{\theta}_k$

- It is not assumed that $\theta - \hat{\theta}_k$ tends toward zero, nor toward anything else.

- It is not assumed that x_k is bounded.

- It is thus not assumed that e_k tends toward zero, nor even that e_k is bounded.

Existence Theorem :

There is a set $\mathcal{J}$ of identification methods I, such as :

$$HL \implies HI1, \quad HI2, \quad HI3$$

This is particularly true of all algorithms of the following type :

$$\hat{\theta}_{k+1} = \hat{\theta}_k + \Lambda_k x_k R_k^{-1}(y_k - \hat{\theta}_k^T x_k)$$

with $\quad \Lambda_k^{-1} = P_k^{-1} + x_k R_k^{-1} x_k^T$

$\quad\quad P_{k+1} = \Lambda_k + Q_k$

where : $\quad \hat{\theta}_0 P_0^{-1} \hat{\theta}_0$ is bounded,

$\quad\quad R_k$ is any bounded positive sequence,

$\quad\quad Q_k$ is any sequence of defined positive matrices, chosen so that P_k remains

bounded.

The demonstration is given in [3] and [4].

Furthermore, the following hypothesis is defined :

$\widehat{HS}$: when $k \to \infty$, the zeros common to $\hat{A}_k$, $\hat{B}_k$ are strictly interior to the unit

circle.

Comments :

HS does not imply $\widehat{HS}$:

Consider the following counter-example :

Process : $\quad y_k = a\, y_{k-1} + b\, u_{k-1}$

(HS is verified if $b \neq 0$ when $|a| > 1$)

Suppose a control $u_k = p\, y_k$

Then $y_k = (a + bp)\, y_{k-1}$

It is thus possible that the identification algorithm gives

$$\hat{a} = (a + bp)$$

$$\hat{b} = 0$$

and if $|a + bp| > 1$, $\hat{HS}$ will not be verified.

It will then be necessary to go back over the conditions relating to the verification of $\hat{HS}$.

4 - HYPOTHESES ABOUT THE CONTROL METHOD

Consider a state feedback control :

$$u_k = r_k^T x_k$$

which may also be written

$$Q_k(z) u_k = P_k(z) y_k$$

with

$$Q_k(z) = [1 - q_{1k} z^{-1} - \ldots - q_{nk} z^{-n}]$$

$$P_k(z) = [\quad p_{1k} z^{-1} + \ldots + p_{nk} z^{-n}]$$

$$r_k^T = [p_{1k} \ldots p_{nk} q_{1k} \ldots q_{nk}]$$

A more general expression would be :

$$u_k = r_k^T x_k + p_{ok} y_k$$

HI3 permits writing

$$u_k = [r_k + p_{ok} \hat{\theta}_k^T] x_k + p_{ok} \tilde{\theta}_k^T x_k + p_{ok} v_k$$

It will be admit here that, if $p_{ok} \neq 0$, all the further demonstrations may be extended because of the convergence of $\tilde{\theta}_k$ and v_k toward zero.

Let C be an algorithm for calculating r_k :

$$(\hat{\theta}_1 \ldots \hat{\theta}_k) \overset{C}{\longmapsto\!\!\!\longrightarrow} r_k$$

Comments :

r_k could have been defined as a function of $\hat{\theta}_k$ alone. However, imperatives relative to the time available for calculation may require working out r_k from $\hat{\theta}_{k-i}$. More generally, r_k may be a function of the whole $\hat{\theta}_k$ sequence, as in the example below :

Suppose a first order process :

$$y_{k+1} = a y_k + b u_k \qquad (b \neq 0)$$

Define the optimal control so that $y_{k+1} = 0$, then :

$$u_k = -p y_k, \quad \text{with } p = a/b.$$

In the adaptive control loop, (a/b) may be replaced by the identified parameters :
$p_k = \hat{a}_k / \hat{b}_k$.

An other way is to define :

$$p_k = p_{k-1} + \frac{\hat{b}_k}{\hat{b}_k^2 + 1} (\hat{a}_k - \hat{b}_k \, p_{k-1})$$

Then, p_k is now a function of the whole sequence $\{\hat{a}_i, \hat{b}_i\}$, but one can observe that :

- p_k does not becomes infinite when $\hat{b}_k = 0$

- If $\hat{a}_{k+1} - \hat{a}_k \to 0$, $\hat{b}_{k+1} - \hat{b}_k \to 0$, with $\hat{b}_k \neq 0$, then $p_k \to \hat{a}_k / \hat{b}_k$

Generally, the conditions sought concerning C are those which make it possible to ensure that the system

$$x_{k+1} = [F + g \, r_k] \, x_k$$

is asymptotically stable.

Define the matrix $\hat{F}_k$, from the identified model :

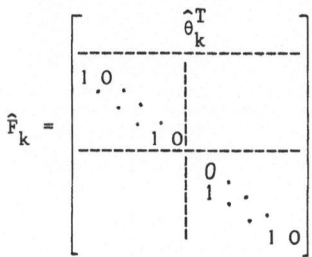

Define the following hypotheses concerning C :

HC1 : r_k remains bounded

HC2 : $\lim\limits_{k \to \infty} [r_{k+1} - r_k] = 0$

HC3 : $[\hat{F}_k + g \, r_k^T]$ has all its eigenvalues strictly interior to the unit circle, when $k \to \infty$.

Existence Theorem

There is a set $\mathcal{C}$ of control methods C, such that :

$$\text{HI1, HI2, HI3, } \hat{HS} \Longrightarrow \text{HC1, HC2, HC3}$$

Generally, all control methods designed for state feedback stabilization of a stabilizable linear system meet the requirements of the above hypothesis.

The hypothesis of the final stabilizability of the model ($\hat{HS}$) is obviously necessary to establish HC3. Certain methods may even require that $(\hat{A}_k, \hat{B}_k)$ have <u>never</u> common zeros outside of the unit circle, at the risk of not satisfying HC1 (cf. the example of the first-order system, given the definition $p_k = \hat{a}_k/\hat{b}_k$). Nevertheless, this same example demonstrates that certain adjustments make it possible to remain limited to $\hat{HS}$.

5 - MAIN THEOREM

If HL, $\hat{HS}$, Ie$\mathcal{J}$, ce$\mathcal{C}$,

the system $\quad x_{k+1} = (F + g\, r_k)\, x_k$

is asymptotically stable.

Demonstration :

Recall that $\quad y_k = \hat{\theta}_k^T x_k + e_k$

$$e_k = \tilde{\theta}_k^T x_k + v_k$$

$$x_k = [y_{k-1} \cdots y_{k-n} \; u_{k-1} \cdots u_{k-n}]$$

$$u_k = r_k^T x_k$$

which thus gives $\quad x_{k+1} = \Phi_k x_k + h\, v_k$

whith

$$\Phi_k = \begin{bmatrix} \tilde{\theta}_k^T + \hat{\theta}_k^T \\ \hline \begin{matrix} 1 & 0 & \\ & \ddots & \\ & & 1 & 0 \end{matrix} \\ \hline r_k \\ \hline \begin{matrix} 1 & 0 & \\ & \ddots & \\ & & 1 & 0 \end{matrix} \end{bmatrix} \qquad h = \begin{bmatrix} 1 \\ \vdots \\ 0 \\ \hline 0 \\ \hline \vdots \\ 0 \end{bmatrix}$$

From the previous set of hypotheses, it is straightforward that :

C1 : Φ_k remains bounded

C2 : $\displaystyle\lim_{k \to \infty} [\Phi_{k+1} - \Phi_k] = 0$

C3 : when $k \to \infty$, $\Phi_k \to [F_k + g\, r_k]$, then all the eigen values of Φ_k are strictly inside the unit circle.

As a result of C1 and C3, $\mu < 1$, K and N bounded exist, so that

$$||\Phi_k^N|| \leqslant \mu, \text{ for } k \geqslant K$$

Consider then the expression of x_{k+N} :

$$x_{k+N} = [\Phi_{k+N-1} \times \ldots \times \Phi_{k+1} \times \Phi_k]\, x_k + [\Phi_{k+N-1} \times \ldots \times \Phi_{k+1}]\, h\; v_k + \ldots + h\; v_{k+N-1}$$

N being bounded gives

$$[\Phi_{k+N-1} \times \ldots \times \Phi_k] - \Phi_k^N \to 0 \quad, \quad \text{because of C2}$$

$$[\Phi_{k+N-1} \times \ldots \times \Phi_{k+i}] \quad \text{is bounded because of C1}$$

Moreover, recall that $h \, v_k \to 0$

Finally, it is clear that $\nu < 1$ and K' may be found bounded, so that

$$||x_{k+N}|| \leqslant \nu ||x_k||, \quad \text{for } k \geqslant K'$$

Therefore, x_k tends exponentially towards zero.

6 - COMMENTS ON $\hat{HS}$

Finally, the only hypothesis that remains to be verified before inferring the overall stability of the system is the hypothesis $\hat{HS}$.

Certain authors sometimes suggest that, if unstable modes appear, the resulting data permit identification and, subsequently, the stabilization of these modes.

This reasoning is erroneous.   Let us return to the example of the first order $bz^{-1}/(1-az^{-1})$, with $|a| > 1$, $b \neq 0$.

Assume that when $\hat{a}_k = a$, $\hat{b}_k = 0$ have been identified, the control is $u_k = 0$, in order to minimize $\Sigma \, u_k^2$, since it seems not possible to minimize $\Sigma \, y_k^2$, because $\hat{b}_k = 0$.

Then, the prediction error $e_k = y_k - \hat{a}_k y_{k-1}$ remains nul, and the situation will show no development.

To remedy this, a modification of the identification algorithm may be proposed, by the addition of an arbitrary variation δ_θ, when the identified model is not stabilizable.

Nevertheless, it should be noted that
1) A test of nonstabilizability is not always easy to define and to implement.
2) The practical conditions entering into the choice of δ_θ must be carefully determined. In the above example, consider :
$$b_{k+1} = \delta_b, \quad \text{when } \hat{b}_k = 0, \quad \text{with } \delta_b > 0.$$
If the unknown parameter b is negative, would there not be a risk of seeing the further estimations of b become nil again and the problem reappear ?
3) More generally speaking, it is necessary to make sure that the modifications introduced do not destroy the properties HI1, HI2 and HI3.

Despite these observations, it may be *conjectured* that there are means of imposing $\hat{HS}$ by direct action on the identification algorithm.

A better solution consists in the choice of a control algorithm C, such that one can have $(r_{k+1} - r_k) \to 0$ *only if* the identified model is stabilizable (and many C algorithms can have this property naturally).

Then, when $r_{k+1} \neq r_k$, the loop is no longer closed by means of a *stationary* linear feedback, and it may be *conjectured* that the identified model will vary, until it becomes stabilizable.

In any event, it should be kept in mind that if the process is stabilizable, there is no reason that a nonstabilizable model should be a point of attraction for the identification algorithm. The question concerning hypothesis $\hat{HS}$, though delicate to deal theoretically does not seem then to be crucial, provided that a minimum of common sense precautions is exercised, such as constraints on the control u_k and on the variations of the coefficients r_k.

7 - <u>CONCLUSION</u>

For a considerable time, and sometimes with success, practitioners have associated real-time identification methods with control methods. The present paper offers a solution to the problem of demonstrating the stability of such systems, in the determinist case. First of all, it would appear that the scope of possibilities is much greater than previous theories would make us think, and that the association of the most classical techniques for identification and control is proved justified from the theoretical point of view.

It would also appear that the identification and control association should be considered as an inseparable whole ; in effect, stabilization arise as a property which is obtained independently of the convergence of identified parameters towards the real parameters. In other terms, a good identification is not a <u>necessary</u> condition for stabilization. And in still other terms, the system is stabilized even if the control system is not adapted, that is, not in conformity with the control system which would be theoretically optimal for the process under consideration. For this reason, we believe that the property of stabilization is even more general than that of adaptability, so that we prefer to speak in this respect of unconditional stabilizers.

REFERENCES

[1] DE LARMINAT, Ph., and JEANNEAU, J.L.
     Metodo de regulacion adaptativa para sistemas de fase no minima.
     3e Congreso nacional informatica y automatica, Madrid, 1975.

[2] DE LARMINAT, Ph.
"Quelques considérations sur certains systèmes autoadaptatifs et leur stabilité globale inconditionnelle"
Ecole d'été d'analyse numérique EDF, IRIA, CEA. Juin-Juillet 1978.

[3] DE LARMINAT, Ph.
On overall stability of certain adaptive control systems.
5th IFAC Symp. on identification and system parameter estimation. Darmstadt, Sept. 24-28, 1979.

[4] DE LARMINAT, Ph.
A theorem for the analysis of the unconditional overall stability of M.R.A.S.
INFO II CONF. PATRAS, 1979.

[5] GOODWIN, G.C., RAMADGE, P.J., CAINES, P.E.
"Discrete time stochastic adaptive control"
Univ. of Newcastle, Australia, Dec. 1978.

[6] NARENDRA, K.S., LIN, Y.M., VALAVANI, L.S.
"Stable adaptive controller design, Part II : Proof of stability"
Report 7904, Yale Univ., April 1979.

[7] LANDAU, I.D., LOZANO, R.
Design and evaluation of discrete time explicit M.R.A.C. for tracking and regulation.
Note interne LA6, 79.14, Grenoble, Juin 1979.

[8] FUCHS, J.J.
Commande adaptative directe des systèmes linéaires discrets.
Thèse D.E. Univ. de Rennes, France, 6 déc. 1979.

[9] DE LARMINAT, Ph.
Systèmes autoadaptatifs avec modèle de référence et régulateurs autoaccordables.
Approche unifiée. Introduction à leur stabilité globale inconditionnelle.
4e Congreso Informatica y Automatica. Madrid 16/19 octobre 1979.

<u>ON SOME ADAPTIVE CONTROLLERS FOR STOCHASTIC SYSTEMS</u>

<u>WITH SLOW OUTPUT SAMPLING</u>

Torsten Söderström
Department of Automatic Control and Systems Analysis
Institute of Technology, Uppsala University

P.O. Box 534, S-751 21 Uppsala, Sweden

ABSTRACT

A discrete time control problem is studied where the output is measured only at every r:th sampling. Adaptive controllers based on recursive parameter estimation are examined and their self-tuning properties are investigated. The explicit analysis is carried out for a general first order system with an additional delay. The adaptive control of non minimum phase systems raises a new problem. Requirements on identifiability and self-tuning are then in conflict.

1. INTRODUCTION

In some digital control systems the output is bounded to be measured with a low frequency. The reason can be that no fast on-line sensor exists but a time consuming laboratory analysis has to take place. Examples include measurements of chemical composition in certain process industries as well as measurements of various biological variables. Despite this constraint on the measuring rate it may for other reasons be beneficial to use a relatively short sampling interval, i.e. to let the input change a number of times between the arrivals of the output measurements.

How should adaptive control be applied in this situation? Adaptive controllers can be designed in many ways. The model reference technique is one possibility. Here we will consider recursive parameter estimation (identification) and a time varying control law based on the last available parameter estimates. Such regulators have become very popular and are often called self-tuning regulators since quite often the controller converges to the optimal regulator. Åström et al (1977) give a good survey of the field, while Clarke and Gawthrop (1979) discuss more recent extensions.

The purpose of this paper is to discuss and illustrate some difficulties that arise when self-tuning regulators are used for the mentioned problem. The regulators will be examined using the ODE analysis as given by Ljung (1977). Alternative approaches for analysis are also given in Ljung (1979). The topic is treated in some more detail in the report Söderström (1980) where e.g. proofs of the propositions can be found.

2. MINIMUM VARIANCE CONTROLLERS

Minimization of the output variance is often chosen as the objective in adaptive control. In the usual case this objective requires only simple polynomial operations, cf
Åström (1970), which is far less than solving two Ricatti equations in the full LQG
problem. When the output is measured only every r:th sampling interval (r being an
integer > 1), i.e. the situation to be treated in this paper, the problem is more
difficult. The minimum variance strategy can of course be obtained from the solution
of a full LQG problem. Söderström (1979), Söderström and Lennartson (1980) give the
modification of the LQG problem to r > 1. However, for some simple systems the explicit minimum variance strategy is known.

Proposition 2.1 Consider the system

$$y(t)-ay(t-1)=bu(t-k)+e(t)+ce(t-1) \tag{2.1}$$

where $1 \le k \le r$, $|c| < 1$ and $e(t)$ is zero mean white noise
Consider also the criteria

$$V=\lim_{N\to\infty} \frac{1}{N} \sum_{t=1}^{N} Ey(t)^2 \tag{2.2}$$

The optimal controller is then given by (m being an   arbitrary integer)

$$u(mr)=- \frac{a^{k-1}}{b} (a+c/p)y(mr)$$
$$\tag{2.3}$$
$$u(mr+i)=0 \quad 1 \le i \le r-1$$

where p is the positive solution of the equation

$$p=1+(c+a)^2 \frac{1-a^{2r-2}}{1-a^2} + a^{2r-2}c^2(1-1/p) \tag{2.4}$$

□

Proposition 2.2 Consider the system

$$y(t)-ay(t-1)=bu(t-1)+b\beta u(t-2)+e(t) \tag{2.5}$$

where $a+\beta \neq 0$ if $\beta^2 \ge 1$ and $e(t)$ is white noise with zero mean. The controller minimizing the criteria V, (2.2) is then given by

$$u(mr+i)=\alpha u(mr+i-r)-\delta\lambda^i y(mr) \quad 0 \le i \le r-1 \tag{2.6}$$

When $\beta^2 \leq 1$ the involved parameters (α, δ and λ) are given by

$$\alpha=(-\beta)^r \qquad \delta=\frac{a+c/p}{b} \qquad \lambda=-\beta \tag{2.7}$$

When $\beta^2 \geq 1$ the parameters are instead given by

$$\alpha=\left(-\frac{1}{\beta}\right)^{r-1}\left[-\frac{c}{p}\frac{\beta^2-1}{\beta(a+\beta)}-\frac{1+a\beta}{a+\beta}\right] \qquad \delta=\frac{1+a\beta}{\beta(a+\beta)}\frac{(a+c/p)}{b} \qquad \lambda=-\frac{1}{\beta} \tag{2.8}$$

The parameter p is still the positive solution of (2.4).

□

Note that simplified expressions are obtained when c=0. It is easy to verify that expected and well-known results take place when r=1.

A suboptimal controller can be derived if the output variance is penalized only at the measuring times t=0, r, 2r ... . It is then appropriate to model the system as a difference equation in the measured output $\{y(mr)\}_m$ and the inputs $\{u(t)\}_t$. Such a model can be viewed as a multiple input single output system with r sampling intervals as time unit and the following inputs

$$u_1(mr)=u(mr) \qquad u_2(mr)=u(mr+1) \quad \ldots \quad u_r(mr)=u(mr+r-1) \tag{2.9}$$

The multiple inputs give a lack of uniqueness since the stationary variance $Ey(mr)^2$ can be minimized by many regulators. The extra degrees of freedom can be used to achieve further objectives. The situation is illustrated in the following example.

Example 2.1   Consider the system

$$y(t)-ay(t-1)=bu(t-k)+e(t) \tag{2.10}$$

where $1 \leq k \leq r$ and e(t) is zero mean white noise with variance λ^2. The system can be rewritten as

$$y(mr)=a^r y(mr-r)+ \sum_{i=o}^{r-1} a^i bu(mr-k-i)+\varepsilon_o(mr) \tag{2.11}$$

where the disturbance term $\varepsilon_o(mr)$ is white noise (with mr as time unit) fulfilling

$$\varepsilon_o(mr)= \sum_{i=o}^{r-1} a^i e(mr-i) \qquad \sigma^2 \triangleq E\varepsilon_o(mr)^2=\lambda^2\frac{1-a^{2r}}{1-a^2} \tag{2.12}$$

Every regulator will then fulfil

$$Ey(mr)^2 \geq \sigma^2 \tag{2.13}$$

Equality is obtained if and only if

$$a^r y(mr) + \sum_{i=o}^{r-1} a^i bu(mr+r-k-i)=0 \qquad (2.14)$$

Equation (2.14) shows precisely how the lack of uniqueness acts in this example. Consider now some different regulators fulfilling (2.14)

$$R_1 \qquad u(mr)=-\frac{a^k}{b} y(mr)$$

$$u(mr+i)=0 \qquad 1 \le i \le r-1 \qquad (2.15)$$

$$R_2 \qquad u(mr-k)=-\frac{a^r}{b} y(mr-r)$$

$$u(t)=0 \qquad t \ne mr-k \qquad (2.16)$$

and assuming especially k=1

$$R_3 \qquad u(mr+i)=-\frac{a^r}{b} \frac{1-a}{1-a^r} y(mr) \qquad 0 \le i \le r-1 \qquad (2.17)$$

$$R_4 \qquad u(mr+i)=-\frac{a^{2r-1-i}}{b} \frac{1-a^2}{1-a^{2r}} y(mr) \qquad 0 \le i \le r-1 \qquad (2.18)$$

Regulator R_1 is optimal in the sense stated in Prop 2.1. In regulator R_2 only the latest input in (2.14) is used. Regulator R_3 is obtained by resampling the process to an r times longer sampling period. Finally regulator R_4 is constructed to minimize

$$V_u = \lim_{N \to \infty} \frac{1}{N} \sum_{t=1}^{N} Eu(t)^2 \qquad (2.19)$$

subject to the constraint (2.14).

□

3. ADAPTIVE CONTROLLERS FOR SYSTEM WITH UNKNOWN PARAMETERS

When an adaptive controller based on parameter estimation is to be designed it is of large importance how the model structure (model parameterization) and the identification method are chosen. One of the simplest methods is least squares (LS). The convergence of the parameter estimates is relatively quick although the limiting estimates may be biased. Then they may not give the optimal controller. The model structure is most naturally chosen as a difference equation in the measured outputs and the computed inputs. When the LS method is to be applied the model has the form

$$y(mr)+a_1 y(mr-r)+\ldots +a_p y(mr-pr)=b_1 u(mr-1)+b_2 u(mr-2)+\ldots b_q u(mr-q)+\varepsilon(mr) \qquad (3.1)$$

It may not be possible to estimate all the parameters $a_1 \ldots a_p$, $b_1, \ldots b_q$ for identifiability reasons. Also in the common situation (r=1) it is a good practice to fix one parameter e.g. b_1 and to estimate the remaining ones.

Adaptive controllers based on the above principles will now be examined for first order systems.

Example 3.1 Consider the system cf Example 2.1, with $1 \le k \le r$

$$y(t)-ay(t-1)=bu(t-k)+e(t) \tag{3.2}$$

Consider also some adaptive regulators based on (2.15)-(2.18). All the regulators are related to the model structure

$$y(mr)-\hat{a}y(mr-r)=v(mr-r)+\varepsilon(mr) \tag{3.3}$$

For regulator R_1^* (based on R_1)

$$v(mr)=\beta_1 u(mr)$$
$$\tag{3.4}$$
$$u(mr)=-\frac{\hat{a}}{\beta_1}y(mr) \qquad u(mr+i)=0 \qquad 1 \le i \le r-1$$

where β_1 is an arbitrary fixed nonzero constant. For R_2^*

$$v(mr)=\beta_2 u(mr+r-k)$$
$$\tag{3.5}$$
$$u(mr+r-k)=-\frac{\hat{a}}{\beta_2}y(mr) \qquad u(t)=0 \text{ otherwise}$$

where again β_2 is fixed. Regulator R_3^* (applicable for k=1) corresponds to

$$v(mr)=\beta_3 u(mr)$$
$$\tag{3.6}$$
$$u(mr+i)=-\frac{\hat{a}}{\beta_3}y(mr) \qquad 0 \le i \le r-1$$

with a fixed value of β_3. Finally also regulator R_4^* can be applied with k=1 and

$$v(mr)=\gamma_0 u(mr)+\ldots +\gamma_{r-1}u(mr+r-1)$$
$$\tag{3.7}$$
$$u(mr+i)=-\frac{\hat{a}\gamma_i}{\gamma_0^2+\ldots +\gamma_{r-1}^2}y(mr) \qquad 0 \le i \le r-1$$

where all the parameters $\gamma_0 \ldots \gamma_{r-1}$ are kept fixed in order to guarantee identifiability.

The convergence properties of the regulators are given in the following proposition.

Proposition 3.1

i) The regulator R_1^* (3.4) converges to R_1 (2.15) if and only if β_1 and $a^{r-k}b$ have the same sign.

ii) The regulator R_2^* (3.5) converges to R_2 (2.16) if and only if β_2 and b have the same sign.

iii) The regulator R_3^* (3.6) converges to R_3 (2.17) if and only if β_3 and b have the same sign.

iv) The regulator R_4^* (3.7) converges to R_4 (2.18) if and only if $\gamma_i = ca^{r-1-i}$, $0 \leq i \leq r-1$ for some constant c of the same sign as b.

□

Remark: The regulator R_1 (2.15) is known to be the optimal output variance controller, see Proposition 2.1. The adaptive regulator R_1^* has in fact a stronger self-tuning property. Assume that R_1^* is applied as given by (3.3), (3.4) but for a system with coloured noise

$$y(t)-ay(t-1)=bu(t-k)+e(t)+ce(t-1) \tag{3.8}$$

Then the regulator R_1^* converges to the optimal controller (2.3) if and only if β_1 and $a^{r-k}b$ have the same sign.

□

The regulators $R_1^* \ldots R_3^*$ thus possess the self-tuning property under quite weak conditions for the treated example. The conditions for R_4^* to be self-tuning is considerably more restrictive.

Example 3.2 Consider the system $(r > 1)$

$$y(t)-ay(t-1)=bu(t-1)+b\beta u(t-2)+e(t)+ce(t-1) \tag{3.9}$$

Since the deterministic part fulfils

$$y(mr+r)=a^r y(mr)+b[u(mr+r-1)+\ldots +a^{r-1}u(mr)]+b\beta[u(mr+r-2)+\ldots +a^{r-1}u(mr-1)] \tag{3.10}$$

it seems reasonable to use the model structure

$$y(mr+r)=\hat{a}^r y(mr)+\hat{b}u(mr+r-1)+(\hat{\hat{b}}\hat{a}+\hat{b}\hat{\beta})u(mr+r-2)+\ldots +(\hat{\hat{b}}\hat{a}^{r-1}+\hat{\hat{b}}\hat{\hat{\beta}}\hat{a}^{r-2})u(mr)+\hat{\hat{b}}\hat{\hat{\beta}}\hat{a}^{r-1}u(mr-1)+$$
$$+\varepsilon(mr+r) \tag{3.11}$$

Assume that a minimum variance regulator as given in Prop 2.2 is applied but with the regulator parameters depending adaptively on the estimated model parameters. To investigate identifiability properties examine then

$$\text{rank}[y(mr),u(mr+r-1), \ldots u(mr-1)]$$
$$=\text{rank}[v(mr)-\alpha v(mr-r),-\delta\lambda^{r-1}v(mr), \ldots ,-\delta v(mr), -\delta\lambda^{r-1}v(mr-r)]$$
$$=\text{rank}[v(mr), v(mr-r)]=2 \tag{3.12}$$

where the signal v(mr) is given by

$$v(mr)-\alpha v(mr-r)=y(mr) \tag{3.13}$$

has been used. As a consequence no more than two parameters can be estimated. Thus self-tuning cannot take place for the non minimum phase case ($\beta^2 > 1$) since the optimal regulator (2.6), (2.8) then contains three independent parameters.

For minimum phase systems ($\beta^2 < 1$) such a conflict may not be necessary. One parameter in (3.11) must be fixed. It is most convenient to fix $\hat{a}$ since linear least squares can then applied for recursive estimation of $\hat{b}$ and $\hat{\hat{b\beta}}$.

Proposition 3.2 Consider the minimum phase system (3.9) (assume $\beta^2 < 1$) and an adaptive regulator based on recursive estimation of $\hat{b}$ and $\hat{\hat{b\beta}}$ in (3.11) with $\hat{a}$ arbitrary and fixed and the control function as in Prop 2.2. The necessary conditions for convergence are fulfilled by the following parameter values (p fulfilling (2.4))

1. $\tilde{\beta}=\beta$ $\hat{b}=\dfrac{\hat{a}b}{a+c/p}$

2. $\hat{\beta}=0$ $\hat{b}=\dfrac{\hat{a}b}{a+c/p}\ \dfrac{a+\beta}{a}$ (3.14)

3. $\hat{\beta}=a$ $\hat{b}=\dfrac{\hat{a}b}{a+c/p}\ \dfrac{\beta}{a}$

The third solution is applicable if and only if r is an even integer.

$\square$

The first solution of (3.14) means that the optimal regulator (2.6), (2.7) is obtained. Thus there seems to be good hope for self-tuning under certain conditions although a complete proof presently is lacking. Sufficient conditions for convergence could be derived by examining the stability properties of the associated ODE:s. They have the stationary solutions (3.14).

4. A CONTROLLER BASED ON ADAPTIVE INTERPOLATION

The control law (2.3) coincides with the usual minimum variance controller for t=mr provided c=0. It is then appealing to try an adaptive scheme where the usual minimum variance controller is used and where the missing output measurements are substituted with interpolated and predicted values. This interpolation/prediction can be performed in an adaptive way. The output measurements are to be fitted recursively to an auto-regressive model using the least squares method. Based on the latest model parameters interpolation and prediction can take place. Since these parameters thus influence the controller and the controlled system an adaptive mechanism is reached. The indicated way of interpolation implies in particular that the measurements are used unchanged for t=0, r, 2r etc. A consequence is that the adaptive method cannot generally give the optimal control if $c \neq 0$. However, for some cases this occurs.

Example 4.1 Consider the following system, cf Example 2.1.

$$y(t)-ay(t-1)=bu(t-1)+e(t) \tag{4.1}$$

Let the controller be

$$u(t)=- \frac{a}{b} \hat{y}(t) \tag{4.2}$$

Note that (4.2) is the usual minimum variance controller if $\hat{y}(t)$ is substituted by $y(t)$. Assume further that the measured outputs of the controlled system is modelled as a first order autoregression

$$y(mr)-\hat{a}y(mr-r)=\epsilon(mr) \tag{4.3}$$

The parameter $\hat{a}$ is estimated using recursive least squares. The model (4.3) implies that the output for t=1, 2 ... follows an autoregression as well

$$y(t)-\alpha y(t-1)=\tilde{\epsilon}(t) \qquad \alpha= \sqrt[r]{\hat{a}} \tag{4.4}$$

Due to the non-dynamic feedback (4.2) it is for this example sufficient to predict future unmeasurable outputs. Eqs. (4.2), (4.4) imply that the total, adaptive regulator will fulfil

$$u(mr+i)=- \frac{a}{b} \alpha^i y(mr) \qquad 0 \leq i \leq r-1 \tag{4.5}$$

To be more precise, α in (4.5) is to be based on (4.4) and the estimate $\hat{a}$ where measurements up to time t=mr are used. Note that the solution of (4.4) wrt α can give complications if r is even. No solution exists if $\hat{a} < 0$ while $\hat{a} > 0$ implies two possible solutions.

Assuming convergence takes place the closed loop system will fulfil

$$y(mr)+a\alpha \frac{\alpha^{r-1}-a^{r-1}}{\alpha-a} y(mr-r)= \epsilon_o(mr) \tag{4.6}$$

The convergence properties are then determined by the stability properties of the differential equation, cf Ljung (1977)

$$\overset{*}{a}=E\epsilon(mr)y(mr-r)=Ey(mr)y(mr-r)-\hat{a}Ey(mr)^2 \tag{4.7}$$

Since multiplication of the RHS with a positive factor does not effect the stability properties it is equivalent to examine

$$\dot{\hat{a}} = -a\alpha \frac{\alpha^{r-1} - a^{r-1}}{\alpha - a} \quad -\hat{a} = \frac{\alpha(a^r - \hat{a})}{\alpha - a} \tag{4.8}$$

The stationary solutions of (4.8) are easily found to be

$$\hat{a}_1 = 0 \qquad \hat{a}_2 = a^r (\alpha = -a) \tag{4.9}$$

The first solution represents the optimal controller since then the regulators (2.3) and (4.5) become identical. Also the stability properties of the solutions are of essential importance for the convergence properties of the adaptive scheme. The local stability properties are often investigated from linearization. However, the ODE (4.8) <u>cannot</u> be linearized around $\hat{a}_1$. The second stationary solution, $\hat{a}_2$, is valid only if r is even. A simply analysis using linearization shows that this solution is (locally) stable. Assume that a > 0 (which is most natural) and let $\alpha = +\sqrt[r]{a}$ be chosen in (4.4). Then $\hat{a}_1 = 0$ is the only stationary solution. It can then in fact also be seen that $\hat{a}(0) > 0$ implies $\hat{a}(t) \to 0$ in (4.8).

□

5. CONCLUSIONS

The control problem when the output is measured at every r:th sampling instant is treated. Explicit expressions for the output minimum variance controller is given for general first order systems. Some adaptive controllers based on recursive identification of the model parameters have been discussed. Although the explicit analysis is limited to general first order systems with an additional delay it has been shown that self-tuning takes place in many cases. The control of non minimum phase systems offers new difficulties since then good identifiability properties are in conflict with possibilities for self-tuning.

A controller based on adaptive prediction has also been discussed. The adaptive mechanism is then to recursively identify the output as an autoregressive process. This model is used for prediction and interpolation of lacking output measurements. Also such a controller can be self-tuning under certain circumstances.

REFERENCES

K.J. Åström (1970)
Introduction to Stochastic Control Theory. Academic Press.

K.J. Åström, U. Borisson, L. Ljung and B. Wittenmark (1977)
Theory and application of self tuning regulators. Automatica, vol 13, pp 457-476.

D.W. Clarke and P.J. Gawthrop (1979)
Self-tuning control. Proc IEE, vol 126, pp 633-640.

L. Ljung (1977)
Analysis of recursive stochastic algorithms. IEEE Transactions on Automatic Control,
vol AC-22, pp 551-575.

L. Ljung (1979)
Convergence of recursive estimators. Preprints 5th IFAC Symp. on Identification and
System Parameter Estimation, pp 131-144.

T. Söderström (1979)
Linear optimal control of systems with slow output sampling. Report UPTEC 7926R,
Institute of Technology, Uppsala University, Sweden.

T. Söderström (1980)
On adaptive control of stochastic systems with infrequent output sampling. UPTEC
report 8004 R, Institute of Technology, Uppsala University, Sweden.

T. Söderström and B. Lennartson (1980)
On linear optimal control with infrequent output sampling. UPTEC report in prepara-
tion, Institute of Technology, Uppsala University, Sweden.

REALIZATION AND APPLICATION OF A SELF-TUNING ON-OFF CONTROLLER

R. Breddermann

Institut für Regelungstechnik, RWTH Aachen, F.R.G.

Abstract. A concept for the realization of a self-
tuning on-off controller is presented. The controller
contains a predictor which is used to estimate the
future system output sequences taking into account
the whole set of future on-off input sequences over
a specified number of time intervals. For every time
interval the optimal input is selected by minimizing
a suitable cost-function, measuring the deviation
between the predicted output sequence and the desired
set-point sequence. The prediction is based on a parametric system model, whose parameters are estimated
using the on-line recursive-least-squares algorithm.
As an example the proposed on-off controller is applied
to the wall temperature control of a water cooled metal
cylinder and the results are presented.

Keywords. Adaptive control; direct digital control;
predictive control; on-off control; recursive parameter
estimation; temperature control.

1. INTRODUCTION

Within the last few years self-tuning control has been applied successfully to solve various control problems. The basic type of a self-tuning
regulator was developed by Åström and Wittenmark /1/ for the minimum
variance control of randomly-disturbed discrete-time systems with constant but unknown parameters. However, this strategy applied to continuous-time systems tends to produce excessive control actions, and non-
minimum-phase systems can only be controlled in a stable manner with
additional computational efforts. To overcome these disadvantages Clarke
and Gawthrop /2/ introduced an extended control law which makes the control effort adjustable and moreover allows the incorporation of set-
points into the self-tuning concept.

The different self-tuning controllers published up to now are based on
the assumption that any value of the control signal within certain limits
can be processed by the given controlling element. However, in some prac-

tical applications e.g. in heating devices controlling elements are used which produce two different system input signals (e.g. on and off) only.

In this paper a concept for realization of a self-tuning on-off control-ler is developed and its efficiency is investigated in an application. The statement of the problem to be solved is given in section 2. Based on a heuristic approach the structure of the regulator is described in section 3. The problem of modeling and parameter estimation is dealt with in section 4. Section 5 shows the realization of a predictive on-off controller using a parametric model of the plant. Finally the re-sults of an application of this strategy to temperature control are pre-sented in section 6.

2. STATEMENT OF THE PROBLEM

The block diagram in Fig.1 shows a continuous-time plant which has to be controlled with a discrete-time self-tuning regulator. It is assumed, that the relay, used as a controlling element, is switched on or off by the regulator at equidistant time instants only. The regulator has to adapt his behaviour to that of the plant in such a way that optimal pro-cess control is reached for disturbances and set-point variations. The transient output deviations from the desired set-point, caused by the nonlinear controlling element, should be reduced to a minimal value. The sampling period has to be small enough to ensure that all significant transient deviations can be observed.

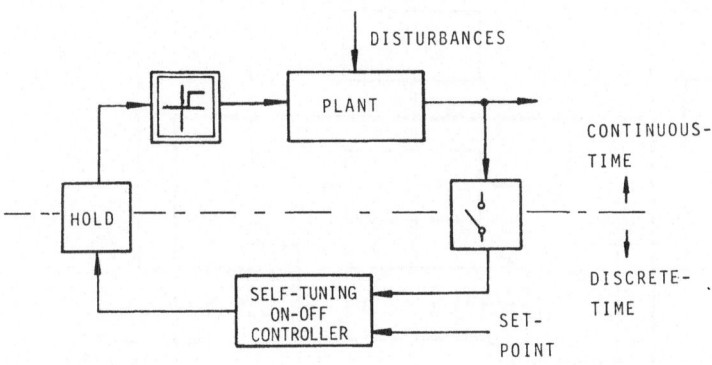

Fig. 1   Block diagram of control problem

The application of the well-known concepts of self-tuning regulators is not possible in this case, for the controller in connection with the switching controlling element cannot be considered as a linear filter.

3. HEURISTIC APPROACH

The presented problem is solved in a heuristic way by modifying the minimum variance strategy, which forms the basis of the well-known self-
tuning regulators. The fundamentals of the solution are given by Åström's
separation theorem /3/. It says that a minimum variance regulator can be
separated into a predictor and a dead-beat regulator. The predictor is
used to predict the influence of the disturbances on the output and the
dead-beat regulator has to calculate the value of the following control
variable in order to make the predicted output equal to the desired set-
point. With the given controlling element the dead-beat strategy cannot
be realized, as the relay allows only two different values of the control
variable. For this reason the prediction is extended to a greater number
of time intervals and the dead-beat strategy is approximated by an optimal on-off control strategy. The structure of such a self-tuning on-
off regulator is shown in Fig. 2.

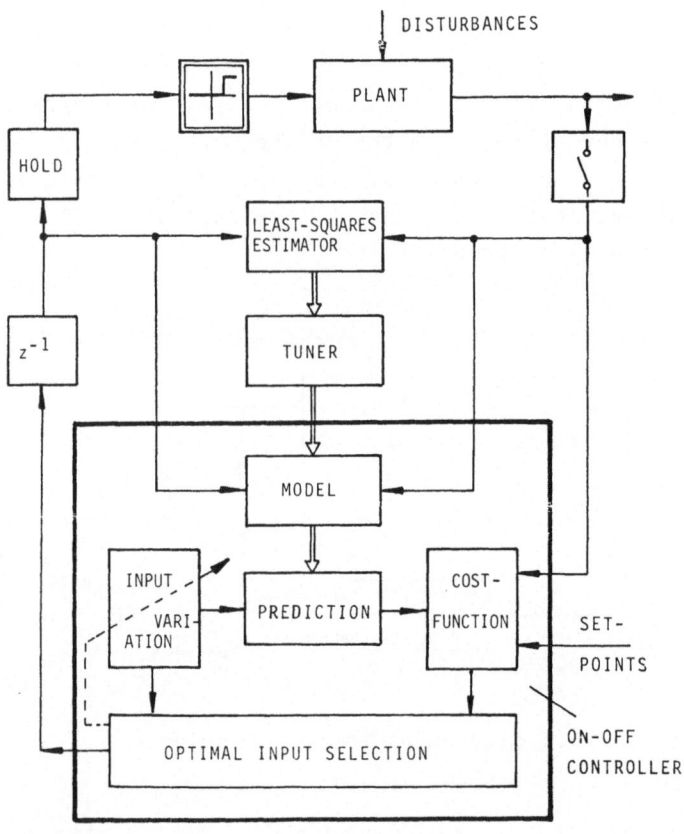

Fig. 2   Structure of the self-tuning on-off regulator

The prediction is carried out on the basis of a parametric model of the plant, whose parameters are determined by a suitable parameter estimation method. The parameters of the model are adjusted by the tuner and permanently updated if necessary. By minimization of a cost-function, which evaluates the predicted system deviations, it is decided whether the controlling element has to turn on or off in the following sample instant. Therefore all possible on-off input sequences within a defined number of prediction steps are evaluated.

4. MODELING AND PARAMETER ESTIMATION

The system to be controlled is described by a linear parametric model whose parameters are estimated by the least-squares method for reasons of computational simplicity. The bias is tolerable if the signal to noise ratio is high enough. Fig. 3 shows the system model and the equation error of the least-squares method.

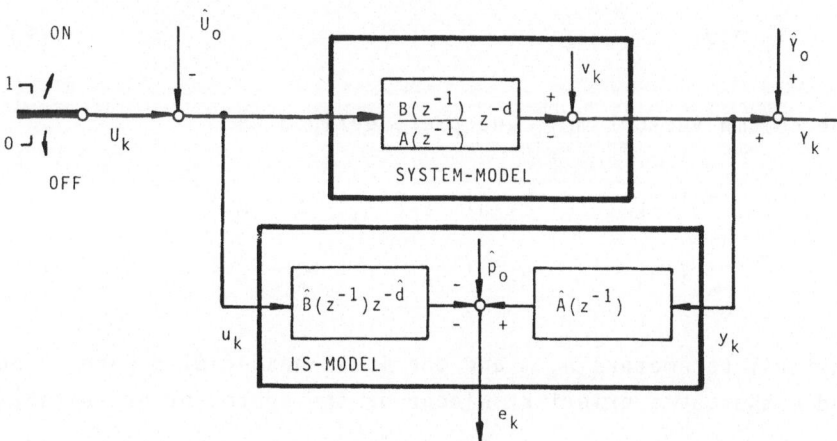

Fig. 3 System model and equation error representation

The system model and the least-squares model are characterized by polynomials in the backward-shift operator z^{-1} of the following form:

$$A(z^{-1}) = 1 + a_1 z^{-1} + \ldots + a_n z^{-n}$$

$$B(z^{-1}) = b_0 + b_1 z^{-1} + \ldots + b_m z^{-m}$$

and

$$\hat{A}(z^{-1}) = 1 + \hat{a}_1 z^{-1} + \ldots + \hat{a}_{\hat{n}} z^{-\hat{n}}$$

$$\hat{B}(z^{-1}) = \hat{b}_0 + \hat{b}_1 z^{-1} + \ldots + \hat{b}_{\hat{m}} z^{-\hat{m}} .$$

The switch, representing the controlling element, allows only the two values 0 and 1 for the input signal U_k. The influence of steady-state values is considered in two ways:

The a priori knowledge about the steady-state values $\hat{U}_0$ and $\hat{Y}_0$ is taken into account at the summing points placed at the input and output of the system-model. Missing or inaccurate estimates of the steady-state values are compensated by an additional steady-state parameter $\hat{p}_0$ in the least-squares model.

The equation error of the least-squares model is described by the following difference equation

$$\hat{e}_k = y_k + \sum_{i=1}^{\hat{n}} \hat{a}_i \, y_{k-i} - \sum_{i=0}^{\hat{m}} \hat{b}_i \, u_{k-\hat{d}-i} - \hat{p}_0 \tag{1}$$

or shorter

$$\hat{e}_k = y_k - \underline{m}_{k-1}^T \, \hat{\underline{p}} \tag{2}$$

where the column vectors $\underline{m}_{k-1}$ and $\hat{\underline{p}}$ are defined as

$$\underline{m}_{k-1}^T = (\, y_{k-1} \, \cdots \, y_{k-\hat{n}} \mid u_{k-\hat{d}} \, \cdots \, u_{k-\hat{d}-\hat{m}} \mid 1 \,) \tag{3}$$

$$\hat{\underline{p}}^T = (-\hat{a}_1 \, \cdots -\hat{a}_{\hat{n}} \mid \hat{b}_0 \, \cdots \, \hat{b}_{\hat{m}} \mid \hat{p}_0) \, . \tag{4}$$

The structural parameters $\hat{n}$, $\hat{m}$ and the model dead-time $\hat{d}$ have to be pre-specified either by a priori knowledge of the system or by suitable tests.

The standard-recursive algorithm of the time-weighted least-squares method

$$\hat{\underline{p}}_{k+1} = \hat{\underline{p}}_k + \underline{k}_k \, (\, y_{k+1} - \underline{m}_k^T \, \hat{\underline{p}}_k \,) \tag{5}$$

$$\underline{k}_k = \underline{P}_k \, \underline{m}_k \, (\, \rho_k + \underline{m}_k^T \, \underline{P}_k \, \underline{m}_k \,)^{-1} \tag{6}$$

$$\underline{P}_{k+1} = \frac{1}{\rho_k} \, (\, \underline{P}_k - \underline{k}_k \, \underline{m}_k^T \, \underline{P}_k \,) \tag{7}$$

is used for on-line updating of the parameter vector $\hat{\underline{p}}$.

By an appropriate choice of the forgetting factor ρ_k past data are weighed out in order to follow slowly changing parameters.

5. PREDICTIVE ON-OFF CONTROLLER

The structure of the predictive on-off controller is presented in Fig.4. To realize the optimal on-off control strategy, that has been proposed in section 3, a certain number of future values of the controlled variable has to be predicted. Because of the system dead-time one has to distinguish between predictions based on already determined input signals and predictions based on possible future input sequences of given length.

The general solution of a prediction problem has to take into account the deterministic input signal and stochastic disturbances. In this case, however, the influence of the disturbances is considered to be negligible compared to the input signal, which is permanently switched between its two extreme levels. With this assumption the prediction problem can be solved using only the deterministic part of the process model whose parameters are estimated by the least-squares method.

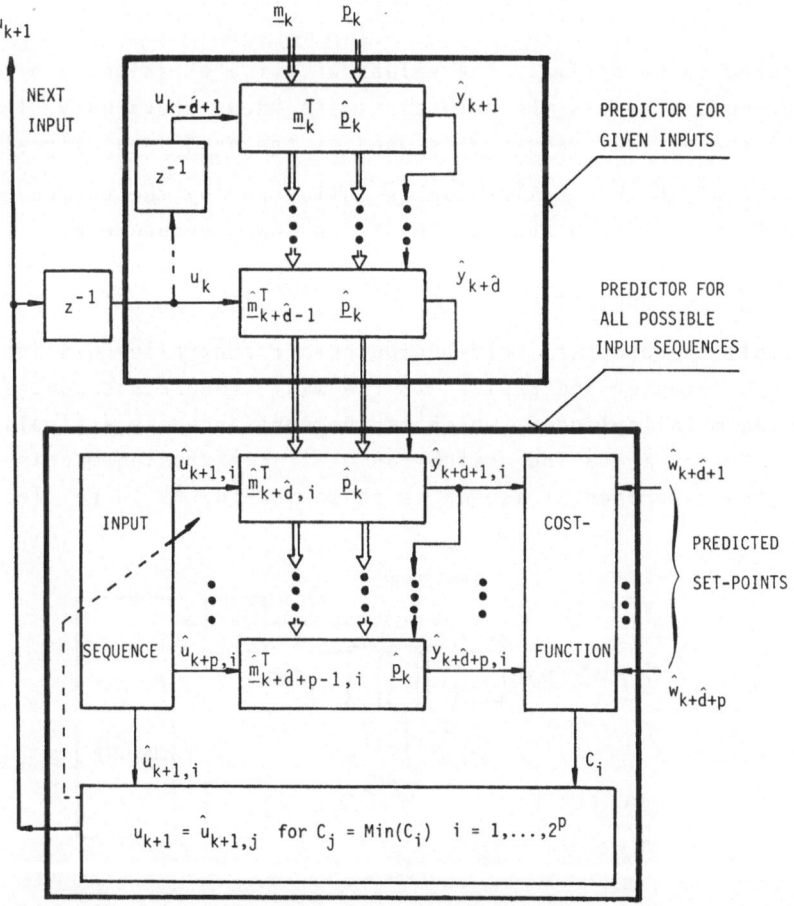

Fig. 4   Structure of the on-off controller

Based on the parameter vector $\hat{\underline{p}}_k$ at the time k, described in eq. 4, and the signal vector

$$\underline{m}_k^T = (\ y_k \ \cdots \ y_{k+1-\hat{n}} \ | \ u_{k+1-\hat{d}} \ \cdots \ u_{k+1-\hat{d}-\hat{m}} \ | \ 1 \ ) \qquad (8)$$

the future output values are estimated recursively by the equation

$$\hat{y}_{k+1} = \underline{m}_{k+1-1}^T \ \hat{\underline{p}}_k \qquad (9)$$

for all numbers of prediction steps 1 within the range $1 \leq l \leq \hat{d}+p$, where p is the previously determined length of the investigated input sequence. The input sequence within the whole set of 2^p possible input sequences which minimizes the cost-function

$$c_i = \sum_{\nu=1}^{p} g_\nu \ |\hat{w}_{k+\hat{d}+\nu} - \hat{y}_{k+\hat{d}+\nu,i}| \qquad (10)$$

is considered to be optimal. The values $\hat{w}_{k+\hat{d}+1}$ are estimates of the future set-points unless the set-points are known previously. The first element of the optimal sequence is used as the next input signal u_{k+1}.

The behavior of the controller can be influenced by the choice of the weighting factors g_ν and the length of the input sequence p.

6. RESULTS OF AN APPLICATION

As an example the proposed self-tuning on-off controller was implemented on a process computer and applied to the wall temperature control of a water cooled metal cylinder, which may be considered as a simplified physical model of a heating device for plastic extrusion machines. The scheme of the experimental set-up is shown in Fig. 5. Three electrical

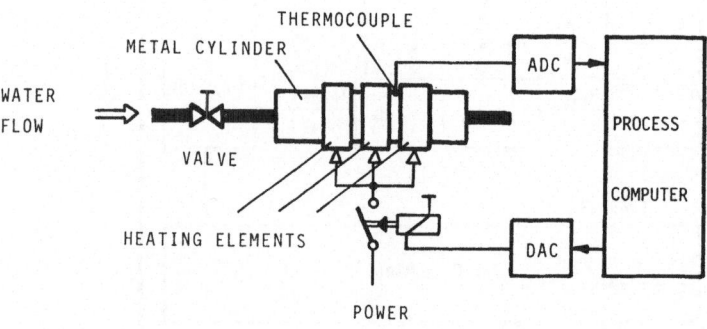

Fig. 5   Scheme of experimental set-up

resistance heaters were used which could only be switched on or off
simultanously by the controlling element. The wall temperature, representing the controlled variable, is measured with a thermocouple.

Fig. 6 shows the results of the self-tuning on-off control, when the
system model is estimated in open-loop and is not updated in closed-
loop. The set-point is composed of triangular and rectangular pro-
files. Fig. 6a represents the controlled variable. The system deviation
in Fig. 6b shows that, due to the missing updating of the system model,
the set-point variations lead to mean values of system deviations that
cannot be reduced to zero. It is plain to see that the variance of the
system deviation is affected by the input sequence length p. Figures
6c-d show the system deviation and the control variable for two enlarged
sections of Fig. 6b for p=4 and p=2. It is evident that the variance
of the output can be reduced to a certain extend by an appropriate choice
of p (which in this case is p=3 or 4).

Fig. 7 shows the results of the self-tuning on-off control when the system
model is permantently updated in every sample interval. Compared to the case previously discussed, the mean value of the system deviation is kept
near zero in spite of setpoint variations, whereas slight deteriorations of
the output variance are observed. Due to the diminished dead-time
estimate $\hat{d}$ increasing overshooting is observed with decreasing length of
the evaluated input sequence p.

7. CONCLUSIONS

The concept for the realization of a self-tuning on-off regulator, that was
developed by suitable modification of well-known self-tuners, has been presented. The proposed self-tuning controller was implemented on a process
computer and applied to solve a temperature control problem under different
operating conditions. The results have shown that acceptable temperature
control can be achieved with this control scheme. Due to the nonlinear
feedback law, updating of the model parameters under closed-loop conditions
is possible as long as the on-off control signal is persistently exciting.

REFERENCES

/1/ Åström, K.J., and Wittenmark, B.: 'On self-tuning regulators',
     Automatica, Vol.9, 1973, pp. 189-199

/2/ Clarke, D.W., and Gawthrop, P.J.: 'Self-tuning controller',
     Proc. IEE, Vol. 122, No. 9, 1975, pp. 929-934

/3/ Åström, K.J.: 'Introduction to Stochastic Control', New York:
     Academic Press, 1970

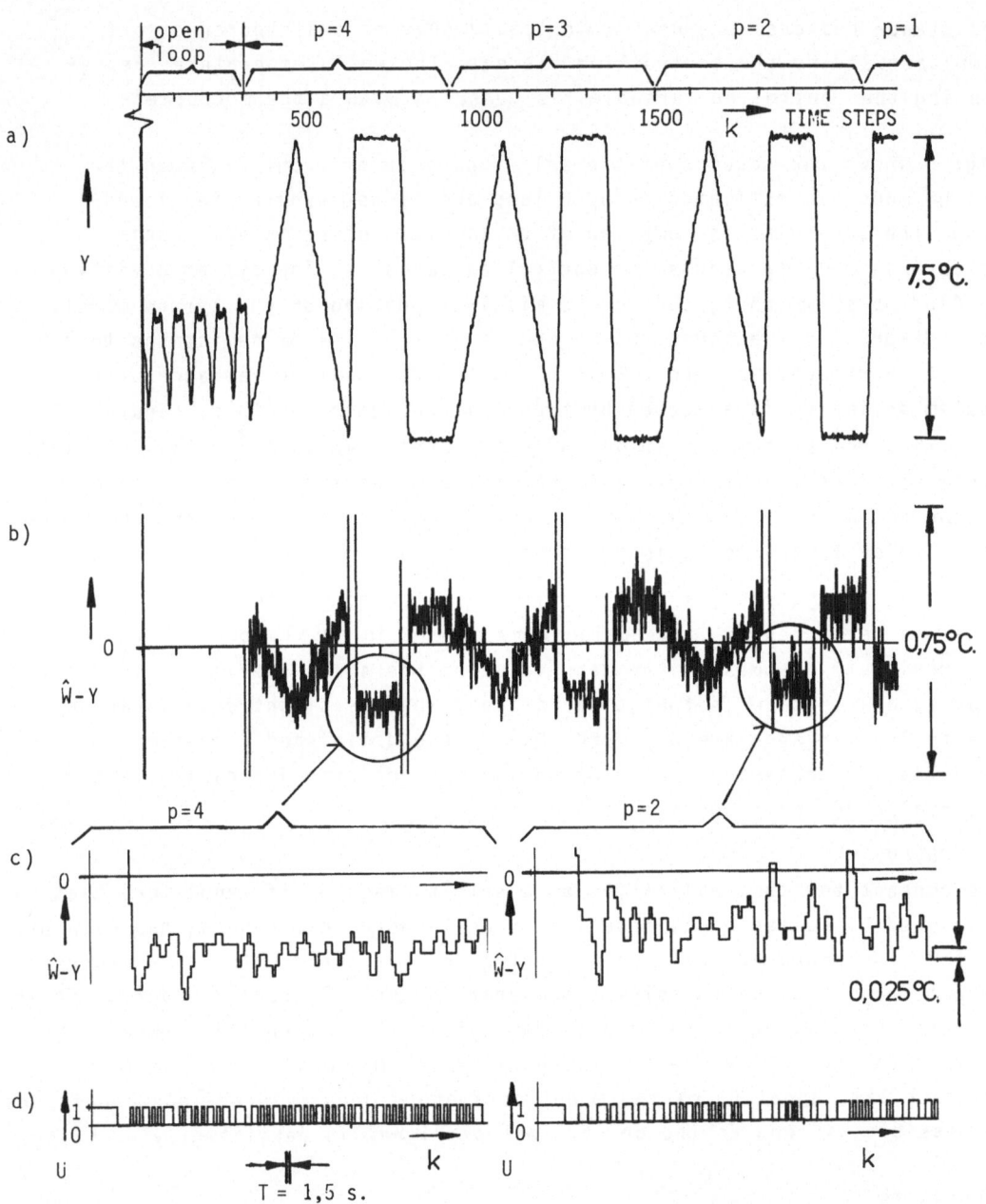

Fig. 6 Self-tuning on-off control without system model
updating ($\hat{n}=3$; $\hat{m}=2$; $\hat{d}=3$; $g_\nu=\nu$; sample interval= 1,5 s.)

a) Real output Y

b) System deviation $\hat{W}-Y$

c) System deviation $\hat{W}-Y$ for two enlarged sections

d) Control variable U for two enlarged sections

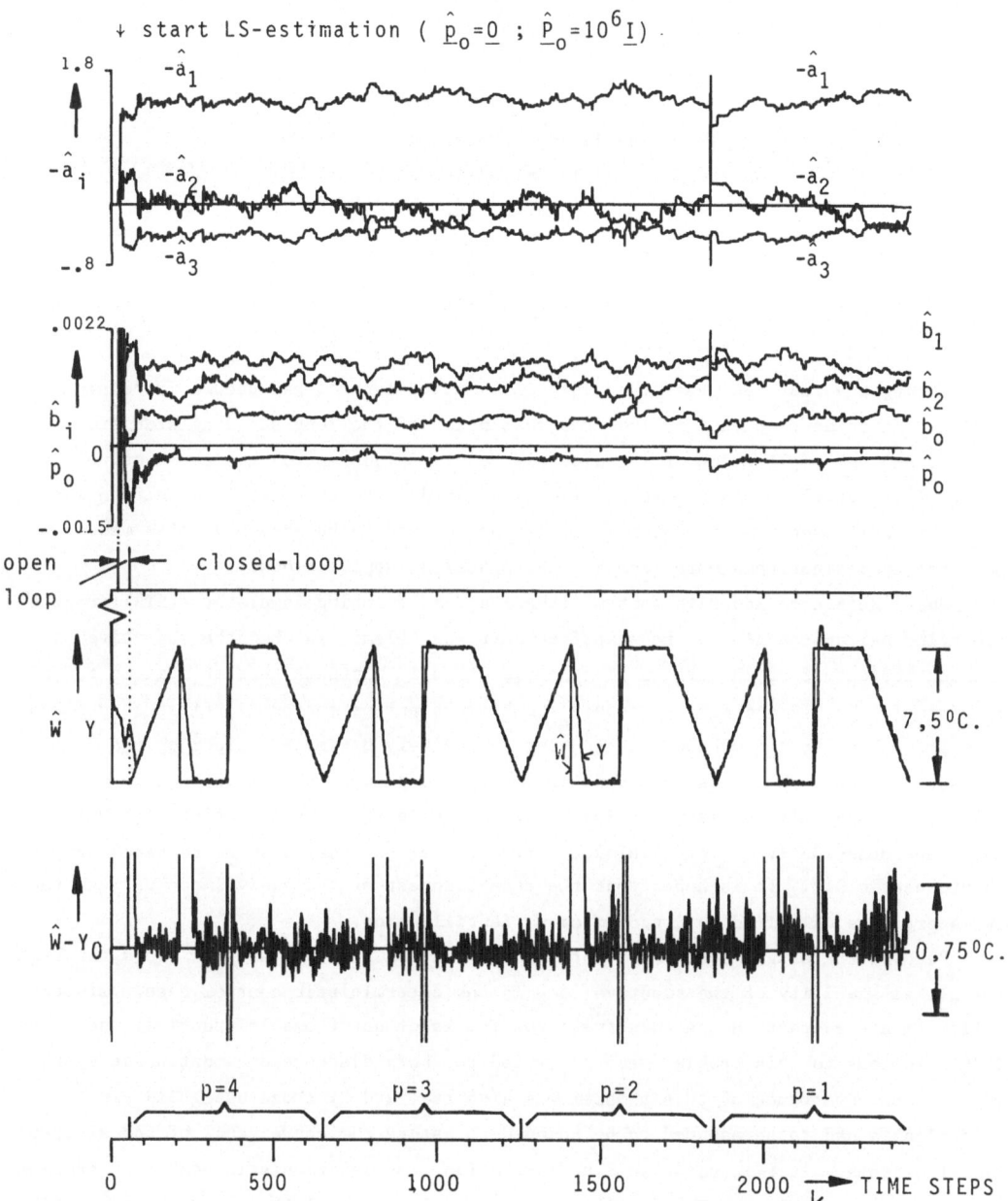

Fig. 7 Self-tuning on-off control with permanent
system model updating
(ρ_k=0.98; $\hat{n}$=3; $\hat{m}$=2; $\hat{d}$=2; g_ν=ν; sample interval= 1,5 s.)

a) Recursive estimates of the $\hat{a}_i$-parameters
b) Recursive estimates of the parameters $\hat{b}_i$ and $\hat{p}_o$
c) Real output Y versus set-point $\hat{W}$
d) System deviation $\hat{W}$-Y

RECENT DEVELOPMENTS IN ADAPTIVE CONTROL

Kumpati S. Narendra

Benjamin B. Peterson

Yale University

New Haven, Ct.

I. Introduction

During the past few years the field of adaptive control has become increasingly active. Major developments include the resolution of the long standing stability problem and the realization of the equivalence of various approaches to adaptive control. These developments, in turn, allow the field to be examined in a unified manner. Further, recent major advances in microprocessor technology have also made sophisticated algorithms more feasible for practical applications.

Model Reference Adaptive Control (MRAC) and Self Tuning Regulators (STR) are the two principal approaches to the adaptive control problem. In MRAC the objective is to make the output of an unknown plant asymptotically approach that of a given reference model. In STR a design procedure for known plant parameters is first chosen and this is applied to the unknown plant using recursively estimated values of these parameters. While MRAC has been applied to both continuous and discrete time systems, STR is analyzed only in discrete time. However while thus far the efforts in MRAC have been principally in deterministic systems, a significant body of research exists on stochastic STR. It is hoped that the establishment of the equivalence between the two approaches will lead to improved cross fertilization.

Perhaps the single most important question in MRAC in the past few years has been the global stability of the adaptive loop in the deterministic control case. If stability is not guaranteed the adaptive algorithm is of questionable practical use. In 1979 solutions to this problem were suggested for both discrete and continuous systems [1-3]. The importance of this problem was also realized by those using STR since some signals had to be assumed to be bounded to assure the convergence of the stochastic algorithms. In [4], using an STR formulation, the deterministic stability problem was completely resolved. The significant feature is that all the above methods, which have been shown to be globally stable, are equivalent.

For some time it has been suggested by many authors that STR and MRAC have many common features. Recently in [5-9] the similarities have been more precisely examined. Egardt [5] has presented a unified view of a number of MRAC schemes while Narendra and Valavani [6] demonstrate that for a particular parametrization of the plant direct and indirect control lead to identical equations. In [7] Johnson has shown that the same basic algorithms result using four different adaptive controllers. However the stability of the various equivalent schemes cannot always be shown. In light of the recent stability results the equivalence relationships become more important since it is now possible to demonstrate, with but minor modifications, how these schemes can also be made stable.

The objective of this paper is to review briefly some of the recent major advances in adaptive control and indicate possible directions for future research. The adaptive control problem as well as the prior information that is required to find a unique solution to it are stated in section II. Since error models provide a convenient basis for comparing different schemes they are presented in some detail in section III. The application of the error models in the analysis of adaptive observers and identifiers as well as some of the stability questions that arise in the control problem are treated briefly in section IV. The proof of global stability of the adaptive loop for both discrete and continuous systems is outlined in section V and the conditions for the equivalence of different schemes are stated in section VI. Section VII contains some of the results that are currently available in stochastic adaptive control. Finally, some of the questions of theoretical and practical interest that remain unresolved are presented in section VIII.

II. The Adaptive Control Problem

In spite of the many attempts that have been made during the past two decades to define an adaptive system, there is no universally accepted definition at the present time. Since adaptation is the regulation or control of a system in the presence of uncertainties the difficulty in defining the term precisely may be attributed to the myriad forms the uncertainty regarding the system can take. Most theoretical work on adaptive control has tended to concentrate on the control of a linear time-invariant system with unknown parameters.

Figures 1a and 1b represent the adaptive control problem using MRAC and STR approaches. In MRAC (Fig. 1.a) the input and output of a linear time-invariant plant are $u(\cdot)$ and $y_p(\cdot)$ respectively. A linear time-invariant model and reference input $r(\cdot)$ are specified which result in a model output $y_m(.)$. From all the available data it is desired to determine a control input $u(\cdot)$ such that the error $e_1(t)$ between $y_p(t)$ and $y_m(t)$ tends to zero asymptotically.

In the STR approach the first step is the selection of any known procedure for the design of a controller when the plant parameters are known. In the absence of such knowledge the plant parameters are estimated on line and the control input is generated using these estimates.

Two philosophically different approaches exist for MRAC of an unknown plant. In

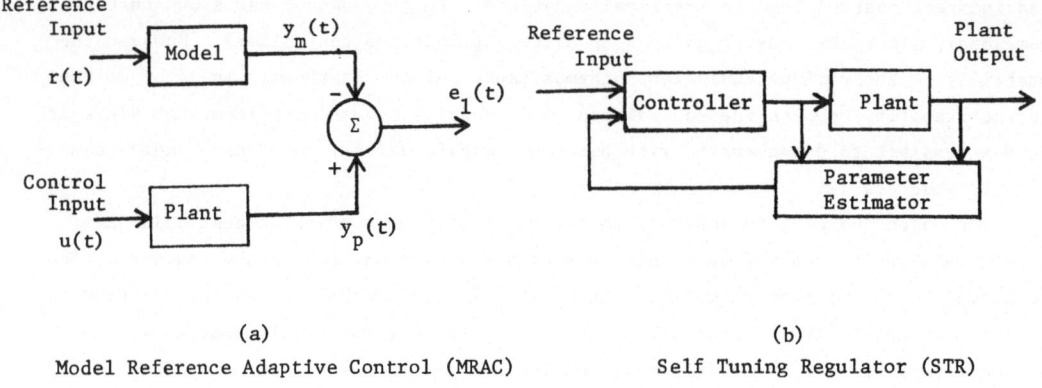

(a)   (b)

Model Reference Adaptive Control (MRAC)   Self Tuning Regulator (STR)

FIGURE 1

Indirect (explicit) Control the plant parameters are estimated and the control parameters are adjusted on the basis of these estimates [obviously this corresponds to one
form of STR]. In Direct (implicit) Control the parameters of the controller are directly adjusted to minimize the error between model and plant outputs. As shown in
section VI stable adaptive controllers obtained using the two methods have recently
been shown to be equivalent [6].

     In sections III and V the stability problem is discussed in detail. The follow-
ing assumptions regarding the plant have to be made in all the approaches which have
successfully resolved the stability problem.

Assumptions Regarding the Plant: Let the plant transfer function be $W_p(s)$ and that
of the reference model $W_M(s)$. Let $W_p(s)$ and $W_M(s)$ have n_1 and n_2 poles and m_1 and
m_2 zeros respectively. $n_1^* \stackrel{\Delta}{=} n_1 - m_1$, $n_2^* \stackrel{\Delta}{=} n_2 - m_2$ are defined as the relative degrees
of the plant and model and play an important role in the stability problem. The as-
sumptions may be stated in terms of n_i, m_i and n_i^* $(i = 1,2)$ as follows:

    (i) n_1^* should be known exactly.

   (ii) $n_2^* \geqslant n_1^*$

  (iii) An upperbound on n_1 should be known.   (1)

   (iv) If $W_p(s) = k_p \dfrac{p(s)}{q(s)}$, where $p(s)$ and $q(s)$ are monic polynomials, the sign
        of k_p must be known.

   (v) The zeros of the plant must lie in the open left half of the complex plane.

     For the discrete case n_1^* and n_2^* represent the effective delays of plant and model
and the same conditions (i)-(iv) have to be satisfied. In (v) the zeros of $W_p(z)$ must

lie in the interior of the unit circle. Finally an upperbound on the gain k_p must also be known.

III. Error Models

The stability of adaptive control systems can best be analyzed using error models. As a result of the work done during the past few years it is now possible to describe error models for both discrete and continuous systems in a unified manner. The state error vector e(t) between plant and model can in general be represented by a nonlinear time varying differential equation of the form

$$\dot{e}(t) = f_1[e(t), \phi(t), t] \tag{2a}$$

or difference equation

$$e(k+1) = f_1[e(k), \phi(k), k] \tag{2b}$$

where $\phi(\cdot)$ is the parameter error vector. The error depends explicitly on t (or k) due to the reference input $r(\cdot)$. If θ^* represents the desired (but unknown) controller parameter vector and $\theta(t)$ its true value $\theta(t) - \theta^* \overset{\Delta}{=} \phi(t)$. The aim of adaptive control is to generate an adaptive law of the form

$$\dot{\phi}(t) = f_2[e(t), t] \tag{3a}$$

or

$$\Delta\phi(k) = f_2[e(k), k] \tag{3b}$$

using all available data such that all signals are uniformly bounded and $\lim_{t \to \infty} e(t) = 0$. In the following discussion we shall confine our attention to continuous systems but equivalent error models for the discrete case can be achieved by replacing $\dot{\phi}$ by $\Delta\phi$ and the continuous time index t by the discrete index k.

Error Model A: If $x_p(t)$ and $x_m(t)$ are respectively the state vectors of the plant and model

$$x_p(t) - x_m(t) \overset{\Delta}{=} e(t)$$

If $e_1(t) = c^T e(t)$ is a scalar output error, the error model is described by

$$\phi^T(t)u(t) = e_1(t) \qquad \text{(output error equation)}$$

$$\dot{\phi}(t) = \frac{-\alpha\Gamma e_1(t)u(t)}{1 + u^T(t)\Gamma u(t)} \qquad \text{(adaptive law)} \tag{4}$$

$$\alpha > 0 \qquad \Gamma = \Gamma^T > 0.$$

where $\phi(t)$ and $u(t)$ are (mxl) vectors. This error model has been analyzed extensively for various classes of inputs $u(\cdot)$.

For any input $u(\cdot)$ it can be shown that $v(t) = \phi^T(t)\Gamma^{-1}\phi(t)$ is a Lyapunov function so that $\|\phi\|$ is bounded and $\lim_{t\to\infty} \phi^T(t)\Gamma^{-1}\phi(t) = v_\infty < \infty$. However, very little can be said about the convergence of $\phi(t)$ as $t \to \infty$.

When $u(t)$ is uniformly bounded, $e_1(t)$ is also bounded and tends to zero as $t \to \infty$. From equation (4) it also follows that $\dot{\phi}(t) \to 0$ as $t \to \infty$ but $\phi(t)$ may not tend to a constant.

When $u(t)$ is "sufficiently rich" (persistently exciting or satisfies a mixing condition) it has been shown by many authors [15-17] that $\phi(t) \to 0$ as $t \to \infty$ and the system (4) is uniformly asymptotically stable. The analysis of this error model with stochastic instead of deterministic inputs was recently reported in [18]. As one might expect it is shown here that ergodic inputs under fairly weak conditions are "sufficiently rich" and hence result in the parameter error vector converging to zero.

Error Model B: In this case the state error equation is given by

$$W(s)\phi^T(t)u(t) = e(t) \tag{5}$$

where $W(s)$ is a known stable linear operator. [Note that 's' in all the equations is used to denote both the differential operator $\frac{d}{dt}$ as well as the Laplace transform variable.] If $W(s) = (sI-A)^{-1}b$ it can be shown that the adaptive law

$$\dot{\phi}(t) = -e^T(t)Pbu(t) \qquad\qquad P = P^T > 0 \qquad\left.\right\} \tag{6}$$
$$A^TP + PA = -Q \qquad Q > 0$$

results in uniformly bounded parameter and state error vectors. The comments made earlier for error model A are also applicable in this case when $u(\cdot)$ is uniformly bounded or sufficiently rich.

Error Model C: A third error model which has found wide application in the adaptive control problem can be described by the equations

$$W(s)v(t) = e_1(t) \; ; \; \phi^T(t)u(t) - u^T(t)\Gamma u(t)e_1(t) = v(t)$$
$$\dot{\phi}(t) = -\Gamma \, e_1(t)u(t) \tag{7}$$

where $W(s)$ is a strictly positive real transfer function.

When $W(s) \equiv 1$ error model C is identical to model A.

When $u(t)$ is uniformly bounded it has been shown in [12] that $e_1(t) \to 0$ as $t \to \infty$, $\phi(t)$ is uniformly bounded and $\dot{\phi}(t) \to 0$ as $t \to \infty$. Even when $u(t)$ grows in an unbounded fashion with time, i.e. $u(\cdot) \in L_e^\infty$ but $u(\cdot) \notin L^\infty$, it has been shown that $\dot{\phi}(\cdot) \in L^2$. It is this fact that has been used in the proofs of stability given in [3] and [2].

Error Model D: If in equation (7) $W(s)$ is not strictly positive real {SPR}, the adaptive law no longer assures global stability. Since such an error equation occurs frequently in adaptive control, the following error augmentation method is used to generate stable adaptive control laws.

If the error equation is of the form

$$W(s)\phi^T(t)u(t) = e_1(t) \tag{8}$$

an auxiliary error signal

$$[W(s)\theta^T(t) - \theta(t)^T W(s)]u(t) = [W(s)\phi^T(t) - \phi^T(t)W(s)]u(t) = e_2(t) \tag{9}$$

is generated and added to $e_1(t)$. If

$$\varepsilon(t) = e_1(t) + e_2(t) \tag{10}$$

$\varepsilon(t)$ is called an augmented error signal [14]. This yields the error equation in the form

$$\varepsilon(t) = \phi^T(t)W(s)u(t) = \phi^T(t)\zeta(t) \tag{11}$$

From error model A the adaptive law can be determined by inspection as

$$\dot{\phi}(t) = \frac{-\Gamma \ \varepsilon(t)\zeta(t)}{1 + \zeta^T(t)\Gamma\zeta(t)} \tag{12}$$

The adaptive law (12) assures that $\varepsilon(t) \to 0$ as $t \to \infty$ and $\dot{\phi}(t) \ \varepsilon \ L^2$ if $u(t)$ is uniformly bounded. This in turn assures that the true error $e_1(t)$ also tends to zero.

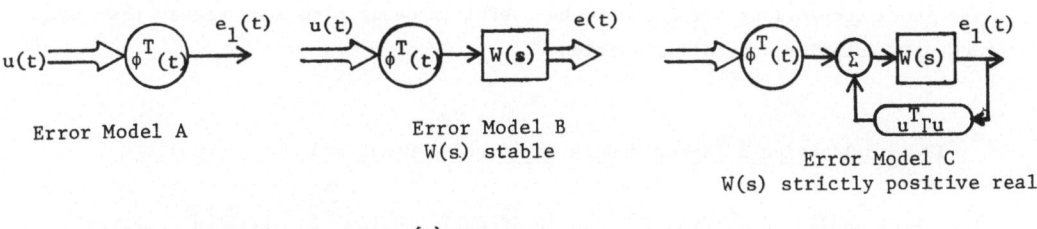

Error Model A

Error Model B
W(s) stable

Error Model C
W(s) strictly positive real

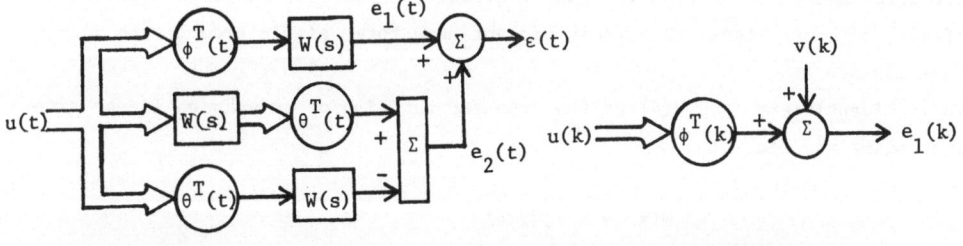

Error Model D

Error Model E

FIGURE 2

Error Model E: Error models A-D can be represented by homogeneous differential or difference equations. When external disturbances are present the corresponding non-homogeneous equations have to be analyzed. We describe here briefly the first error model with an additional input $v(k)$.

Let

$$\phi^T(k)u(k) + v(k) = e_1(k) \tag{13}$$

be the discrete error equation with $v(\cdot)$ an external scalar disturbance. If the same adaptive law (4) as in the noise free case is used here the resulting equations have the form

$$\phi(k+1) = \left[I - \frac{\alpha\Gamma u(k)u(k)^T}{1+u(k)^T\Gamma u(k)} \right] \phi(k) + \frac{\alpha\Gamma u(k)v(k)}{1 + u^T(k)\Gamma u(k)} \tag{14}$$

A complete analysis of equation (14) provides considerable insight into the effect of an external disturbance on the adaptive system.

(i)   If $u(\cdot)$ and $v(\cdot)$ are uniformly bounded and the homogeneous equation is uniformly asymptotically stable (i.e. input $u(\cdot)$ is sufficiently rich), the parameter error vector ϕ will also be uniformly bounded.

   In equation (14) it is seen that the term $\dfrac{\alpha\Gamma u(k)v(k)}{1 + u^T(k)\Gamma u(k)}$ is uniformly bounded provided $v(\cdot)$ is uniformly bounded.

(ii)   If the input $u(\cdot)$ is not sufficiently rich and the inputs $u(\cdot)$ and $v(\cdot)$ are correlated, the parameter error vector can grow in an unbounded fashion.

(iii)   If in (i), $u(\cdot)$ and $v(\cdot)$ are uncorrelated the expected value of $\phi(\cdot)$ will evolve asymptotically to zero with a finite variance. However if the parameter α is made time-varying such that $\sum_{k=0}^{\infty} \alpha(k) = \infty$ $\sum_{k=0}^{\infty} \alpha^2(k) < \infty$, $\phi(k)$ will tend to zero in the mean square sense.

The above conclusions are found to be useful in describing some recent results in stochastic adaptive control in section VII.

IV.   Application of Error Models to Adaptive Observers and Controllers

The error models of section III can be directly applied to adaptive observers, identifiers and controllers. We merely present here three simple examples to illustrate the ideas.

Example 1: Identifier. Let $W(s)$ be the transfer function of an unknown linear time-invariant plant and let

$$W(s) = \sum_{i=1}^{N} \alpha_i^* W_i(s)$$

where $W_i(s)$ are specified but the constants α_i^* are unknown. A model of the plant is constructed as $\sum\limits_{i=1}^{N} W_i(s)\hat{\alpha}_i$ where $\hat{\alpha}_i$ are estimates of α_i^* ($i = 1,..N$). If the plant and model are subject to the same input $u(t)$, the output error $e_1(t) \stackrel{\Delta}{=} y_p(t) - y_m(t)$ is given by

$$[\sum_{i=1}^{N} \hat{\alpha}_i W_i(s)]u(t) - [\sum_{i=1}^{N} \alpha_i^* W_i(s)]u(t) = e_1(t) \tag{15}$$

or

$$\phi^T(t)\zeta(t) = e_1(t) \tag{16}$$

where $\zeta^T(t) = [\zeta_1(t),...,\zeta_N(t)]$, $W_i(s)u(t) = \zeta_i(t)$
and $\hat{\alpha}_i(t) - \alpha_i^* = \phi_i(t)$.

Since equation (16) corresponds to the error model A the adaptive laws for updating $\hat{\alpha}_i(t)$ can be written by inspection as

$$\dot{\hat{\alpha}} = \frac{-\Gamma e_1(t)\zeta(t)}{1 + \zeta^T(t)\Gamma\zeta(t)} \qquad \Gamma = \Gamma^T > 0 \tag{17}$$

If $u(t)$ is uniformly bounded and $W_i(s)$ are asymptotically stable $e_1(t) \to 0$ as $t \to \infty$.

Example 2: In example 1 the poles of $W(s)$ are known and only the zeros of the model are adjusted by varying $\hat{\alpha}_i(t)$. In this example we consider a different parametrization of the plant so that both poles and zeros of the model can be adjusted.

Any transfer function $W(s)$ with n poles and m ($\leq$n-1) zeros can be represented in the form

$$W(s) = \frac{A(s)}{R(s)} \left[\frac{1}{1 + \frac{B(s)}{R(s)}} \right] \tag{18}$$

where $R(s)$ is a known Hurwitz polynomial of degree n and $A(s)$ and $B(s)$ are (n-1) degree polynomials in 's'. The identification of $W(s)$ now reduces to the identification of the coefficients of the polynomials $A(s)$ and $B(s)$.

If $u(t)$ is the input to the plant and $y_p(t)$ the corresponding output, we have

$$y_p(t) = \frac{-B(s)}{R(s)} y_p(t) + \frac{A(s)}{R(s)} u(t) \tag{19}$$

Let the model be such that its output $\hat{y}_p(t)$ is defined by

$$\hat{y}_p(t) = \frac{-\hat{B}(s)}{R(s)} y_p(t) + \frac{\hat{A}(s)}{R(s)} u(t) \tag{20}$$

where $\hat{A}(s)$ and $\hat{B}(s)$ are (n-1) degree polynomials in 's' whose coefficients can be adjusted. Once again the output error $e_1(t)$ can be expressed as

$$\phi^T(t)\zeta(t) = \begin{bmatrix} \phi_1(t) \\ \phi_2(t) \end{bmatrix}^T \begin{bmatrix} \zeta_1(t) \\ \zeta_2(t) \end{bmatrix} = e_1(t) \tag{21}$$

where ϕ_i are the parameter error vectors whose elements are the coefficients of the polynomials $\hat{A}(s) - A(s)$ and $\hat{B}(s) - B(s)$ respectively and $\zeta_i(t)$ ($i = 1,2$) are the states of filters with input $u(t)$ and $y_p(t)$ respectively.

Once again equation (21) is in the form of error model A and hence the adaptive law follows directly.

Example 3: ARMA Model. The discrete version of example 2 follows along very similar lines and has been used extensively in identification of discrete systems.

If a plant is described by the ARMA model

$$y(k) = \sum_{i=1}^{n-1} a_i^* y(k-i) + \sum_{i=1}^{m-1} b_i^* u(k-i) \tag{22}$$

and the estimates of a_i^* and b_i^* are $a_i(k)$ and $b_i(k)$ respectively, a model is constructed such that its output $\hat{y}(k)$ is given by

$$\hat{y}(k) = \sum_{i=1}^{n-1} a_i(k) y(k-i) + \sum_{i=1}^{m-1} b_i(k) u(k-i). \tag{23}$$

From equations (22) and (23) the error equation may be written as

$$e_1(k) = \phi^T(k) w(k) \tag{24}$$

where $\hat{y}(k) - y(k) = e_1(k)$; $\phi^T(k) = [a_1(k) - a_1^*, \ldots, b_{m-1}(k) - b_{m-1}^*]$ and $w^T(k) = [y(k-1), \ldots y(k-n+1), u(k-1), \ldots u(k-m+1)]$.

Again in this case equation (24) is in the form of error model A and the adaptive law for updating $\phi(k)$ (and hence $a_i(k)$ and $b_i(k)$) follows directly.

Comments on the Adaptive Control Problem: In all the above examples it was assumed that the input to the plant (and hence the model) was uniformly bounded. Such an assumption is quite reasonable in identification problems where the input can be chosen by the designer.

In the adaptive control problem, as pointed out in [19] this assumption can no longer be made. The plant input is generated as a feedback signal from a loop whose stability is under investigation. Hence the resulting error equations, even though they have the form of error model A have to be analyzed for unbounded inputs as well. It is this that leads to the stability problem of the adaptive control loop which is discussed in the next section.

V. Global Stability of Adaptive Control

While many schemes have been suggested in the literature during the past years for the adaptive control problem it is only recently that the global stability of some of these schemes was established. In view of the importance of this result we shall attempt in this section to outline briefly its highlights. In particular, we shall indicate how the stability problem arises, the difficulties encountered in resolving it and how the recent solutions suggested overcome these difficulties. Only a simplified version of the controller is discussed here to illustrate the principal ideas involved; for further details regarding the general problem the reader is referred to [11].

In Figure 3 $W_M(s)$ and $W_p(s)$ are the transfer functions of a reference model and an unknown plant, F is a known filter, $\zeta(t)$ an (mx1) vector output of F and $\theta^T(t)\zeta(t)$ a feedback control signal. It is known that if $\theta(t) \equiv \theta^*$, a constant vector, the transfer function of the plant together with the controller is $W_M(s)$ (showing that the controller structure is such that this is indeed the case is the algebraic part of the adaptive control problem).

If $\theta(t) - \theta^* \overset{\Delta}{=} \phi(t)$ is a parameter error vector, the output error $e_1(t) \overset{\Delta}{=} y_p(t) - y_m(t)$ may be related to $\phi(t)$ by

$$W_M(s)\phi^T(t)\zeta(t) = e_1(t) \tag{25}$$

If the additional feedback signal $\zeta(t)^T\Gamma\zeta(t)e_1(t)$ is included (as shown in dotted lines in Figure 3) the resulting error equation has the form

$$W_M(s)v(t) = e_1(t) \quad ; \quad v(t) = \phi^T(t)\zeta(t) - \zeta(t)^T\Gamma\zeta(t)e_1(t) \tag{26}$$

If $W_M(s)$ the model transfer function is strictly positive real and strictly proper, (26) corresponds to error model C. In such a case the adaptive law may be expressed as

$$\dot{\phi}(t) = -\alpha\Gamma e_1(t)\zeta(t) \qquad \alpha > 0 \qquad \Gamma = \Gamma^T > 0$$

and results in $e_1(t)$ and $\phi(t)$ being bounded. Since the output $y_m(t)$ is uniformly bounded and $y_p(t) - y_m(t) \overset{\Delta}{=} e_1(t)$, it follows that $y_p(t)$ and $\zeta(t)$ are also uniformly bounded. Hence $e_1(t) \to 0$ as $t \to \infty$ and all the signals in the plant and controller are uniformly bounded.

If in equation (26) $W_M(s)$ is not strictly positive real the error model C can no longer be used. Instead, the error model D having an augmented error signal is required to prove stability. The auxiliary signal $e_2(t)$ where

$$[\phi^T(t)W_M(s) - W_M(s)\phi^T(t)]\zeta(t) = e_2(t)$$

is now added to the true error signal $e_1(t)$ as shown in Figure 4 to generate the augmented error $\varepsilon(t)$. The corresponding adaptive control law may be expressed as

$$\dot{\phi}(t) = \frac{-\alpha\Gamma\varepsilon(t)\zeta_1(t)}{1 + \zeta_1^T(t)\Gamma\zeta_1(t)} \quad ; \quad W_M(s)\zeta(t) = \zeta_1(t) \tag{27}$$

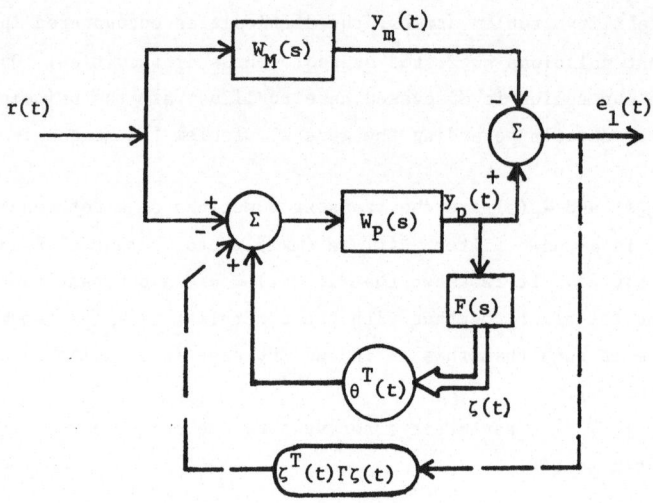

FIGURE 3

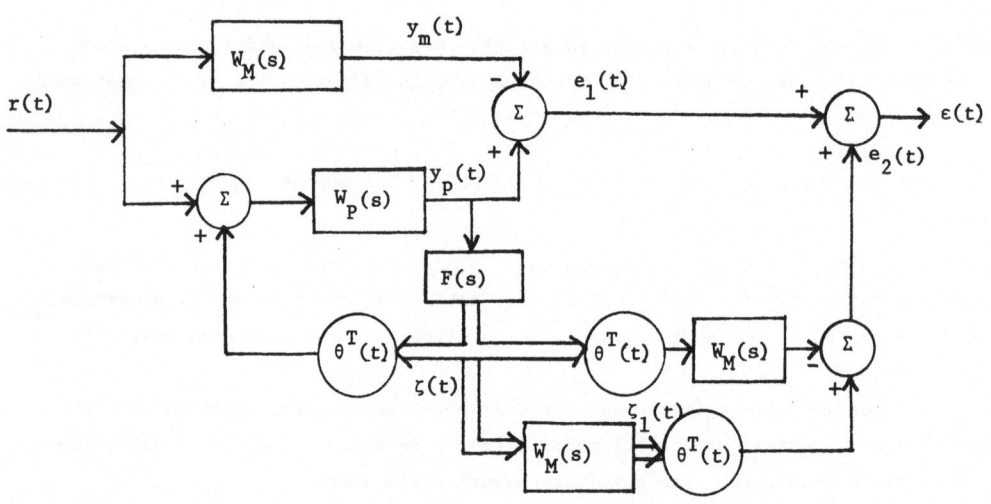

FIGURE 4

This results in a bounded augmented error $\varepsilon(t)$. and $\dot{\phi}(\cdot) \in L^2$.

While the above procedure assures the uniform boundedness of the augmented error $\varepsilon(t)$ it does not directly follow that the true error $e_1(t)$ (Figure 4) is also uniformly bounded. Since $y_p(t) - y_m(t) = e_1(t)$ and $e_1(t) + e_2(t) = \varepsilon(t)$ the possibility exists that both $y_p(t)$ and $e_2(t)$ grow in an unbounded fashion even while $\varepsilon(t)$ is uniformly bounded. To demonstrate that this cannot happen is the heart of the stability problem.

In [3] it is shown that the plant output $y_p(t)$ can be expressed as the sum of the output of a feedback system and other signals which are uniformly bounded. The feedback system consists of a uniformly asymptotically stable linear system in the forward path and $\dot{\phi}(t)$ in the feedback path. Since $\dot{\phi}(\cdot) \in L^2$ this results in a uniformly bounded $y_p(t)$ (and hence $e_2(t)$ and $e_1(t)$). The proof is seen to crucially depend on $\dot{\phi}(\cdot)$ being an L^2 function and this, in turn, emphasizes the importance of the error models A,C and D.

VI. Equivalence of Adaptive Schemes

In MRAC and STR, over the years, a variety of schemes have been developed for the control problem. Recently, it has been recognized [5-9] that many of these schemes are similar either by being special cases of one algorithm or by giving rise to identical error equations. However it is not always known that the resulting systems are globally stable. Since global stability is an important prerequisite for any reliable adaptive controller we shall confine our attention in this section to schemes that have been shown to have this property.

Two adaptive systems A and B will be considered to be equivalent (A∿B) if

(i) the error equations can be made identical by the proper parametrization of
plant and controller

and (ii) the overall system is globally stable.

By this definition of equivalence the schemes in [1,2,3,4 and 13] are equivalent. In [7] it is shown that Input Matching, Error Augmentation (MRAC), STR, and Output Error Identification all lead to updating equations of the form

$$\theta(k+1) = \theta(k) + \frac{\Gamma e_1(k)u(k)}{1 + u^T(k)\Gamma u(k)} \tag{28}$$

or parameter error equations

$$\Delta\phi(k+1) = \frac{\Gamma e_1(k)u(k)}{1 + u^T(k)\Gamma u(k)} \tag{29}$$

where $e_1(k)$ is the output error. This corresponds to the error model A discussed in section II. In those cases where $u(k)$ can be assumed to be uniformly bounded (e.g. identifiers and adaptive observers) $e_1(k) \to 0$ as $k \to \infty$ and $\phi(k)$ is uniformly bounded. Hence such algorithms are equivalent according to our definition. When $u(k)$ cannot be assumed to be bounded (as in the control problem) error augmentation is needed as described in the previous section and the parameter error equations have the same form as (29) with $e_1(k)$ replaced by the augmented error $\varepsilon(k)$. All the equivalent schemes currently known have this feature either explicitly or implicitly and it is the presence of the augmented error which enables the global stability of the schemes to be demonstrated as described in the previous section.

In [6] it was shown that by a particular parametrization of the plant, indirect control and direct control of continuous time systems lead to identical error equations. The same results also carry over to the discrete case. In [1], the proof of global stability of a discrete adaptive system is given and using a similar approach the global stability of continuous direct control systems is shown in [3]. In view of the results presented in [6] it follows that indirect adaptive control systems which result in global stability can also be designed. In [4] such a globally stable STR is derived. We describe briefly below how the use of the error model D enables the equivalence of the schemes in [1] and [4] to be established.

A discrete plant is described by the model

$$A(z^{-1})y(k) = z^{-d}B(z^{-1})u(k)$$

where A and B are polynomials in z^{-1} and d is a specified time delay. The transfer function of a reference model is z^{-d} and it is desired to make the output $y(k)$ of the plant track the output $y^*(k)$ of the model asymptotically so that

$$\lim_{k \to \infty} |y(k) - y^*(k)| = 0$$

Using the identification scheme described in section IV the coefficients of A and B are estimated and the parameters of a feedback controller are adjusted so that the overall transfer function is z^{-d}. If $\phi(k)$ is the parameter error vector and $\zeta(k)^T = \{y(k), y(k-1), \ldots y(k-n+1), u(k), \ldots u(k-m-d+1)\}$ the error equations may be expressed as

$$z^{-d}[\phi^T(k)\zeta(k)] = e_1(k) \overset{\Delta}{=} y(k) - y^*(k) \tag{30}$$

Since z^{-d} is not strictly positive real an auxiliary error signal is generated as described in error model D and added to $e_1(k)$. This results in the error equation

$$\phi^T(k)\zeta(k-d) = \varepsilon(k) \tag{31}$$

and from error model A we have the adaptive control law

$$\Delta\phi(k) = \frac{-\Gamma\varepsilon(k)\zeta(k-d)}{1 + \zeta(k-d)^T\Gamma\zeta(k-d)} \tag{32}$$

which is identical to that obtained in [3].

In a similar manner it can also be shown that all the globally stable schemes known at the present time are equivalent.

VII. Some Results in Stochastic Systems

In sections I-VI we have analyzed several error models and applied them to deterministic adaptive observers and controllers. As mentioned in section I a significant body of research exists on stochastic STR [8,9,20] and in this section we briefly outline how the error model E can be used to describe some of these results. Since STR is formulated only for discrete time systems we shall confine our attention only to such systems in this section.

Let a plant be defined by

$$A(z^{-1})y(k) = z^{-d}B(z^{-1})u(k) + C^{1}(z^{-1})w(k) \tag{33}$$

where

$$A(z^{-1}) = 1 + a_1 z^{-1} + \ldots + a_n z^{-n}$$

$$B(z^{-1}) = b_0 + b_1 z^{-1} + \ldots + b_m z^{-m} \tag{34}$$

$$C(z^{-1}) = c_0 + c_1 z^{-1} + \ldots + c_n z^{-n}$$

$\{u(k)\}$ is the input sequence and $\{w(k)\}$ is a disturbance sequence. Our aim them is to determine the effect of the disturbance on the system when the same adaptive laws are used as in the deterministic case and study the nature of convergence of the output and parameter errors. Figure 5 represents a block diagram of the STR with the disturbance input.

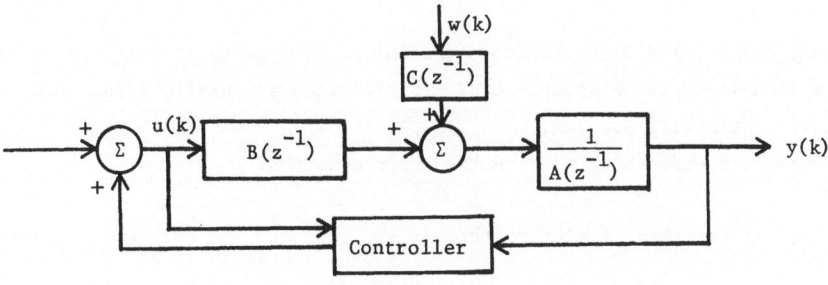

FIGURE 5

In the absence of the disturbance, using the adaptive law (32) the controller
will evolve asymptotically to a steady state such that the transfer function of the
adaptive loop is z^{-d} (where d is the inherent delay in the plant) as described in
section VI. The transfer function to the output as seen by the disturbance $w(\cdot)$ in
this case is then

$$\frac{1}{z^d} \frac{C(z^{-1})}{B(z^{-1})} \cdot z^{-d} = \frac{C(z^{-1})}{B(z^{-1})} \tag{35}$$

The output of the controller $u(k)$ is of the form

$$u(k) = \theta^T(k) \zeta(k) \tag{36}$$

where $\theta(k)$ is a control parameter vector and $\zeta(k)$ is a signal vector generated by the
controller. If it is assumed that a constant desired parameter vector θ^* exists (such
that the loop transfer function is z^{-d}) and $\theta(k) - \theta^* \overset{\Delta}{=} \phi(k)$ the error equations may
be expressed as

$$e_1(k) = z^{-d}[\phi^T(k)\zeta(k)] + v(k) \tag{37}$$

where $\qquad \dfrac{C(z^{-1})}{B(z^{-1})} w(k) = v(k)$

If $v(k) \equiv 0$ the adaptive law (32) using an augmented error signal is

$$\Delta\theta(k) = \Delta\phi(k) = \frac{-\Gamma\varepsilon(k)\zeta(k-d)}{1 + \zeta^T(k-d)\Gamma\zeta(k-d)}$$

where $\varepsilon(k) = e_1(k) + e_2(k)$ and $e_2(k) = \theta^T(k)\zeta(k-d) - \theta^T(k-d)\zeta(k-d)$.

From the results of sections V and VI it follows that $\varepsilon(k)$, $e_1(k)$ and $e_2(k) \to 0$ as
$k \to \infty$.

In the case when $v(k) \not\equiv 0$ the adaptive control problem is to analyze the behavior
of the various signals in the adaptive loop. If $v(k)$ is a stochastic input this becomes a stochastic stability problem.

With an augmented error $\varepsilon(k)$ we have the error equation

$$\phi^T(k)\zeta(k-d) + v(k) = \varepsilon(k) \tag{38}$$

which is of the form given in error model E. Hence the results given in section II
may be applied here directly.

Since the stochastic stability of the adaptive loop is under investigation and
$\zeta(k)$ is a feedback signal it is no longer possible to assume that $\zeta(k-d)$ is bounded.
The adaptive law (32) now yields the non-homogeneous difference equation

$$\phi(k+1) = \left[I - \frac{\alpha \Gamma \zeta(k-d) \zeta^T(k-d)}{1 + \zeta^T(k-d) \Gamma \zeta(k-d)} \right] \phi(k) + \frac{\alpha \Gamma \zeta(k-d) v(k)}{1 + \zeta^T(k-d) \Gamma \zeta(k-d)} \tag{39}$$

Using the results for error model E we conclude that

   (i)   if the input $\zeta(k-d)$ is sufficiently rich and $v(k)$ is uniformly bounded, the error $e_1(k)$ is also uniformly bounded.

  (ii)   if the input $\zeta(k-d)$ is not sufficiently rich and $v(k)$ and $\zeta(k-d)$ are correlated it is possible for the adaptive loop to have unbounded solutions.

 (iii)   if $\zeta(k-d)$ is sufficiently rich and $v(k)$ is a white noise sequence and $v(k)$ and $\zeta(k-d)$ are uncorrelated the expected value of $\phi(k)$ tends to zero asymptotically. By making Γ a time-varying matrix (as shown in section E) $\phi(k)$ can be made to tend to zero in a mean square sense.

 (iv)   if $w(k)$ is a white noise sequence and $\dfrac{C(z^{-1})}{B(z^{-1})} = D(z^{-1})$ is strictly positive real, equation (38) may be rewritten as

$$D^{-1}(z^{-1}) \; [\phi^T(k) \zeta(k-d) + v(k)] =$$

$$= D^{-1}(z^{-1}) \phi^T(k) \zeta(k-d) + w(k) = D^{-1}(z^{-1}) \varepsilon(k) = \varepsilon_1(k) \tag{40}$$

and a similar analysis as before can be carried out.

VIII.   Comments and Conclusions

The design of adaptive observers and controllers can be broadly divided into two parts. The first, which is an algebraic part, is concerned with the choice of a suitable structure for the observer or controller so that a solution to the problem exists in the form of a set of constant parameters. The second part is analytic in nature and deals with the manner in which the observer or controller parameters are to be adjusted so that they evolve to the desired values. The error models described in section III are central to both these aspects of the design process.

In observers and identifiers the input to the plant can be assumed to be uniformly bounded. This simplifies considerably the analysis of the error equations. In adaptive control such an assumption cannot be made and this in turn results in the complex problem of stability of the adaptive loop.

All the globally stable adaptive schemes known at the present time are equivalent in that they have the same error equations and the same adaptive laws. The fact that several different independent analyses led to one common overall scheme suggests that this important result was not entirely fortuitous. Rather, it is quite probable that with the techniques that are currently available these represent the only class of systems for which a globally stable adaptive controller can be designed. It is also worth mentioning that other schemes which have proved successful in practice may be stable even though their global stability has not so far been demonstrated.

The assumptions in section II regarding the plant are required by all the known stable adaptive schemes. Even in the single input - single output case these assumptions are much too restrictive. n_1^*, the relative degree of a plant is rarely known exactly and assumption (iii) implies that the method cannot be extended to distributed parameter systems. Further, the corresponding assumptions for multivariable systems are likely to be even more restrictive regarding a priori information about the plant that may be needed. In view of this, it is the authors' opinion that the adaptive control problem should be reformulated so that less is demanded of the controller allowing less restrictive assumptions to be made. Instead of requiring perfect model matching so that the output error goes to zero at $t \to \infty$ emphasis in the future should be on globally stable schemes in which the output error is merely bounded.

With the advent of sophisticated microprocessors and the recent interest in adaptive control, attempts will be made in many areas to apply the new techniques. Inputs from such real applications would be most helpful at this stage.

References

[1] K. S. Narendra and Y. H. Lin, "Stable Discrete Adaptive Control," to appear in IEEE Trans. Auto. Cont., June 1980.

[2] A. S. Morse, "Global Stability of Parameter-Adaptive Control Systems," to appear in IEEE Trans. Auto. Cont., June 1980.

[3] K. S. Narendra, Y. H. Lin, and L. S. Valavani, "Stable Adaptive Controller Design - Part II, Proof of Stability," to appear in IEEE Trans. Auto. Contr., June 1980.

[4] G. C. Goodwin, P. J. Ramadge, and P. E. Caines, "Discrete Time Multi-Variable Adaptive Control," to appear in IEEE Trans. Auto. Control, June 1980.

[5] B. Egardt, "Unification of Some Continuous-Time Adaptive Control Schemes," IEEE Trans. Auto. Cont., Vol. AC-24, No. 4, pp. 588-592, Aug. 1979.

[6] K. S. Narendra and L. S. Valavani, "Direct and Indirect Model Reference Adaptive Control," Automatica, Vol. 15, pp. 653-654, Nov. 1979.

[7] C. R. Johnson, Jr., "Input Matching, Error Augmentation, Self-Tuning, and Output Error Identification: Algorithmic Similarities in Discrete Adaptive Model Following," to appear in IEEE Trans. Auto. Cont.

[8] L. Ljung and I. D. Landau, "Model Reference Adaptive Systems and Self-Tuning Regulators - Some Connections," Proc. 7th IFAC Congress, Vol. 3, pp. 1973-1980, June 1978.

[9] I. D. Landau, "Model Reference Adaptive Control and Stochastic Self-Tuning Regulators - Towards Cross-Fertilization," Presented at AFOSR Workshop on Adaptive Control, Univ. of Illinois, Urbana-Champaign, May 1979.

[10] K. S. Narendra, "Stable Identification Schemes," System Identification: Advances and Case Studies, Academic Press, New York, 1976.

[11] K. S. Narendra and Y. H. Lin, "Design of Stable Model Reference Adaptive Controllers," Proceedings of the Workshop on Applications of Adaptive Control, Yale University, August 1979.

[12] Y. H. Lin and K. S. Narendra, "A New Error Model for Adaptive Systems," to appear in IEEE Trans. Auto. Cont., June 1980.

[13] B. Egardt, "Stability Analysis of Discrete Time Adaptive Control Schemes," to appear in IEEE Trans. Auto. Cont.

[14] R. V. Monopoli, "Model Reference Adaptive Control with an Augmented Error Signal," IEEE Trans. Auto. Cont., Vol. AC-19, pp. 474-484, Oct. 1974.

[15] J. S-C. Yuan and W. M. Wonham, "Probing Signals for Model Reference Identifica-tion," IEEE Trans. Auto. Control, Vol. AC-22, No. 4, pp. 530-538, Aug. 1977.

[16] A. P. Morgan and K. S. Narendra, "On the Uniform Asymptotic Stability of Certain Linear Nonautonomous Differential Equations," SIAM J. Contr. and Opt., Vol. 15, No. 1, pp. 5-24, Jan. 1977.

[17] M. M. Sondhi and D. Mitra, "New Results on the Performance of a Well-Known Class of Adaptive Filters," Proc. IEEE, Vol. 64, No. 11, pp. 1583-1597, 1976.

[18] R. R. Bitmead and B.D.O. Anderson, "Exponentially Convergent Behavior of Simple Stochastic Estimation Algorithms," Proc. 17th IEEE Conf. on Decision and Control, San Diego, 1979.

[19] K. S. Narendra and L. S. Valavani, "Stable Adaptive Observers and Controllers," Proc. IEEE, Vol. 64, No. 8, pp. 1198-1208, Aug. 1976.

[20] L. Ljung, "On Positive Real Transfer Functions and the Convergence of Some Re-cursive Schemes," IEEE Trans. Auto. Contr., Vol. AC-22, No. 4, pp. 539-551, Aug. 1977.

DESIGN OF MULTIVARIABLE ADAPTIVE CONTROL SYSTEMS WITHOUT THE NEED FOR PARAMETER IDENTIFICATION[*]

by

K. Sobel, H. Kaufman and O. Yekutiel
Electrical and Systems Engineering Department
Rensselaer Polytechnic Insitute
Troy, New York 12181

ABSTRACT

Implicit model reference adaptive control algorithms for multi-input multi-output plants are reviewed, and the procedures for satisfying the constraints sufficient for stability are presented. Examples showing the applicability of the theory are discussed.

1. INTRODUCTION

A general approach for designing an implicit adaptive controller (i.e. without a parameter identifier) for multivariable linear systems which do not have to satisfy the conditions of perfect model following (PMF) was proposed by Mabius and Kaufman (in 1976).[1] By making direct use of Lyapunov procedures, they were able to develop a control algorithm with the following characteristics:

- Asymptotically stable when PMF is valid.
- Stable (in the sense of a bounded error) when PMF does not hold, provided that certain inequality constraints independent of the model can be satisfied for all possible plant parameter variations.
- Independent of having to have explicit estimates for plant parameters.
- Easily implementable in a microprocessor configuration.

Subsequent efforts reported in 1979[2] showed that for a step command, the error between plant and model outputs goes to zero provided that the number of controls is equal to the number of outputs and also provided that the plant input-output transfer matrix is positive real for some feedback gain matrix. Although somewhat restrictive in application because of this latter constraint, it should be noted that the implementation of such an adaptive system, once designed, can easily be achieved with microprocessor circuitry. Thus further research has been devoted towards increasing the applicability of multivariable implicit adaptive controllers.

To this effect, the following results are discussed:

- Replacement of the step command constraint by any command that can be generated by a linear system;
- Steady state behavior;
- Procedures for determining if a system is positive real over a given range of parameter variations and procedures for compensating accordingly when necessary;

- The use of interactive computer graphics as an aid in tuning the adaptive control parameters;
- A design study using the F-8 aircraft.

2. PROBLEM FORMULATION

The continuous linear model reference control problem will be solved for the linear process equations:

$$\dot{x}_p(t) = A_p\, x_p(t) + B_p\, u_p(t) \tag{2-1}$$

$$y_p(t) = H_p\, x_p(t) \tag{2-2}$$

where

$x_p(t)$ is the (n x 1) plant state vector

$u_p(t)$ is the (m x 1) control vector

$y_p(t)$ is the (m x 1) plant output vector

and A_p, B_p are matrices with the appropriate dimensions. The range of the plant parameters is assumed to be bounded and all possible (A_p, B_p) pairs are assumed controllable and output stabilizable. The objective is to find, without explicit knowledge
of A_p and B_p, the control $u_p(t)$ such that the plant output vector $y_p(t)$ approximates
"reasonably well" the output of the following model forced by a step command:

$$\dot{x}_m(t) = A_m\, x_m(t) + B_m\, u_m(t) \tag{2-3}$$

$$y_m(t) = H_m\, x_m(t) \tag{2-4}$$

where

$x_m(t)$ is the (n x 1) model state vector

$u_m(t)$ is the (m x 1) model step input or command

$y_m(t)$ is the (m x 1) model output vector

and A_m, B_m are matrices with the appropriate dimensions. Furthermore, the controller
structure is to be such that as time approaches infinity, the error $y_p - y_m$ approaches
zero. This is to be valid regardless of model structural characteristics, i.e., even
if the conditions for perfect model following do not hold. Thus, it is being assumed
that it is not necessary for matrices G, H to exist such that

$$(A_m - A_p) = B_p\, G \quad \text{and} \quad B_m = B_p\, H.$$

To facilitate the controller development, it is useful to incorporate the command
generator tracker concept developed by Broussard at TASC.[3] When $y_p = y_m$ for $t \geq 0$,
(i.e. perfect tracking occurs), the corresponding plant state and control trajectories
will be denoted as $x_p^{*}(t)$ and $u_p^{*}(t)$ respectively. By definition then the ideal plant
response x_p^{*} is such that

$$H_p\, x_p^{*}(t) = H_m\, x_m(t) \tag{2-5}$$

and furthermore

$$\dot{x}_p^{*}(t) = A_p\, x_p^{*}(t) + B_p\, u_p^{*}(t) \quad \text{for all } t \geq 0 \tag{2-6}$$

In addition, the ideal plant response will be assumed to satisfy the following equation[3]

$$\begin{bmatrix} x_p^*(t) \\ u_p^*(t) \end{bmatrix} = \begin{bmatrix} S_{11} & S_{12} \\ S_{21} & S_{22} \end{bmatrix} \begin{bmatrix} x_m(t) \\ u_m \end{bmatrix}$$

where S_{11}, S_{12}, S_{21} and S_{22} are matrices (with the appropriate dimensions) whose existence can be shown under rather mild restrictions provided that u_m is a step and that the number of controls, m is not less than the number of outputs, q. Thus, we define a new error $e = x_p^*(t) - x_p(t)$ and seek a controller which guarantees that $e \rightarrow 0$ as $t \rightarrow \infty$. We observe that when $x_p(t) = x_p^*(t)$, we have $H_p x_p(t) = H_p x_p^*(t)$. By definition we know that $H_p x_p^*(t) = H_m x_m(t)$ and therefore $H_p x_p(t) = H_m x_m(t)$ or $y_p = y_m$ which is the result we require. Introducing this new error into the model reference adaptive control problem results in an asymptotically stable control law. This result is in contrast to the bounded error stability achieved with the error being defined as in Ref. 1 ($e = x_m - x_p$).

It should be noted here at the outset that even though the subsequent CGT based analysis is valid only when u_m is a step command, any command signal which can be described as the solution of a differential equation forced by a step input (or zero) can be used, provided it is augmented to the model state and not to the model output.

3. ADAPTIVE CONTROL ALGORITHMS

The resulting adaptive control algorithm is of the form:

$$u_p = K_e(y_m - y_p) + K_m x_m + K_u u_m + K_r r \qquad (3-1)$$

where

$$r = \begin{array}{c} y_m - y_p \\ x_m \\ u_m \end{array}$$

and where the gains are such that:

$$K_r = K_I + K_p \qquad (3-2)$$

$$K_p = (y_m - y_p)r^T \bar{T} \qquad (3-3)$$

$$\dot{K}_I = (y_m - y_p)r^T T \qquad (3-4)$$

The closed loop system which results from this algorithm gives rise to an asymptotically stable error provided that the matrices T and $\bar{T}$ are positive definite and also provided that the plant input-output transfer function matrix $Z(s) = H_p(SI - A_p + B_p \tilde{K}_e H_p)^{-1} B_p$ is positive real for some feedback gain matrix $\tilde{K}_e$.[2] Equivalently, this implies that there must exist a positive definite matrix $P(A_p, B_p)$, a positive definite matrix $Q(A_p, B_p)$ and a gain matrix $\tilde{K}e(A_p, B_p)$ such that

$$(1) \quad P(A_p - B_p \tilde{K}_e H_p) + (A_p - B_p K_e H_p)^T P < 0 \qquad (3-5)$$

(2) $H_p = Q \, B_p^T \, P$ (3-6)

4. STABILITY DISCUSSION

As described in Ref. 2, stability under constraints (3-5) and (3-6) can be shown using the Lyapunov function

$$V(e, K_I) = e^T It)P \, e(t) + TR \left[S(K_I - \tilde{K}) \, T^{-1}(K_I - \tilde{K})^T \, S^T \right] \qquad (4-1)$$

where

P is an n x n positive definite symmetric matrix

$\tilde{K}$ is an m x n_r matrix (unspecified)

S is an m x m nonsingular matrix.

and e is the error $(x_p^*(t) - x_p(t))$ which satisfies the equation

$$\begin{aligned} \dot{e}(t) &= \dot{x}_p^*(t) - \dot{x}_p(t) \\ &= A_p \left[x_p^*(t) - x_p(t) \right] + B_p \left[u_p^*(t) - u_p(t) \right] \end{aligned} \qquad (4-2)$$

The corresponding time derivative of V thus (see Ref. 2) becomes:

$$\begin{aligned} \dot{V} &= e^T \left[P(A_p - B_p \, \tilde{K}_e \, H_p) + (A_p - B_p \, K_e \, H_p)^T \, P \right] e \\ &\quad -2 \, e^T \, P \, B_p(S^T S)^{-1} \, B_p^T \, Pe \, r^T \, \bar{T} \, r \end{aligned} \qquad (4-3)$$

Thus, provided that the constraints (3-5, 3-6) are satisfied, $\dot{V}$ will be a negative definite function of e. This implies that e and $\dot{e} \to 0$. Furthermore from Eq. (4-2) it can be seen that if B_p has full rank, then this steady state condition on e will in turn force $u_p \to u_p^*$.

5. CONSTRAINT SATISFACTION

5.1 Introduction

Given the bounds on (A_p, B_p) an implementable procedure is needed in order to determine the satisfaction of the previous constraints. To this effect section 5.2 discusses two procedures for demonstrating positive realness, while section 5.3, discusses possible design procedures to follow if these conditions are not satisfied.

5.2 Constraint Satisfaction

5.2.1 Constraint Satisfaction Using Frequency Domain Considerations

From a result due to Mabius[1] it can be shown that (3-5) and (3-6) will be satisfied if $Z(s) = H_p(sI - A_p + B_p \, \tilde{K}_e \, H_p)^{-1} B_p$ is strictly positive real. A modification of a procedure proposed by Mabius[1] is presented for selecting Q, and $\tilde{K}_e$ such that (5-1) and (5-2) are satisfied for H_p.

Step 1. Choose the matrix product $\tilde{K}_e H_p$ such that the eigenvalues of $A_p - B_p$ $\tilde{K}_e H_p$ have negative real parts.

Step 2. Define $Z(s) = H_p(sI - A_p + B_p \, \tilde{K}_e \, H_p)^{-1}$ and define: (5-1)

$$F(w) = Z(jw) + Z^T(-jw) \qquad (5-2)$$

Step 3. Validate that H_p is such that

(a) $[(A_p - B_p \tilde{K}_e H_p), H_p]$ is observable for all admissible A_p and B_p.

(b) $F(w)$ is positive definite for all w.

This step is perhaps best carried out by checking that all m principal minors of $F(w)$ are positive.[4,5]

5.2.2 Constraint Satisfaction Based upon Time Domain Considerations

A time domain approach for showing strict positive realness of the transfer function

$$Z(s) = J + H(sI - F)^{-1} G \tag{5-3}$$

is based upon the following result.

Assume $Z(\infty) < \infty$ and that (F, G, H, J) is a minimal realization of $Z(s)$ with F having all of its eigenvalues in the left half plane. Then $Z(s)$ is a strictly positive real matrix of rational functions of s if and only if there exists a negative definite matrix π satisfying the equation:[16]

$$\pi(F - GR^{-1} H) + (F^T - H^T R^{-1} G^T)\pi - \pi GR^{-1} G^T\pi - H^T R^{-1} H = 0 \tag{5-4}$$

where $J + J^T = R$.

However since in the problem of interest $J=0$, R^{-1} will not exist, an alternate approach is suggested based upon a test for the discrete positive realness of a transformed system.[6] To this effect define the following quantities:

$$A = (I + F)(I - F)^{-1} \tag{5-5a}$$

$$B = \frac{1}{\sqrt{2}} (A + I)G \tag{5-5b}$$

$$C = \frac{1}{\sqrt{2}} (A^T + I)H^T \tag{5-5c}$$

$$U = R + C^T (A + I)^{-1} B + B^T(A^T + I)^{-1} C \tag{5-5d}$$

Then $Z(s)$ as defined in Eq. (5-3) will be strictly positive real (for any J including $J=0$) if and only if the following recursive difference equation has a negative definite steady state solution.[6]

$$\pi(n + 1) = A^T \pi(n) A$$
$$- [A^T \pi(n) B + C][U + B^T \pi(n) B]^{-1} [B^T\pi(n) A + C^T] \tag{5-6}$$
$$\pi(0) = 0$$

Thus to apply this test to the model reference adaptive control problem, A, B, C, U would be computed using the following equivalences:

$$F = A_p - B_p \tilde{K}_e H_p \tag{5-7a}$$

$$G = B_p \tag{5-7b}$$

$$H = H_p \tag{5-7c}$$

$$J = 0 \tag{5-7d}$$

5.3 Design of Suitable Output Configurations

5.3.1 Full State Availability ($H_p = I$)

If measurements for all states are available, then $H_p = I$, and it is possible to

find an output matrix that will result in positive realness by solving the following
linear quadratic regulator problem.[7]

$$\text{MIN} \int_0^\infty (x^T Q x + u^T R u)\, dt \tag{5-8}$$

Subject to:

$$\dot{x} = A_p x + B_p u \tag{5-9}$$

The well known solution (when it exists) to this problem is

$$u = -K x \tag{5-10}$$

where

$$K = +R^{-1} B_p^T P \tag{5-11}$$

$$\text{Re } \lambda(A_p - B_p K) < 0 \tag{5-12}$$

and

$$A_p^T P + P A_p - P B_p R^{-1} B_p^T P + Q = 0 \tag{5-13}$$

Then selection of both H_p and $\tilde{K}_e H_p$ as K will result in the positive realness of
the matrix $H_p(sI - A_p + B_p \tilde{K}_e H_p)^{-1} B_p$.

Since such a design of H_p requires apriori knowledge of A_p and B_p, its use is
contingent upon the availability of nominal A_p and B_p matrices. The robustness of
this output matrix in the sense of retaining positive realness for deviations in A_p
and B_p must then be examined.

5.3.2   Dynamic Compensation

If somehow it can be determined that the transfer matrix

$$Z(s) = C(s) H_p(sI - A_p + B_p \tilde{K}_e H_p)^{-1} B_p \tag{5-14}$$

is positive real for some stabilizing $\tilde{K}_e$, then an adaptive model following controller
exists such that

$$\lim_{t \to \infty} L^{-1}[C(s) Y_p(s)] = \lim_{t \to \infty} L^{-1}[C(s) Y_m(s)]$$

Note that if $C(s)$ is strictly Hurwitz, then y_p itself will approach y_m. However implementation of such compensation presents several problems. First $C(s)$ must be robust enough such that (5-16) is positive real for wide enough variations in A_p and
B_p. Second, $C(s)$ will most probably require differentation of the output vector y_p.

This differentation could possibly be alleviated using state variable filters
as suggested by Courtiol provided that the rate of adaptation is much slower than the
system transients.[8] To illustrate this concept, consider the following definitions:

$$Y_{p_1}(s) = C(s) Y_p(s) \tag{5-15a}$$

$$Y_{p_{1_F}}(s) = C^{-1}(s) Y_{p_1}(s) = Y_p(s) \tag{5-15b}$$

$$U_{pF}(s) = C^{-1}(s) U_p(s) \tag{5-15c}$$

$$Y_{m_1}(s) = C(s) Y_m(s) \tag{5-15d}$$

$$y_{m_{1F}}(s) = C^{-1}(s) \, Y_{m_1}(s)(s) = Y_m(s) \tag{5-15e}$$

Note that since the transfer function between $y_{p_{1F}}$ and U_{p_F} is the same as eq. (5-14), (i.e. positive real), a model reference adaptive controller exists such that $y_{p_{1_F}}$ will asymptotically approach $Y_{m_{1_F}}$. In particular

$$U_{p_F}(t) = (K_{I_F} + K_{p_F}) \, r_F \tag{5-16}$$

where the filtered subscript on K_{I_F}, K_{p_F}, r_F denotes dependence upon the corresponding filtered inputs, filtered outputs, and filtered model states. The actual control $U_p(t)$ to be applied to the system is:

$$U_p(t) = L^{-1}[C(s) \, U_{p_F}(s)] \tag{5-17}$$

Assuming that the adaptation is slow enough (i.e. T and $\overline{T}$ sufficiently small):[8]

$$U_p(t) \underset{\sim}{\sim} (K_{I_F} + K_{p_F}) \, r(t) \tag{5-18}$$

5.3.2.2 Feedforward of the Control

If equations (3-3), (3-4) are modified to be

$$K_p = v \, r^T \, \overline{T} \tag{5-19a}$$

$$\dot{K}_I = v \, r^T \, T$$

where

$$v = Q \, H_p \, e + G(U_p^* - U_p + \tilde{K}_e \, H_p \, e) \tag{5-20}$$

Then asymptotic stability can be proven provided that:[9]

$$J + H_p(SI - A_p + B_p \tilde{K}_e H_p)^{-1} B_p$$

is positive real and

$$Q^{-1}G > J.$$

Note that this is not as severe as the previous constraint since it allows the addition of the matrix J to possibly compensate for any negativeness in $H_p(SI - A_p + B_p \tilde{K}_e H_p)^{-1} B_p$. However implementation does require

- Apriori knowledge of U_p^*.
- Apriori knowledge of a gain matrix $\tilde{K}_e$ that is stabilizing over all (A_p, B_p).

Since computation of the correct U_p^* requires values for A_p and B_p, it is proposed that a nominal value of U_p^* be used in (5-20). In fact it is shown in Ref. 9 that such use of a nominal value for U_p^* will at worst result in stability with respect to a bounded error.

The requirement that a value for $\tilde{K}_e$ be available would clearly not be a problem, if the plant to begin with, were open loop stable; in this case $\tilde{K}_e = 0$. Otherwise some means of apriori designing a sufficiently robust feedback gain is necessary.

Finally one might then say that if reasonable values for $\tilde{K}_e$ and U_p^* were available, then an appropriate controller would be:

$$U_p(t) = U_p^*(t) + \tilde{K}_e \, H_p \, e \qquad\qquad (5\text{-}21)$$

To this effect reference 9 shows that in terms of the final error, the adaptive controller will be better than this non-adaptive controller if

$$M^T M - P^T M^T M P > 0 \qquad\qquad (5\text{-}22a)$$

where

$$M = -H_p A_p^{-1} B_p \qquad\qquad (5\text{-}22b)$$

$$P = [Q^{-1} \, G - H_p A_p^{-1} B_p]^{-1} \, Q^{-1} G \qquad\qquad (5\text{-}22c)$$

6. EXAMPLES

Example 1 - Non-positive Real Second Order System

Consider the roll and roll rate dynamics of an F-8 aircraft as given in Ref. (2):

$$\begin{bmatrix} \dot{x}_1 \\ \dot{x}_2 \end{bmatrix}_{p\,p} = \begin{bmatrix} -3.598 & 0 \\ 0.99.47 & 0 \end{bmatrix} \begin{bmatrix} x_1 \\ x_2 \end{bmatrix}_p + \begin{bmatrix} 14.65 \\ 0 \end{bmatrix} U_p \qquad\qquad (6\text{-}1)$$

The corresponding model is:

$$\begin{bmatrix} \dot{x}_1 \\ x_2 \end{bmatrix}_m = \begin{bmatrix} -10 & 0 \\ 1 & 0 \end{bmatrix} \begin{bmatrix} x_1 \\ x_2 \end{bmatrix}_m + \begin{bmatrix} 20 \\ 0 \end{bmatrix} U_m \qquad\qquad (6\text{-}2)$$

where U_m is a unit step, x_1 is the roll rate, and x_2 is the roll angle.

If the output is chosen as $H = H_m = H_p = (0, 0.1)$ then the plant transfer function is given by

$$Z_p(s) = \frac{a_{21} \, b_1 \, h_2}{s^2 - a_{11}s + a_{21} \, b_1 \, \tilde{k}_e \, h_2}$$

which does not satisfy the positive real property. Note also that the PMF conditions are also not satisfied.

Therefore, we choose a dynamic compensator represented by $C(s) = s+1$. The plant cascaded with compensator has a transfer function given by

$$C(s) \, Z_p(s) = \frac{a_{21} \, b_1 \, h_2 \, s + a_{21} \, b_1 \, h_2}{s^2 - a_{11}s + a_{21} \, b_1 \, \tilde{k}_e \, h_2}$$

which can be shown to be positive real. Then, as shown in Section 5.3.2, we attempt the use of state variable filters with transfer functions, $C^{-1}(s) = \frac{1}{s+1}$ to avoid explicit differentation. Interactive graphics computer simulations for this system with $T = \bar{T} = .5I$ and $20I$ are shown in Figures 1 and 2 respectively. These indicate that the output error $H_p e = 0.1(x_{m2} - x_{p2})$ does approach zero, and that for $T = 2I$ the error in X_{p_1} also goes to zero. Clearly the values for T and $\bar{T}$ are very influential on the response characteristics and thus the use of interactive computer graphics is an asset to the design process.

7. CONCLUSION

Design procedures for satisfying the constraints characteristic of multivariable implicit adaptive controllers have been discussed, and an illustrative example has been presented. The utility of interactive computer graphics as a design aid was noted in view of the choice of parameters that the designer has to select.

8. REFERENCES

1. Mabius, L., and Kaufman, H., "An Adaptive Flight Controller for the F-8 Without Explicit Parameter Identification", 1976 IEEE Conference on Decision and Control, FL, December 1976.

2. Sobel, K., Kaufman, H., and Mabius, L., "Model Reference Output Adaptive Control Systems without Parameter Identification", 18th IEEE Conf. on Decision and Control, Ft. Lauderdale, FL, December 1979.

3. Broussard, J. R. and O'Brien, M. J., "Feedforward Control to Track the Output of a Forced Model", the 17th IEEE Conference on Decision and Control, January 1979, pp. 1149-1154.

4. Anderson, B. D. O., "A System Theory Criterion for Positive Real Matrices", Journal SIAM Control, Vol. 5, No. 2, 1967.

5. Siljak, D., "New Algebraic Criteria for Positive Realness", Proc. of 4th Annual Princeton Conf. on Info. Science and System, March 1970, pp. 329-335.

6. Anderson, B. D. O., and Vongpanitlerd, S., Network Analysis and Synthesis: A Modern Systems Theory Approach, Prentice Hall, Englewood Cliffs, NJ, 1973.

7. Molinari, B., "The Stable Regulator Problem and its Inverse", IEEE Trans. Auto. Cont., Vol. AC-18, No. 5, October 1973, pp. 454-459.

8. Courtiol, B., "On a Multidimensional Systems Identification Method", IEEE Trans. Auto. Cont., June 1972, pp. 390-394.

9. Sobel, K., "Model Reference Adaptive Control of Multi-Input Multi-Output Plants Without the Need for Explicit Parameter Identification", Ph.D Thesis, RPI, Troy, NY, May 1980.

9. ACKNOWLEDGMENT

This research was supported by NSF Grant No. ENG 77-07446 to Rensselaer Polytechnic Institute.

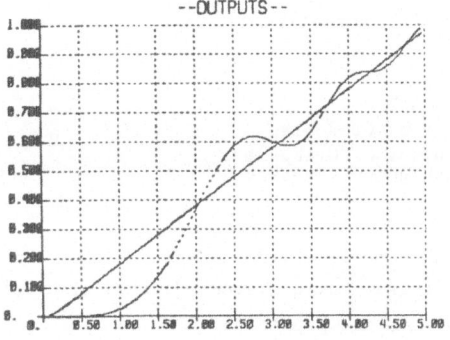

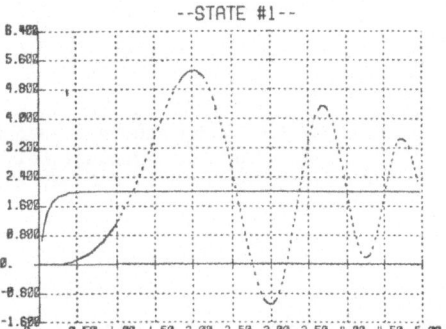

Figure 1 $T = \overline{T} = .5I$
Solid Line Denotes Model

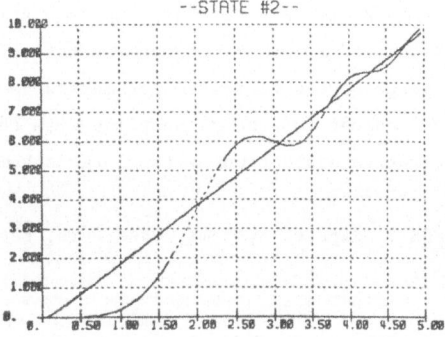

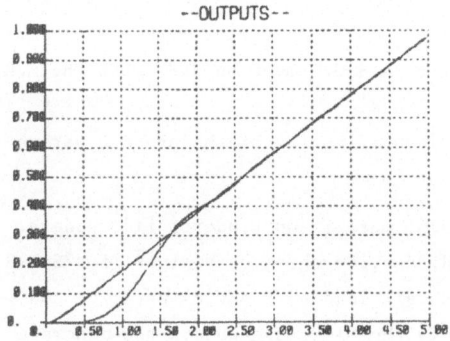

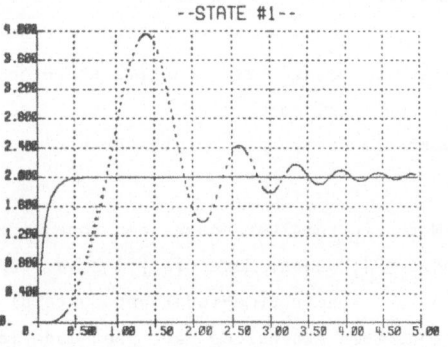

Figure 2 $T = \overline{T} = 2I$
Solid Line Denotes Model

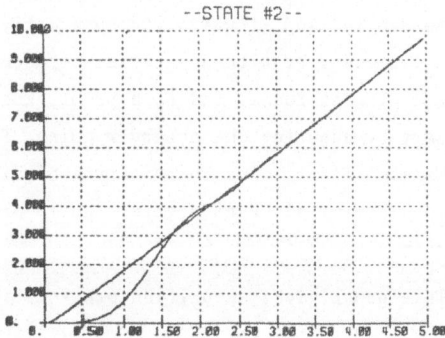

CONVERGENCE ANALYSIS OF M.R.A.S. SCHEMES USED FOR

ADAPTIVE STATE ESTIMATION

L. DUGARD - I.D. LANDAU
Laboratoire d'Automatique de Grenoble
Institut National Polytechnique de Grenoble
B.P. 46 - 38402 ST MARTIN D'HERES - FRANCE

ABSTRACT : Three adaptive state observers for discrete time systems derived from Model Reference Adaptive System (M.R.A.S.) techniques are presented.

The schemes are analysed both in the deterministic and stochastic environments using the "Equivalent Feedback Representation" (E.F.R.) and the "Ordinary Differential Equation" (O.D.E.) methods respectively. From this analysis, conditions for the convergence of the estimated parameters to the desired ones are given in the stochastic environment and the connections with the adaptive Kalman Filter are discussed. A comparative evaluation of the three schemes concludes the paper.

KEYWORDS : Adaptive Systems, Observers, Estimators, Convergence Analysis, Stochastic Systems.

INTRODUCTION

In order to reconstruct inaccessible states of a linear system with unknown or time-varying parameters, adaptive state observers (A.S.O.) must be used with the M.R.A.S. techniques, one can derive A.S.O. from the linear asymptotic observer. Which allow both the observation of the states and the identification of the process parameters.

Works in this aera have been done since 1970. Caroll and Lindorff [1], Lüders, Narendra [2], Kudva, Narendra [3] have proposed adaptive observer designs (for S.I.S.O continuous, linear time-invariant systems) which all lead to a series-parallel configuration. Some extensions have been made later in the MIMO case and in the discrete time case.

Landau [4] and Silveira [5] have derived parallel A.S.O. starting from a linear asymptotic observer with a parallel structure. They offer a better robustness in the presence of disturbance and no a priory knowledge of the system to be observed is necessary by adapting the observer gains. This last solution corresponds to an adaptive Kalman filter for a certain type of disturbance. The corresponding algorithms use time-varying adaptation gains, while previous designs have considered only constant gains.

This paper gives an unified presentation of three A.S.O. (series-parallel, pa-

rallel, parallel with adaptive observer gains) with the analysis in the deterministic
and stochastic environments. The paper is organized as follows. In section II, the
three observers and the corresponding algorithms are given. The convergence analysis
in the deterministic case is made in section III using the EFR method. In section IV,
conditions for the convergence of the algorithms in the stochastic environment are
given, using a theorem derived from the ODE method. Connections with the Kalman Filter are examined in section V.

II - PRESENTATION OF THE SYSTEM AND OBSERVERS

We consider a single input single output system, observable and controllable, represented under the canonical observability form.

$$x(t+1) = \begin{bmatrix} a_1 & I \\ \vdots & \\ a_n & 0 \end{bmatrix} x(t) + \begin{bmatrix} b_1 \\ \vdots \\ b_n \end{bmatrix} u(t) \qquad (II.1)$$

$$= A\, x(t) + b\, u(t)$$

$$y(t) = [1,\ 0 \quad 0]\, x(t) + \frac{M(q^{-1})}{N(q^{-1})} \quad w(t) = c^T x(t) + W(t) \qquad (II.2)$$

$$= y_1(t) + W(t)$$

with : $x^T(t) = [x_1(t),\ \ldots,\ x_n(t)]$

$$M(q^{-1}) = 1 + \sum_{i=1}^{n} m_i\, q^{-i}$$

$$N(q^{-1}) = 1 + \sum_{i=1}^{n} n_i\, q^{-i}$$

$$q^{-1}\, y(t) = y(t-1)$$

$x(t)$ is the state vector, $u(t)$ the input, $y(t)$ the output of the system. $W(t)$,
the disturbance added at the noiseless output of the system, is represented by an
ARMA model driven by a gaussian white sequence $w(t) = N(0, \sigma_w)$. $\{u(t)\}$ is supposed
to be a stationary process and $\{u(t)\}$, $\{w(t)\}$ have rational spectral densities such
that all moments exist. M and N are supposed to be relatively prime polynomials.

For the three A.S.O. considered, the adaptation algorithms used for up-dating
$\hat{\theta}(t)$ are given by the general formula :

$$\hat{\theta}(t+1) = \hat{\theta}(t) + F(t+1)\ \Phi(t+1)\ \varepsilon(t+1)$$

$$\qquad (II.3\ a)$$

$$= \hat{\theta}(t) + \frac{F(t)\ \Phi(t+1)}{1 + \Phi^T(t+1)\ F(t)\ \Phi(t+1)}\ \varepsilon_o(t+1)$$

$$F^{-1}(t+1) = \lambda_1(t) \ F^{-1}(t) + \lambda_2(t) \ \Phi(t+1) \ \Phi^T(t+1) \tag{II.3 b}$$

$$0 < \lambda_1(t) \leqslant 1 \ ; \ 0 \leqslant \lambda_2(t) \leqslant \max \lambda_2(t) = \lambda < 2$$

where $\hat{\theta}(t)$, $F(t)$, $\Phi(t)$ differ from one algorithm to another. How $\varepsilon_o(t)$ and $\Phi(t)$ are generated and the structures of the observation models are summarized in Table II.1.

II.1   ALGORITHM 1 : Series-parallel A.S.O. with constant observer gains

This algorithm is derived from the observer developped by Kudva, Narendra [3] for a continuous S.I.S.O. system with a constant adaptation gain. The adaptive state observer is given by :

$$x_M(t+1) = \begin{bmatrix} \ell_1' & \\ \vdots & I \\ \ell_n' & 0 \end{bmatrix} x_M(t) + \begin{bmatrix} \hat{a}_1(t+1)-\ell_1' \\ \vdots \\ \hat{a}_n(t+1)-\ell_n' \end{bmatrix} y(t) + \begin{bmatrix} \hat{b}_1(t+1) \\ \vdots \\ \hat{b}_n(t+1) \end{bmatrix} u(t) + \begin{bmatrix} u_a^1(t+1) \\ \vdots \\ u_a^{n-1}(t+1) \\ 0 \end{bmatrix} \begin{bmatrix} u_b^1(t+1) \\ \vdots \\ u_b^{n-1}(t+1) \\ 0 \end{bmatrix}$$

$$= L' \ x_M(t) + (\hat{a}(t+1) - \ell') \ y(t) + \hat{b}(t+1) \ u(t) + u_a(t+1) + u_b(t+1) \tag{II.4}$$

where $\hat{a}(t+1)$ and $\hat{b}(t+1)$ are the adjustable parameter vectors, ℓ' the observation coefficient vector, $u_a(t+1)$ and $u_b(t+1)$ are auxiliary transient signals ensuring the stability of the overall adaptive scheme.

$$\hat{\theta}^T(t) = [\hat{a}^T(t), \ \hat{b}^T(t)] = [\hat{a}_1(t), \ \ldots, \ \hat{b}_1(t), \ \ldots]$$

$$\ell'^T = [\ell_1', \ \ldots, \ell_n']$$

The state variable filter :

$$h_F(q^{-1}) = \frac{q^{-1}}{1 + \sum_{i=1}^{n-1} d_i \ q^{-i}} = \frac{q^{-1}}{D(q^{-1})}$$

is introduced for the generation of the auxiliary variables used in the adaptation algorithm. The a priori and a posteriori errors, generalized errors, auxiliary variables, observation vector and adjustable parameter vector of the A.S.O. are given in Table II.2.

II.2   ALGORITHM 2 : Parallel A.S.O. with constant observer gains

This algorithm has been proposed by Landau [4]. The observer is made, starting from an equivalent representation of the linear asymptotic parallel observer. This observer is given by :

$$x_M(t+1) = \begin{bmatrix} \hat{a}_1(t+1)-\ell_1 & \\ \vdots & I \\ \hat{a}_n(t+1)-\ell_n & 0 \end{bmatrix} x_M(t) + \begin{bmatrix} \ell_1 \\ \vdots \\ \ell_n \end{bmatrix} y(t) + \begin{bmatrix} b_1(t+1) \\ \vdots \\ b_n(t+1) \end{bmatrix} u(t) + \begin{bmatrix} u_a^1(t+1) \\ \vdots \\ u_a^{n-1}(t+1) \\ 0 \end{bmatrix} + \begin{bmatrix} u_b^1(t+1) \\ \vdots \\ u_b^{n-1}(t+1) \\ 0 \end{bmatrix}$$

$$= [\hat{A}(t+1) - \ell c^T] \ x_M(t) + \ell \ y(t) + \hat{b}(t+1) \ u(t) + u_a(t+1) + u_b(t+1) \tag{II.5}$$

All the corresponding definitions are given in Table II.1. For more details, see Landau [4], Silveira [5].

II.3 ALGORITHM 3 : Parallel A.S.O. with adjustable observer gains

Silveira [5] has removed the need for prior knowledge upon the process to be observed, required by the previous observer, by adapting the observer gains simultaneously with the parameters of the model. This observer is given by :

$$x_M(t+1) = [\hat{A}(t+1) - \hat{\ell}(t+1)\ c^T]\ x_M(t) + \hat{\ell}(t+1)\ y(t) + \hat{B}(t+1)\ u(t) +$$

$$+ u_a(t+1) + u_b(t+1) + u_c(t+1) \tag{II.6}$$

which is similar to the previous observer. The other definitions can be found in Table II.1.

These thred observers are more precisely detailed and explained in Dugard, Landau, Silveira [6].

III - CONVERGENCE ANALYSIS - DETERMINISTIC CASE (W(t) ≡ O)

The MRAS can be put under the form of a "Standard Feedback System" with a feed-forward linear part and a feedback non-linear part with time-varying parameters, see [4], [5]. The observers presented here can be designed using this method and the convergence problem is studied as a stability problem of the "Equivalent Feedback Representation" of the system, which can be solved by applying hyperstability and positivity concepts or by using the Lyapunov functions [3], [4], [5].

From [9], we can derive the following theorem used to analyse the A.S.O. in the deterministic environment.

Theorem III.1 : Assume that :

a) The adaptation algorithm used for updating $\hat{\theta}(t)$ is given by the Equations (II.3).

b) There is the following relation between $\varepsilon(t)$ and $\Phi(t)$:

$$\varepsilon(t) = H(q^{-1})[\theta - \hat{\theta}(t)]^T\ \Phi(t) \tag{III.1}$$

where $H(z^{-1})$ is a normalized rational transfer function.

Then :

$$\lim \varepsilon(t) = 0 \quad \forall\ \hat{\theta}(0) - \theta,\ \forall \varepsilon(0) \tag{III.2}$$

if the transfer function $H'(z^{-1}) = H(z^{-1}) - \frac{\lambda}{2}$ is strictly positive real (s.p.r.)

In addition, if the input is sufficiently rich, the convergence of $\varepsilon(t)$ implies the convergence of $\hat{\theta}(t)$ to θ and the norm of the state error vector $e(t) = x(t) - x_M(t)$ goes asymptotically to zero.

For the three observers, we have the following results, after some calculations.

ALGORITHM 1 : Series parallel A.S.O. - Fixed observer gains

$$\varepsilon(t) = H(q^{-1}) \ [\theta - \hat{\theta}(t)]^T \ \Phi(t) \qquad \text{(III.3)}$$

where :

$$H(q^{-1}) = \frac{1 + \sum\limits_{i=1}^{n-1} d_i \ q^{-i}}{1 - \sum\limits_{i=1}^{n} \ell'_i \ q^{-i}} = \frac{D(q^{-1})}{L'(q^{-1})} \qquad \text{(III.4)}$$

$$\theta^T = [a^T, \ b^T] = [a_1, \ \ldots, \ a_n, \ b_1, \ \ldots, \ b_n] \qquad \text{(III.5)}$$

The coefficients d_i and ℓ'_i are chosen by the designer such that $H'(z^{-1})$ is s.p.r.

Then : $\lim\limits_{t \to \infty} \varepsilon(t) = 0$; $\lim\limits_{t \to \infty} e(t) = 0$, $\lim\limits_{t \to \infty} \hat{\theta}(t) = \theta$ $\qquad \text{(III.6)}$

when the conditions of the Theorem III.1 are fullfilled.

Remark : No prior knowledge of the process is necessary to ensure the convergence of the observer to the process.

ALGORITHM 2 : Parallel ASO - Fixed observer gains

$$\varepsilon(t) = H(q^{-1}) \ [\theta - \hat{\theta}(t)]^T \ \Phi(t) \qquad \text{(III.7)}$$

where :

$$H(q^{-1}) = \frac{1 + \sum\limits_{i=1}^{n-1} d_i \ q^{-i}}{1 - \sum\limits_{i=1}^{n} (a_i - \ell_i) q^{-i}} = \frac{D(q^{-1})}{(A+L')(q^{-1})} \qquad \text{(III.8)}$$

$$\theta^T = [a^T, \ b^T]$$

The equations (III.6) are satisfied if d_i and ℓ_i are chosen such that $H'(z^{-1})$ is s.p.r.

Remark : A prior knowledge of the process is necessary to ensure the convergence of the observer.

ALGORITHM 3 : Parallel ASO-Adjustable observation gains

$$\varepsilon(t) = H(q^{-1}) \ [\theta - \hat{\theta}(t)]^T \ \Phi(t) \qquad \text{(III.9)}$$

where :

$$H(q^{-1}) = 1 + \sum\limits_{i=1}^{n-1} d_i \ q^{-i} = D(q^{-1}) \qquad \text{(III.10)}$$

$$\theta^T = [a^T, \ b^T, \ a^T] \qquad \text{(III.11)}$$

The equations (III.6) are satisfied if d_i are chosen such that $H'(z^{-1})$ is s.p.r.

Remark : No prior knowledge of the process is necessary to ensure the convergence of the observer.

The speed of convergence depends on the dynamic of the linear forward block defined by $H(z^{-1})$.

IV - CONVERGENCE ANALYSIS - STOCHASTIC CASE

In the presence of disturbance, the A.S.O. presented in section II act as adaptive state estimators. It is therefore important to establish their convergence properties in this context in order to determine their asymptotic behaviour and the conditions upon the disturbances which allow to obtain unbiased estimates. This analysis is carried on, using the O.D.E. method of Ljung [7], [8]. From this method, we can derive the following theorem, directly applicable to the schemes discussed in the paper.

Theorem IV.1 : Consider the adaptation algorithm given by the Equations (II.3) with $\lambda_1(t) \equiv 1$ or $\lambda_1(t) \to 1$. Suppose that the stationary processes $\{\bar{\Phi}(t,\theta)\}$, $\{\bar{\varepsilon}(t, \theta)\}$ can be defined for $\theta(t) = \hat{\theta}$ and that $\hat{\theta}(t)$ belongs infinitely often to the domain for which these stationary processes can be defined.

Assume that :

1) $\bar{\varepsilon}(t, \hat{\theta}) = H(q^{-1}) \ \bar{\Phi}^T(t, \hat{\theta}) \ [\theta^* - \hat{\theta}] + v(t)$ (IV.1)

where $\{v(t)\}$ is a white sequence or uncorrelated with $\bar{\Phi}(t, \hat{\theta})$, $\forall$ t.

2) $E \{\bar{\Phi}(t, \hat{\theta}) \ \bar{\Phi}^T(t, \hat{\theta})\} \ [\theta^* - \hat{\theta}] = 0$ (IV.2)

has a unique solution $\hat{\theta} = \theta^*$ (the expectation is taken over $\{u(t)\}$ and $\{w(t)\}$) where $\bar{\Phi}(t, \hat{\theta}) = H(q^{-1}) \ \bar{\Phi} \ (t,\theta)$.

Then :

Prob. $\{\lim_{t \to \infty} \hat{\theta}(t) = \theta^*\} = 1$ (IV.3)

if the transfer function $H'(z^{-1}) = H(z^{-1}) - \frac{\lambda}{2}$ is s.p.r.

For the study, we introduce the stationary processes defined in the Theorem by keeping $\hat{\theta}(t)$ fixed to a constant value $\hat{\theta}^T$. The auxiliary transient signal disappear and we have the following results, obtained after several manipulations of equations.

ALGORITHM 1 : The expression of $\bar{\varepsilon}(t, \hat{\theta})$ is given by :

$$\bar{\varepsilon}(t, \hat{\theta}) = H(q^{-1}) \ \bar{\Phi}^T(t, \hat{\theta}) \ [\theta^* - \hat{\theta}] + \frac{A(q^{-1})}{L'(q^{-1})} \cdot \frac{M(q^{-1})}{N(q^{-1})} \ w(t)$$ (IV.4)

where :

$$H(q^{-1}) = \frac{D(q^{-1})}{L'(q^{-1})}$$

$$\theta^{T*} = [a^T, b^T]$$

The convergence is ensured if $\frac{A}{L'} \cdot \frac{M}{N} \ w(t)$ is a white sequence, i.e. :

$$N(q^{-1}) \equiv A(q^{-1}) \text{ and } M(q^{-1}) \equiv L'(q^{-1})$$ (IV.5)

and if $H'(z^{-1})$ is s.p.r.

ALGORITHM 2 : The expression of $\bar{\varepsilon}(t, \hat{\theta})$ is :

$$\bar{\varepsilon}(t, \hat{\theta}) = H(q^{-1}) \, \bar{\Phi}^T(t, \hat{\theta}) \, [\theta^* - \hat{\theta}] + \frac{A(q^{-1})}{(A+L')(q^{-1})} \cdot \frac{M(q^{-1})}{N(q^{-1})} \, w(t) \qquad (IV.6)$$

where :

$$H(q^{-1}) = \frac{D(q^{-1})}{(A+L')(q^{-1})}$$

$$\theta^{*T} = [a^T, b^T]$$

The convergence is ensured if :

$$N(q^{-1}) \equiv A(q^{-1}) \text{ and } M(q^{-1}) \equiv (A+L')(q^{-1}) \qquad (IV.7)$$

and if $H'(z^{-1})$ is s.p.r.

ALGORITHM 3 : Let us define :

$$\ell^{*T} = [\ell_1^*, \ldots, \ell_n^*] \qquad (IV.8)$$

where : $\ell_i^* = m_i + a_i$, $i = 1, \ldots, n$

The expression of $\bar{\varepsilon}(t, \hat{\theta})$ is then given by :

$$\bar{\varepsilon}(t, \hat{\theta}) = H(q^{-1}) \, \bar{\Phi}^T(t, \hat{\theta}) \, [\theta^* - \hat{\theta}] + \frac{A(q^{-1})}{(A+L^*)(q^{-1})} \cdot \frac{M(q^{-1})}{N(q^{-1})} \, w(t) \qquad (IV.9)$$

where :

$$H(q^{-1}) = \frac{D(q^{-1})}{(A+L^*)(q^{-1})}$$

$$\theta^{*T} = [a^T, b^T, \ell^{*T}]$$

$$(A+L^*)(q^{-1}) = 1 - \sum_{i=1}^{n} (a_i - \ell_i^*) \, q^{-i}$$

The convergence is ensured if :

$$N(q^{-1}) \equiv A(q^{-1}) \text{ and } M(q^{-1}) \equiv (A+L^*)(q^{-1})$$

and if $H'(z^{-1})$ is s.p.r.

For the three algorithms, two essential conditions are required to have convergence w.p.1 of the estimates to the true parameters. Firstly, the disturbance has to be modelized by an ARMA process having the dynamic of the system $(N(q^{-1}) = A(q^{-1}))$. This means that the system (II.1), (II.2) can be reformulated such that the noise appears in the state vector (as seen in the next section). Secondly, a certain transfer function, involving the dynamic of the state variable filter and that of the disturbance, has to be s.p.r. It is interesting to remark that the algorithm 3, built in the deterministic case to remove the prior knowledge of the process, gives an estimation of the disturbance parameters in the stochastic environment.

V - CONNECTIONS WITH THE ADAPTIVE KALMAN FILTER

When the disturbance $W(t)$ is modelized by $\dfrac{M(q^{-1})}{A(q^{-1})} \, w(t)$, the Equations (II.1), (II.2) can be reformulated under the innovation representation form :

$$x(t+1) = A \, x(t) + b \, u(t) + m' \, w(t) \qquad (V.1)$$

$$y(t) = c^T x(t) + w(t) \tag{V.2}$$

where : $m'^T = (m+a)^T = [m_1+a_1, \ldots, m_n+a_n]$

If the parameters are known, we can use the adaptive Kalman filter given by :

$$\hat{x}(t+1) = A \hat{x}(t) + b u(t) + K(t+1) (y(t) - \hat{y}(t)) \tag{V.3}$$

$$\hat{y}(t) = c^T \hat{x}(t) \tag{V.4}$$

in order to obtain an optimal state estimation. The Kalman gain $K(t+1)$ is computed in order to minimize the covariance matrix $P(t) = E\{\tilde{x}(t) \, \tilde{x}^T(t)\}$ where $\tilde{x}(t)$ is the state error vector. The Kalman gain which minimizes $P(t)$ converges to the optimal asymptotic Kalman gain $K = m' = m+a$ (this follows directly from Eq. (V.1).

For the parallel A.S.O. with adjustable observer gains, the $\hat{\ell}(t)$ vector-part converges to the constant vector $m' = m+a$ which is the optimal asymptotic Kalman gain, when the convergence conditions of the theorem IV.1 are satisfied. In this sense, we can say that this adaptive observer is an adaptive Kalman filter for the systems described by Eq. (V.1), (V.5), since $(\theta(t) \rightarrow \theta^*)$ and its tends to the optimal asymptotic Kalman filter. The two other observers can be also considered as adaptive Kalman filters but both process and disturbance must be known to obtain the optimal estimation.

CONCLUSION

We have studied types of A.S.O. derived from the linear basic asymptotic observer, using M.R.A.S. techniques. All these schemes are globally asymptotically stable. Simulation studies have shown that the speed of convergence was similar for the two first observers (depending on the choice of d_i, ℓ_i and ℓ_i') and that the third one provided the fastest convergence and tended to a linear observer with all the poles at the origin, in the deterministic case. In a stochastic environment, however their performances are very different and the convergence conditions have been determined using the O.D.E. method. The theoretical analysis as well as the simulation have clearly shown that the most interesting scheme for the adaptive state estimation is the A.S.O. with adjustable observer gains which can be considered as an adaptive Kalman Filter. It can be also noticed that the schemes having a parallel structure offer a better robustness with respect to the disturbances. At the last, one should mention that the real positivity condition of a certain transfer function plays a crucial role, both in the deterministic and stochastic environments.

REFERENCES

[1] R.L. CAROLL, D.P. LINDORFF
     "An adaptive observer for single input - single output linear systems"
     I.E.E.E. Trans. on Aut. Contr., Vol. AC 18, pp. 428-435, Oct. 1973

[2] G. LÜDERS, K.S. NARENDRA
     "An adaptive observer and identifier for a linear system"
     I.E.E.E. Trans. on Aut. Contr., Vol. AC 18, pp. 496-499, Oct. 1973

[3] P. KUDVA, K.S. NARENDRA
"Synthesis of an adaptive observer using Lyapunov's direct method"
Int. Journ. of Contr., Vol. 18, pp. 1201-1210, Déc. 1973

[4] I.D. LANDAU
"Adaptive control - The model reference approach"
Dekker, March 1979

[5] H.M. SILVEIRA
"Contributions à la synthèse des systèmes adaptatifs avec modèle dans accès aux
variables d'état"
Thèse de Docteur ès Science, Université de Grenoble, Mars 1978

[6] L. DUGARD, I.D. LANDAU, H.M. SILVEIRA
"Adaptive state estimation using M.R.A.S. techniques - Convergence analysis and
evaluation"
Note LAG n° 79-06, 1979

[7] L. LJUNG
"On positive real transfer functions and the convergence of some recursive sche-
mes"
I.E.E.E. Trans. on Aut. Contr., Vol. AC 22, pp. 539-550, Aug. 1977

[8] L. LJUNG
"Analysis of recursive stochastic algorithms"
I.E.E.E. Trans. on Aut. Contr., Vol. AC 22, pp. 551-575, Aug. 1977

[9] I.D. LANDAU, H.M. SILVEIRA
"A stability theorem with application to adaptive control"
I.E.E.E. Trans. on Aut. Contr., Vol. n° 2, April 1979

This work was supported by C.N.R.S. through the A.T.P. n° 31-80

	Algorithm 1 Series-parallel observer constant gains	Algorithm 2 Parallel observer - Constant gains	Algorithm 3 Parallel observer-Adjustable gains
Adjustable Parameter vector $\hat{\theta}^T(t)$	$[\hat{a}^T(t), \hat{b}^T(t)]$	$[\hat{a}^T(t), \hat{b}^T(t)]$	$[\hat{a}^T(t), \hat{b}^T(t), \hat{\ell}^T(t)]$
State vector of the observer $x_M(t+1)$	$L'x_M(t) + (\hat{a}-\ell) y(t) + \hat{b} u(t) + u_a(t+1) + u_b(t+1)$	$(\hat{A}-\ell) x_M(t) + \ell\, y(t) + \hat{b} u(t) + u_a(t+1) + u_b(t+1)$	$(\hat{A}-\ell) x_M(t) - \ell\, y(t) + \hat{b} u(t) + u_a(t+1) + u_b(t+1) + u_c(t+1)$
Output of the observer a posteriori $y_M(t)$	$c^T x_M(t) = x_{M1}(t)$	$x_{M1}(t)$	$x_{M1}(t)$
A priori $y_M^0(t)$	$(\hat{a}_1(t-1)-\ell_1')y(t-1)+x_{M2}(t-1)+\ell_1'\, y_{M1}(t-1) + \hat{b}_1(t-1) u(t-1)$	$(\hat{a}_1(t-1)-\ell_1)y_M(t-1)+x_{M2}(t-1)+\ell_1\, y(t-1) + \hat{b}_1(t-1) u(t-1)$	$(\hat{a}_1(t-1)-\hat{\ell}_1(t-1))y_M(t-1)+x_{M2}(t-1) + \hat{\ell}_1(t-1)\, y(t-1) + \hat{b}_1(t)\, u(t-1)$
Error — A priori $\varepsilon_0(t)$	$y(t) - y_M^0(t)$	$y(t) - y_M^0(t)$	$y(t) - y_M^0(t)$
Error — A posteriori $\varepsilon(t)$	$y(t) - y_M(t)$	$y(t) - y_M(t)$	$y(t) - y_M(t)$
Auxiliary variables	$\tilde{y}_j(t)=\dfrac{1}{D(q^{-1})} y(t-j),\ \tilde{u}_j(t)=\dfrac{1}{D(q^{-1})}\, u(t-j)$	$\tilde{y}_{Mj}(t)=\dfrac{1}{D(q^{-1})}y_M(t-j);\ \tilde{u}_j(t)=\dfrac{1}{D(q^{-1})}\, u(t-j)$	$\tilde{y}_{Mj}(t)=\dfrac{1}{D(q^{-1})}y_M(t-j);\ \tilde{u}_j(t)=\dfrac{1}{D(q^{-1})}\, u(t-j)$ $\tilde{\varepsilon}_j(t) = 1/D(q^{-1})\, \varepsilon(t-j)$
Observation vector $\phi(t)$	$[\tilde{y}_1(t), \ldots, \tilde{u}_1(t), \ldots]$	$[\tilde{y}_{M1}(t), \ldots, \tilde{u}_1(t), \ldots]$	$[\tilde{y}_{M1}(t), \ldots, \tilde{u}_1(t), \ldots, \tilde{\varepsilon}_1(t),\ldots]$
Stability condition - Deterministic case	$\dfrac{1+\sum_{i=1}^{n-1} d_i z^{-i}}{1-\sum_{1}^{n} \ell'_i z^{-i}} - \dfrac{\lambda}{2}$ = s.p.r.	$\dfrac{1+\sum_{i=1}^{n-1} d_i z^{-i}}{1-\sum_{i=1}^{n} (a_i-\ell_i)z^{-i}} - \dfrac{\lambda}{2}$ = s.p.r.	$1+\sum_{i=1}^{n-1} d_i z^{-i} - \dfrac{\lambda}{2}$ = s.p.r.

TABLE II.1

NON MODEL REFERENCE ADAPTIVE MODEL MATCHING[†]

H. Elliott
Department of Electrical Engineering
Colorado State University
Fort Collins, Colorado 80523/USA

I. Introduction

Recent investigations [9], [10] brought considerable insight into similarities
which exist among the many parameter adaptive control schemes proposed in the literature for unknown single-input single-output linear systems. In these reports and
others, parameter adaptive controllers were split into two categories based upon the
type of incorporated adaptation algorithm. Although a number of authors coined
names for these categories, following the terminology introduced in [9], they will
be denoted as "indirect" and "direct". Indirect controllers explicitly identify a
parameterized model of the unknown system (plant) to be controlled and then function-
ally relate the controller parameters to those of the model by use of known control
methodologies. Most generally, direct controllers are defined as those which adjust
the controller parameters directly from measurement of input-output data. However,
to this point direct controllers have usually been synonomous with model reference
adaptive controllers where the controller parameters are adjusted so as to minimize
the error between output of the plant and that of a desired model. If the plant is
time-invariant, model reference adaptive controllers can be shown to converge to
linear model matching control structures. This report further highlights the similarities between direct and indirect adaptive controllers for the case of model
matching by presenting a new design where the controller parameters are directly
adjusted using filtered input-output data without explicit implementation of a reference model.

The proposed controller incorporates an adaptation mechanism which involves
implementation of a linear parameter estimator similar to the type used for online
identification [4], [6] and [8]. In particular the parameters are adjusted so as
to minimize an error signal that serves as a measure of the controller parameter
errors. This error signal consists of a linear combination of the parameter estimates and a set of sensitivity functions generated by filtering input-output data.
Because of the form of this error measure, the controller is similar in many respects
to indirect controllers which make use of linear parameter estimation schemes to
identify a model of the plant. On the other hand, the controller also has many of
the attributes associated with model reference schemes. The design requires knowledge of an upperbound on plant order, the degree difference between numerator and
denominator of the plant transfer function, and the sign and a bound on the high
frequency transfer function gain.

[†]This work was supported by the National Science Foundation under grant ENG-7908014.

Recently, a number of reports have appeared which prove global stability for model reference schemes, e.g. [12]-[15]. Although no global stability proof is presently available for this scheme it is conjectured that it will be shown to be globally stable. The key similarity between this approach and the model reference approach, and the reason one might expect the scheme to be stable, is the generation of a control law that adaptively cancels plant zeros with closed loop poles. Although one might propose the use of scalar error adaptation laws similar to those proposed in [12]-[15], in this report we also discuss use of a multiple equation error adjustment mechanism similar to the identifier proposed by Lion [6]. It can improve convergence speed and hence transient system performance.

In addition to presenting the design for single-input single-output systems, an extension to the multivariable case is briefly outlined. More extensive details on the multivariable case can be found in [16] and [17].

II. Controller Structure for Scalar Systems

To begin, let the dynamical behavior of the unknown plant (to be controlled) be defined by a differential operator representation of the form:

$$p_p(D)x(t) = u(t) \tag{1a}$$

$$y(t) = g_p r_p(D)x(t) \tag{1b}$$

where $y(t)$ and $u(t)$ are the plant output and input signals, respectively, $x(t)$ and its derivatives represent the system state, and $p_p(D)$ and $r_p(D)$ are monic, relatively prime polynomials in the differential operator $D \equiv d/dt$. Let the degrees of $p_p(D)$ and $r_p(D)$ equal m_p and z_p respectively, where $d_p \equiv m_p - z_p \geq 1$. We assume knowledge of d_p, the degree difference between numerator and denominator of the plant transfer function, $n_p \geq m_p$, an upperbound on the plant order, the sign of g_p, and $g_u \geq |g_p|$, an upperbound on the magnitude of g_p. Without loss of generality we take $g_p > 0$. It is also required that $r_p(D)$ be Hurwitz.

Similarly, assume that it is desired to have the output $y(t)$ of (1) track that of a model system,

$$p_m(D)y_m(t) = g_m r_m(D)v(t) \tag{2}$$

where $y_m(t)$ and $v(t)$ are the model output, and external reference input, respectively. The polynomials $p_m(D)$ and $r_m(D)$ are monic, relatively prime, and of degrees m_m and z_m respectively where $d_m \equiv m_m - z_m \geq d_p$. In addition $p_m(D)$ is assumed Hurwitz, and factorable as $p_m(D) = p_{m1}(D)p_{m2}(D)$, where the degree of $p_{m1}(D)$ equals d_p.

Consider the control structure depicted in Figure 1, with control law

$$p_{m2}(D)\tilde{v}(t) = g_m r_m(D)v(t) \tag{3a}$$

$$q(D)s(t) = k*(D)u(t) + h*(D)y_p(t) \tag{3b}$$

$$u(t) = s(t) + g*\tilde{v}(t) \tag{3c}$$

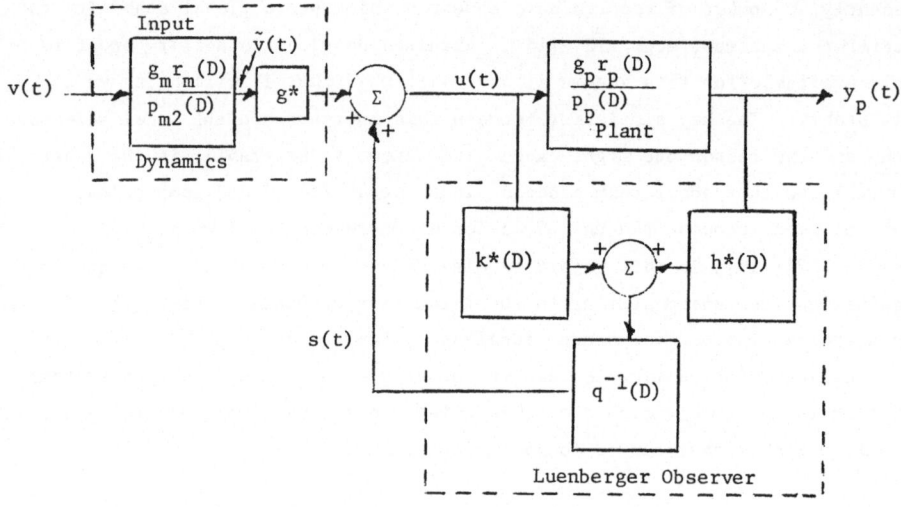

Figure 1. Model Matching by Use of a Differential Operator
Luenberger Observer in Conjunction with Input Dynamics.

In (3), let $q(D)$ be an arbitrary Hurwitz polynomial of degree n_p-1 and let

$$h*(D) = \sum_{i=0}^{n_p-1} h_i^* D^i \tag{4a}$$

$$k*(D) = \sum_{i=0}^{n_p-2} k_i^* D^i \tag{4b}$$

satisfy the polynomial equation

$$h*(D)g_p r_p(D) + k*(D)p_p(D) = q(D)[p_p(D) - p_{ml}(D)r_p(D)] \quad . \tag{5}$$

If $n_p = m_p$ solutions $h*(D)$ and $k*(D)$ to (5) exist and are unique, while if $n_p > m_p$
solutions exist but are nonunique [3]. When (5) holds, control law (3) represents
a combination of input dynamics (3a), and a Luenberger Observer (3b) which produces
closed loop poles corresponding to the stable polynomial $q(D)p_{ml}(D)r_p(D)$ (where the
zeros of $q(D)$ represent the uncontrollable observer poles) [3]. In this case, if $g*$
of equation (3c) equals $1/g_p$, then $y_p(t)$ will converge exponentially to $y_m(t)$ as
desired.

Therefore, adaptive model matching can be implemented by producing a control
law which converges to (3). In particular, consider the following parameter adaptive
control law

$$q(D)\tilde{u}(t) = u(t) \tag{6a}$$

$$q(D)\tilde{y}(t) = y_p(t) \tag{6b}$$

$$p_{m2}(D)\tilde{v}(t) = g_m r_m(D)v(t) \tag{6c}$$

$$u(t) = \sum_{i=0}^{n_p-1} h_i(t)D^i\tilde{y}(t) + \sum_{i=0}^{n_p-2} k_i(t)D^i\tilde{u}(t) + g(t)\tilde{v}(t) \quad . \tag{6d}$$

If

$$\lim_{t\to\infty} h_i(t) = h_i^* \qquad i=0,1,\ldots,n_p-1 \tag{7a}$$

$$\lim_{t\to\infty} k_i(t) = k_i^* \qquad i=0,1,\ldots,n_p-2 \tag{7b}$$

$$\lim_{t\to\infty} g(t) = g^* \equiv 1/g_p \tag{7c}$$

then (6) converges to (3). In the following section a new adaptive algorithm will be derived for estimating the parameters $h_i(t)$, $k_i(t)$, and $g(t)$.

III. Adaptation Equations

In this section a new set of adaptation equations are formulated for adjusting the parameters $h_i(t)$, $k_i(t)$ and $g(t)$ contained in control law (7). Specifically, a method is presented for constructing a measure $\varepsilon(t)$, of the controller parameter errors directly from filtered input-output data without explicit implementation of a reference model. Since this error signal is linear in the parameter estimates, the problem of adaptation becomes a problem of linear parameter estimation, and a modified gradient type adaption law is presented.

To begin let us reformulate (5) as follows:

$$h^*(D)g_p r_p(D) + k^*(D)p_p(D) = q(D)[p_p(D) - g^*p_{m1}(D)g_p r_p(D)] \quad . \tag{8}$$

Since the right hand side of (8) will be of higher degree than the left side unless $g^* = \dfrac{1}{g_p}$, (8) will hold if and only if $g^* = \dfrac{1}{g_p}$, and $h^*(D)$ and $k^*(D)$ satisfy the original equation (5).

Let $d(D)$ be an arbitrary Hurwitz polynomial of degree $n_d \geq n_p + d_p$, and define the physically realizable time varying column vector $\bar{z}(t)$ by the relations

$$d(D)\bar{u}(t) = u(t) \tag{9a}$$

$$d(D)\bar{y}(t) = y_p(t) \tag{9b}$$

$$\bar{z}(t) = [\bar{u}(t), D\bar{u}(t), D^2\bar{u}(t), \ldots, D^{n_p-2}\bar{u}(t), \bar{y}(t), D\bar{y}(t), \ldots, D^{n_p-1}\bar{y}(t),$$

$$p_{m1}(D)q(D)\bar{y}(t)]^T \tag{9c}$$

In addition let one possible set of optimal parameters be contained in the row vector

$$\theta^* = [k_0^*, k_1^*, \ldots, k_{n_p-2}^*, h_0^*, h_1^*, \ldots, h_{n_p-1}^*, g^*] \quad . \tag{10}$$

Observe that if $\bar{x}(t)$ is defined by the relation, $d(D)\bar{x}(t) = x(t)$, then

$$P_p(D)\bar{x}(t) = \bar{u}(t) + \gamma_0(t) \tag{11}$$

and

$$\bar{y}(t) = g_p r_p(D)\bar{x}(t) + \gamma_1(t) \tag{12}$$

where $\gamma_0(t)$ and $\gamma_1(t)$ are linear combinations of decaying exponentials. In view of this we state the following:

Proposition 1

If $\bar{x}(t)$ and its first $2n_p-1$ derivatives are linearly independent over some time interval τ, then (8) holds if and only if $\theta*$ satisfies

$$\theta*\bar{z} = q(D)\bar{u} + \gamma_2(t) \tag{13}$$

over the interval τ, where $\gamma_2(t)$ is a linear combination of decaying exponentials.

Proof:

Multiplying (8) by $\bar{x}(t)$, we obtain

$$[h*(D)g_p r_p(D) + k*(D)P_p(D) - q(D)(P_p(D) - g*P_{m1}(D)g_p r_p(D))]\bar{x}(t)$$

$$= \alpha(D)\bar{x}(t) = 0 \quad . \tag{14}$$

Since $\alpha(D)$ is of maximum degree $2n_p-1$, if $\bar{x}(t)$ satisfies the independence condition, (8) holds if and only if (14) holds. Using (11) and (12) then we have

$$h*(D)\bar{y}(t) + k*(D)\bar{u}(t) + g*P_{m1}(D)q(D)\bar{y} = q(D)\bar{u}(t) + \gamma_2(t) \tag{15}$$

which is just a reformulation of (13).

Let

$$\theta(t) = [h_0(t), h_1(t), \ldots, h_{n_p-1}(t), k_0(t), \ldots, k_{n_p-2}(t), g(t)] \tag{16}$$

be a row vector of the adjustable controller parameters in (6). In light of (13) we will define our parameter error measure $\varepsilon(t)$ as

$$\varepsilon(t) = \theta(t)\bar{z}(t) - q(D)\bar{u}(t) \quad . \tag{17}$$

One possible adaptation law based upon the stable designs presented in [12]-[15] would be

$$\dot{\theta}(t) = -\frac{\varepsilon(t)\bar{z}^T(t)\Sigma}{(\beta+\bar{z}^T\bar{z})} + [0,0,\ldots,0,1] \max(0, g_\ell - g(t)) \tag{18}$$

where $\beta > 0$ is an arbitrary constant, Σ is an arbitrary positive definite symmetric gain matrix, and $g_\ell = 1/g_u$, is a lower bound on $|1/g_p|$.

Equation (18) can be thought of as a modified gradient adjustment law since $\frac{1}{2}\nabla_\theta(\varepsilon^2(t)) = \varepsilon(t)\bar{z}^T(t)$. The term $(\beta+\bar{z}^T\bar{z})$ has analagous counterparts in the adjustment equation proposed in [13]-[15], and serves to insure the boundedness of the right hand side of (18). The second term on the right side of (18) is used to restrict g(t) from converging to zero. If one were to allow g(t) to converge to zero, and the adaptive loop was such that the closed loop system were stable, then the entire system would come to rest.

Simulation studies indicate that considerable improvement in performance can be obtained by replacing (18) by a multiple equation error identifier similar to that proposed by Lion [6]. Let the degree of the filter polynomial d(D) be increased to $n_d \geq 3n_p + d_p - 1$ and define

$$\bar{\varepsilon}(t) = [\varepsilon_0(t), \varepsilon(t), \ldots, \varepsilon_{2n_p-1}(t)] \tag{19}$$

$$\varepsilon_i(t) = \theta(t) D^i \bar{z}(t) - q(D) D^i \bar{u}(t) \tag{20}$$

$$Z(t) = [\bar{z}(t), D\bar{z}(t), \ldots, D^{2n_p-1}\bar{z}(t)] \tag{21}$$

and

$$\dot{\theta}(t) = -\bar{\varepsilon}(t) Z^T(t) \Sigma + [0, 0, \ldots, 0, 1] \max(0, g_\ell - g(t)) \quad . \tag{22}$$

IV. Extension to the Multivariable Case

In extending this approach to the multivariable case, we consider m-input, p-output, $p \geq m$, plants with differential operator representation

$$P_p(D) x(t) = u(t) \tag{23a}$$

$$y(t) = R_p(D) x(t) \quad , \tag{23b}$$

where $P_p(D)$ (mxm) and $R_p(D)$ (pxm) are full rank, relatively right prime polynomial matrices [3]. We also assume $P_p(D)$ to be column proper, and the resulting full rank transfer matrix

$$T_p(s) = R_p(s) P_p^{-1}(s) \tag{24}$$

to be strictly proper [3].

In this case we can construct a nonminimal multivariable Luenberger observer which assigns as closed loop poles, the zeros of $|P_{m1}(D) R_p(D)|$ provided $P_{m1}(D)$ (pxp) is chosen properly. Specifically, in order to implement a multivariable version of the observer shown in Figure 1 we must find polynomial matrices, $H^*(D)$ (mxp) and $K^*(D)$ (mxm), and a scalar matrix G^* (mxp) such that

$$H^*(D) R_p(D) + K^*(D) P_p(D) + Q_m(D) [P_p(D) - G^* P_{m1}(D) R_p(D)] = 0 \quad , \tag{25}$$

where $Q_m(D) = \text{diag}[q(D)]$ (mxm), q(D) being an arbitrary Hurwitz polynomial of degree $\nu-1$, and ν being an upperbound on the systems observability index. It is shown in [17] that for (25) to have solutions $P_{m1}(D)$ must be chosen such that

$$\lim_{s \to \infty} P_{m1}(s) T_p(s) \stackrel{\Delta}{=} M \tag{26}$$

is a real full rank constant matrix. This requires some *a priori* information on the relative polynomial degrees in each row of $T_p(s)$ [17].

As was done in the scalar case, one can convert solution of (25) to solution of

$$H*(D)\bar{y}(t) + K*(D)\bar{u}(t) + G*Q_p(D)P_{m1}(D)\bar{y}(t) = Q_m(D)\bar{u}(t) + \bar{\gamma}(t) \qquad (27)$$

where $Q_p(D) = \text{diag}[q(D)]$ (pxp), $\bar{u}(t)$ and $\bar{y}(t)$ are filtered versions of $u(t)$ and $y_p(t)$ and $\bar{\gamma}(t)$ is a vector of decaying exponentials. Hence one can construct an adaptive multivariable observer similar to (6) and obtain the appropriate gain matrices by building a linear estimator based upon (27).

The key points to observe in extension to the multivariable case are the following. Requirement of an upperbound on plant order generalizes to an upperbound on the plant's observability index. Knowledge of the relative transfer function degree, d_p, generalizes to enough information on relative degrees in each row of $T_p(s)$ to enable one to choose $P_{m1}(D)$ such that (26) holds [17]. The requirement of left half plane zeros generalizes to $R_p(D)$ having a greatest common right divisor matrix [3] with Hurwitz determinant. Finally, the matrix $G*$ in (25) plays the role analogous to $g*$ in (8).

V. Examples

We conclude by presenting some simulation results for the proposed controller. For the simulations the upperbound on plant order $n_p=2$ and the desired model system was chosen such that

$$P_{m1}(D)P_{m2}(D) = (D+2)(D+3)$$

$$g_m r_m(D) = 1 \ .$$

The observer polynomial $q(D) = D+1$, and to simplify some of the filter dynamics we chose $d(D) = q(D)P_{m1}(D)(D+1)^3 = (D+2)(D+1)^4$. For estimation we used a multiple equation identifier of the form (22) with $\Sigma = \text{diag}[20,000]$, but without the term $[0,\ldots,0,1]\max(0,g_\ell-g(t))$. To avoid the problems of $g(t)$ converging to zero, no adjustable gain was used in the feedforward path, i.e. $g(t)$ in (6d) was fixed at one. The estimator was still used to identify all four parameters $g(t)$, $h_0(t)$, $h_1(t)$, and $k_0(t)$ although only the latter 3 were used as adjustable gains.

Case 1. For this case we chose to control the unstable plant

$$(D^2-D+1)x(t) = u(t)$$

$$y_p(t) = (D+4)x(t) \ .$$

In this case there is a unique optimal parameter vector

$$\theta* = [-1, -4, -3, 1] \ .$$

The external input was chosen to be

$$v(t) = 5(\sin 3.14t + \sin 6.28t) \ .$$

Since it is persistently exciting one would expect exact parameter estimation.
Figure 2a shows the parameter trajectories while 2b shows the plant and model outputs.

Case 2. For this case we used the same plant but chose $v(t)$ to be a unit step input
which is rich but not persistently exciting. As shown in Figures 3a and 3b the
plant output converges to that of the model but parameter convergence is slower.

Case 3. For this last case we replaced the second order plant with the first order
plant

$$(D-1)x(t) = u(t)$$

$$y_p(t) = x(t) \quad ,$$

and used the input $v(t)$ of case 1.

In this case the controller is overspecified and there is no unique solution to
(5). However as shown in Figures 4a and 4b, the plant output converged to that of
the model and the parameters converged very rapidly to one optimal vector

$$\theta* = [-3, -3, 0, 1] \quad .$$

VI. References

[1] H. Elliott and W. A. Wolovich, "Parameter Adaptive Identification and
     Control, Proc. 1978 Conference on Decision and Control, San Diego, Janu-
     ary, 1979 (also accepted IEEE Trans. on Aut. Control).

[2] K. J. Astrom, U. Brosson, L. Ljung, and B. Wittenmark, "Theory and Appli-
     cation of Self-Tuning Regulators, Automatica, Vol. 13, 1977, p. 457.

[3] W. A. Wolovich, Linear Multivariable Systems, Springer-Verlag, New York,
     1974.

[4] M. M. Sondhi and D. Mitra, "New Results on the Performance of a Well-Known
     Class of Adaptive Filters," Proc. IEEE, Vol. 64, No. 11, November, 1976.

[5] H. Elliott, "A Model Reference Adaptive Controller with Arbitrarily Small
     Error," Proc. 1978 JACC, October 1978.

[6] P. M. Lion, "Rapid Identification of Linear and Nonlinear Systems," Proc.
     7th Annual JACC, 1966, pp. 605-615.

[7] E. Fogel and D. Graupe, "Convergence of Least Squares Identification
     Algorithms Applied to Unstable Stochastic Processes," Int. J. System Sci.,
     Vol. 8, 1977, pp. 611-618.

[8] G. Kreisselmeier, "Adaptive Observers with Exponential Rate of Convergence,"
     IEEE Trans. Aut. Control, Vol AC-22, February, 1977.

[9] K. S. Narendra and L. S. Valevani, "Direct and Indirect Adaptive Control,"
     Proc. 7th IFAC Congress, Vol. 3, June 1978, pp. 1981-1988.

[10] L. Ljung and I. D. Landau, "Model Reference Adaptive Systems and Self-
     tuning Regulators - Some Connections," Proc. 7th IFAC Congress, Vol. 3,
     June 1978, pp. 1973-1980.

[11] K. S. Narendra and L. S. Valevani, "Stable Adaptive Controller Design-
     Direct Control," IEEE Trans. on Aut. Control, Vol. AC-23, August, 1978,
     pp. 570-582.

[12] A. Feuer and A. S. Morse, "Adaptive Control of Single-Input Single-Output
     Linear Systems," IEEE Trans. on Aut. Control, Vol. AC-23, August, 1978,
     pp. 557-569.

[13] G. C. Goodwin, P. J. Ramadge, P. E. Caines, "Discrete Time Multivariable
     Adaptive Control," Unpublished Technical Report, November, 1978.

[14] A. S. Morse, "Global Stability of Parameter-Adaptive Control Systems,"
     Systems and Information Sciences Report No. 7902, Yale University, Mar. 1979.

[15] B. Egardt, "Stability of Model Reference Adaptive and Self-Tuning Regulators," Ph.D. Dissertation, Dept. of Aut. Control, Lund Institute of Technology, 1978.

[16] H. Elliott and W. A. Wolovich, "Parameter Adaptive Control of Linear Multivariable Systems," Proceedings of the 13th Asilomar Conference on Circuits, Systems and Computers, Pacific Grove, November, 1979.

[17] H. Elliott and W. A. Wolovich, "A Parameter Adaptive Control Structure for Linear Multivariable Systems," Colorado State University Tech. Report #NO79-DELENG-1, Nov., 1979 (submitted IEEE Trans. on Aut. Control).

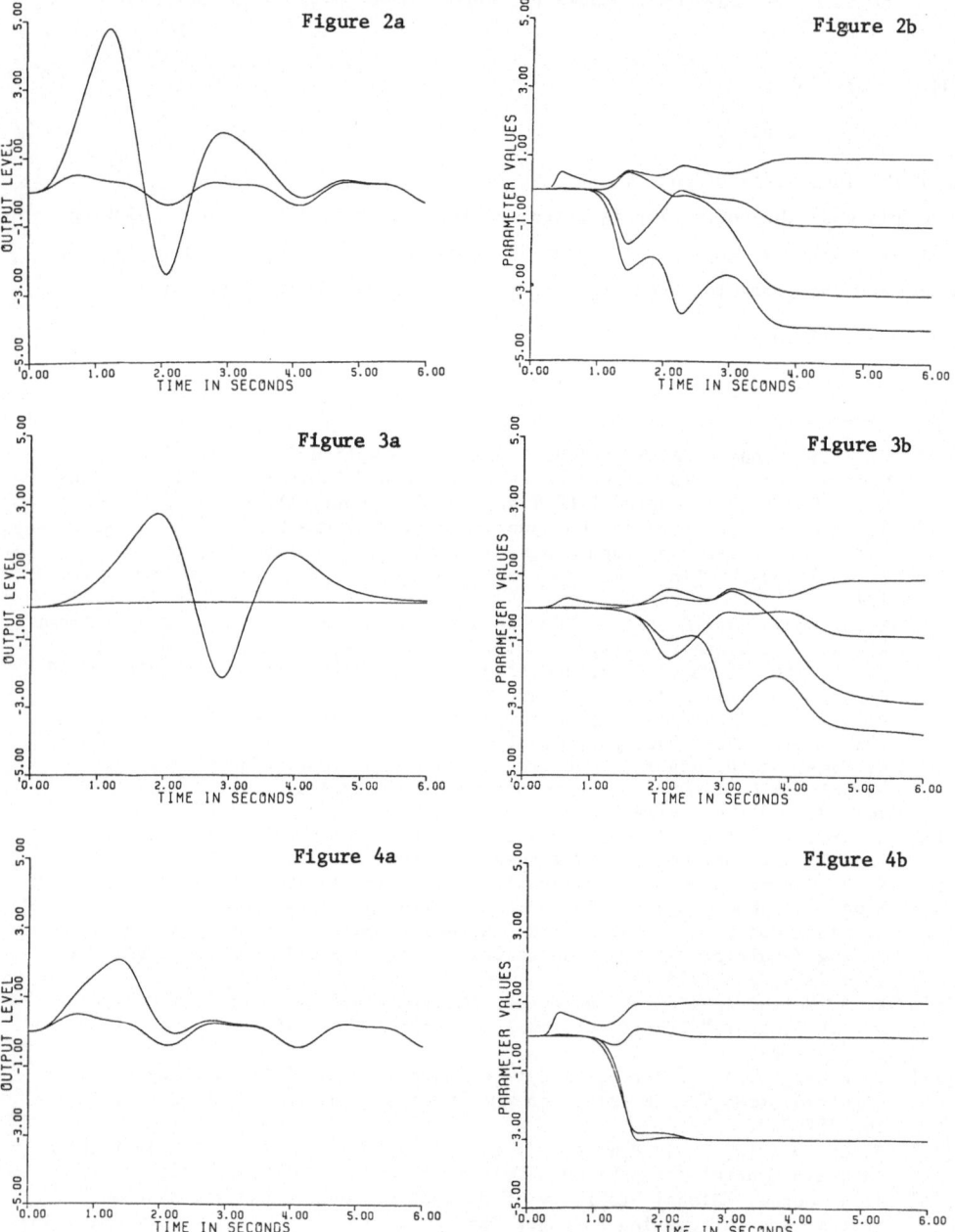

Figure 2a

Figure 2b

Figure 3a

Figure 3b

Figure 4a

Figure 4b

SUBOPTIMAL ADAPTATIVE FEEDBACK CONTROL
OF NONLINEAR SYSTEMS

A. Kuzucu   A. Roch

Institut de Réglage Automatique
Ecole Polytechnique Fédérale de Lausanne
En Vallaire, 1024 Ecublens, Switzerland

1. INTRODUCTION

The problem of suboptimal feedback control for nonlinear systems is studied, and
a method based on linear adaptative feedback is proposed. The nonlinear state equations of the process are approximated by a linear model, and corresponding suboptimal constant feedback gains are computed. Linear model matrices and feedback gains
are updated when the state of the nonlinear system is out of the validity domain of
the model or when a measure of the approximation error exceeds some tolerance value.
Such a *state dependent adaptation mechanism* leads to *corrections made only if necessary*. This feature permits the use of one central ordinator to supervise several processes with priority levels, each process being controlled by microprocessor.

2. SUBOPTIMAL FEEDBACK CONTROL

The practicing engineer often seeks a closed-loop control law, inexpensive and easy
to implement and which achieves a system performance close to optimal with respect
to a particular performance index. Such a suboptimal control law is very attractive
for industrial process control. The proposed method approaches the nonlinear suboptimal regulator and servomechanism problems, via linear constant adaptative feedback.

Consider a dynamical system nonlinear in state and control :

$$\dot{x} = f(x, a, u)$$
$$y = h(x, a)$$

where x is the state vector, a the parameter vector, u the control vector, and the quadratic performance index to be minimized :

$$J = \frac{1}{2} \int_{t_o}^{t_f} \left[(y_d - y)^T Q (y_d - y) + u^T R u \right] dt + \frac{1}{2} (y_d - y)^T S (y_d - y) \Big|_{t_f} \qquad (2)$$

Q and S are constant positive semi-definite output penalization matrices, R is a constant positive definite control penalization matrix.

The optimal control for the nonlinear servomechanism problem formulated by (1) and (2) has the nonlinear feedback structure :

$$u^*(.) = u^*(x, a, t) \qquad t \in (t_o, t_f) \qquad (3)$$

Instead of this feedback control law very hard to be determined and implemented we consider the suboptimal linear feedback :

$$\hat{u}(.) = G_k x(t) + u_{o_k} \qquad t \in (t_k, t_{k+1}) \qquad t_o \leqslant t_k < t_{k+1} < t_f \qquad (4)$$

where G_k is a linear feedback matrix and u_{o_k} an open-loop component of the control vector, both constants between two corrections made at t_k and t_{k+1}. The optimal feedback is approximated by a linear feedback with gains and open-loop components, which are constant by parts.

The quadratic performance index and the linear feedback structure of the suboptimal control law lead to the *repeated linearization* of the nonlinear state equations. The nonlinear state model is approximated by a linear model :

$$\dot{x}_m = A_k(x_k, a_k, u_k) x_m + B_k(x_k, a_k, u_k) u + a_{o_k}(x_k, a_k, u_k)$$

$$\qquad (5)$$

$$y_m = C_k(x_k, a_k) x_m + d_k(x_k, a_k)$$

valid around the measured values x_k, a_k, u_k at the correction instant t_k. Taylor series development of (1) gives such a model. This linear model can also be obtained by a least squares approximation of (1) in some regions of the state space bounded by the limit vectors $\alpha_x, \beta_x, \alpha_u, \beta_u$ such that $\alpha_x \leqslant x_k < \beta_x$ and $\alpha_u \leqslant u_k < \beta_u$ [1]. "Apparent linearization" [2] gives a linear model without the constant term a_{o_k}.

A linear time-invariant servomechanism problem is defined by (5) and (2) if A_k, B_k, C_k, a_{0k} and d_k are considered constants as long as the nonlinear system state stays in the validity domain of the linear stationary model (5). The solution of this problem gives :

$$u(t) = -R^{-1}B_k^T P(t)x(t) - R^{-1}B_k^T p(t) \qquad t \in (t_0, t_f)$$

where $P(t)$ and $p(t)$ are obtained from the adjoint variable via the linear Riccati transformation $\lambda(t) = P(t)x(t) + p(t)$ and satisfy the conventional Riccati differential equations of the linear servomechanism problem.

Further approximation is introduced by taking $t_0 = t_k$, considering only $P_k = P(t_k)$ and $p_k = p(t_k)$ and keeping them constants until another correction. The final time t_f may be kept unchanged or redefined at each correction for a chosen operation time Δ such that $t_f = t_k + \Delta$. We will have in the suboptimal control law (5) :

$$G_k = -R^{-1}B_k^T P_k \qquad \text{and} \qquad u_{0k} = -R^{-1}B^T p_k \qquad t \in (t_k, t_{k+1}) \qquad (6)$$

Our interest is focused on the computation and adaptation of G_k and u_{0k} rather than on the direct generation of the suboptimal control vector. This approach is motivated by the following considerations :

i. The closed-loop structure of the control law is conserved between two corrections

ii. The suboptimal G_k and u_{0k} computed by the use of (5) may be conserved as long as the system state stays in the validity domain of the linear model. This feature leads to a *state dependent adaptation mechanism* and to corrections made *only if necessary,* only if the linear model is no more representative for the controlled system (1). Immediate consequences are less frequent corrections and the possibility to liberate the supervising ordinator for secondary tasks when a correction is not necessary. The control system structure is represented on fig. 1.

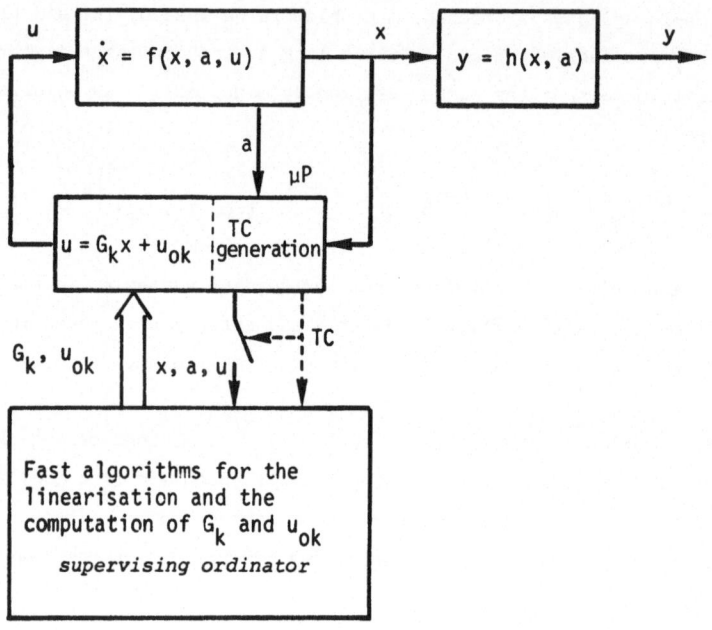

<u>Fig. 1</u> : Control system structure

3. DETERMINATION OF THE FEEDBACK MATRIX G_k AND THE CONTROL U_{ok}

A fast computation of G_k and U_{ok} is essential for the effectiveness of the proposed method. On-line performances expected from the method exclude the backward
integration of the Riccati differential equations with the known numerical technics.
A generalized version of the Davison's method [3] combined to Van Loan's algorithm
[4] is adopted for the fast computation of the Riccati matrix P_k and the vector
p_k.

Canonical equations of the linear servomechanism problem formulated with (5) and (2)
have the form :

$$
\begin{bmatrix} \dot{x} \\ \dot{\lambda} \end{bmatrix} = \begin{bmatrix} A_k & -B_k R^{-1} B_k^T \\ -C_k^T Q C_k & -A_k^T \end{bmatrix} \begin{bmatrix} x \\ \lambda \end{bmatrix} + \begin{bmatrix} V_{1k} \\ V_{2k} \end{bmatrix} \quad \text{with} \quad \begin{bmatrix} V_{1k} \\ V_{2k} \end{bmatrix} = \begin{bmatrix} a_{ok} \\ C_k^T Q(y_d - d_k) \end{bmatrix}
$$

The solution of this linear non-homogeneous equation is given by :

$$
\begin{bmatrix} x(t_2) \\ \lambda(t_2) \end{bmatrix} = \boldsymbol{\Phi}(t_2, t_1) \begin{bmatrix} x(t_1) \\ \lambda(t_1) \end{bmatrix} + \int_{t_1}^{t_2} \boldsymbol{\Phi}(t_2, \tau) \begin{bmatrix} V_{1k} \\ V_{2k} \end{bmatrix} d\tau \tag{7}
$$

where :

$$
\boldsymbol{\Phi}(t_2, t_1) = e^{F_k(t_2 - t_1)} \quad \text{with} \quad F_k = \begin{bmatrix} A_k & -B_k R^{-1} B_k^T \\ -C_k^T Q C_k & -A_k^T \end{bmatrix} \tag{8}
$$

P_k and p_k are obtained from (7) by the use of the linear Riccati transformation
$\lambda = Px + p$ with $t_1 = t_k$, $t_2 = t_f$, and $P(t_f) = S_c$, $p(t_f) = -C_k^T S(y_a - d_k)$

$$
P_k = -(\boldsymbol{\Phi}_{22} - S_c \boldsymbol{\Phi}_{12})^{-1} (\boldsymbol{\Phi}_{21} - S_c \boldsymbol{\Phi}_{11})
$$
$$
\tag{9}
$$
$$
p_k = -(\boldsymbol{\Phi}_{22} - S_c \boldsymbol{\Phi}_{12})^{-1} [g_2 - S_c g_1 + C_k^T S(y_d - d_k)]
$$

where $\boldsymbol{\Phi}_{ij}$ are $n \times n$ submatrices of $\boldsymbol{\Phi}(t_f, t_k)$, $S_c = C_k^T S C_k$ and :

$$
\begin{bmatrix} g_1 \\ g_2 \end{bmatrix} = \int_{t_k}^{t_f} \boldsymbol{\Phi}(t_f, \tau) \begin{bmatrix} V_{1k} \\ V_{2k} \end{bmatrix} d\tau = g \tag{10}
$$

$\boldsymbol{\Phi}(t_f, t_k)$ and the vector g can be directly computed from the exponential of an $(2n+1).(2n+1)$ matrix F_a such that :

$$
F_a = \begin{bmatrix} F_k & \vdots & \begin{matrix} V_{1k} \\ V_{2k} \end{matrix} \\ \cdots & \vdots & \cdots \\ 0 & \vdots & 0 \end{bmatrix} \qquad e^{F_a(t_f - t_k)} = \begin{bmatrix} \boldsymbol{\Phi} & \vdots & \begin{matrix} g_1 \\ g_2 \end{matrix} \\ \cdots & \vdots & \cdots \\ 0 & \vdots & 1 \end{bmatrix}
$$

The computation of this exponential is performed in the following way to improve the precision and the computation time :

- The exponential $e^{F_a h}$ is computed by the use of Padé approximation for a step size such that $t_f - t_k = 2^{kk} h$. The following criterium is adopted to determine kk, iteration number [4] :

$$\frac{\|F_a(t_f - t_k)\|}{2^{kk}} \leq \frac{1}{2} \qquad (11)$$

- $\Phi(h)$ and $g(h)$ are obtained by the appropriate partition of $e^{F_a h}$

- $\Phi(t_f - t_k)$ and $g(t_f - t_k)$ are computed by the use of the "doubling formulas":

$$\Phi_i = \Phi_{i-1} \cdot \Phi_{i-1} \qquad \qquad \Phi_0 = \Phi(h)$$
$$g_i = g_{i-1} + \Phi_{i-1} \cdot g_{i-1} \qquad i = 1, \ldots kk \qquad g_0 = g(h) \qquad (12)$$

Submatrices of Φ_{kk} and g_{kk} are finally replaced in (9) to give P_k and p_k. The feedback matrix G_k and the vector u_{0k} are obtained from (6).

Remarks :

- Generally kk takes values between 5 and 10. The resulting step size is sufficiently small and the Padé approximation of order 2 has given satisfactory results

- The algorithm is approximatively 20 times faster than the numerical integration
  of the Riccati equations by Runge-Kutta IV

- Different approaches may be adopted for the determination of G_k and u_{0k}.
  The problem of the *"regulator with prescribed degree of stability"* [5] or the
  problem of *"receding horizon control"* [6] may be treated with the same formulation by construction of adequately augmented matrices F_a.

4. ADAPTATION MECHANISMS

The computation of the feedback gains and the open-loop components of the control
vector are based on the linear model (5), which has to approximate the nonlinear
state equations of the controlled system. This linear model and corresponding feedback gains have to be corrected when the approximation error introduced by the linearization procedure becomes important with respect to linear terms. Two approaches
are adopted :

i. The first one profits from the a priori knowledge of the nonlinear system characteristics and of the system outputs behaviour. The state space is divided into
   regions so that the maximum modelization error is the same and is within some

tolerance value for each region. Linear models are determined by the least squares identification method and the corresponding feedback gains are computed and
memorized. During the operation, a logical signal is generated when the system
state changes region. Then, the on-line part of the algorithm determines in
which region the state is, picks up the appropriate gains and limit vectors
from the memory and sets them on the regulator. This algorithm is orientated to
the suboptimal feedback control of nonlinear systems which have a prefixed reference value (nonlinear regulator problem). It may be easily implemented on a
microprocessor having sufficient memorization capacity. This approach is illustrated on fig. 2.

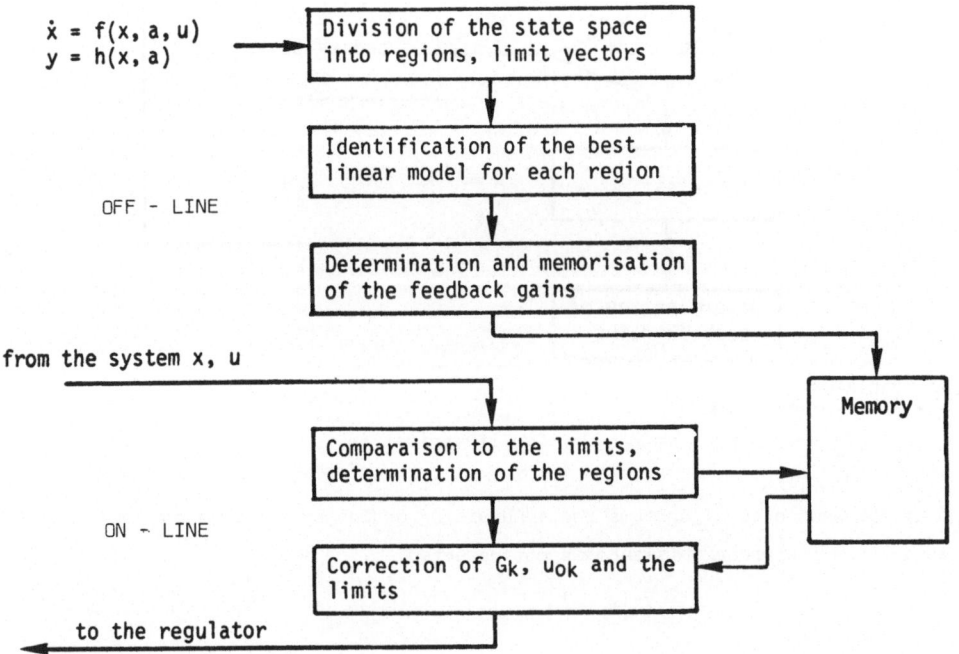

Fig. 2 : Algorithm 1 - Division of the state space into regions

ii. The second approach uses an on-line linearization scheme. A TC signal is gener-
    ated, which represents the approximation error introduced by the linearization.
    When this signal exceeds some heuristically chosen tolerance value, the state x,
    the parameter vector a and the control vector u are measured and transmitted
    to the supervising ordinator. The linear model valid around these values is de-
    termined by one of the available methods, the corresponding feedback gains are
    computed and set on the regulator. This algorithm is illustrated on Fig. 3.

It may easily be implemented by the use of a microprocessor as regulator and a mini-computer as supervising ordinator. Its domain of applications is much wider than in the previous approach. (Nonlinear servomechanism and tracking problems).

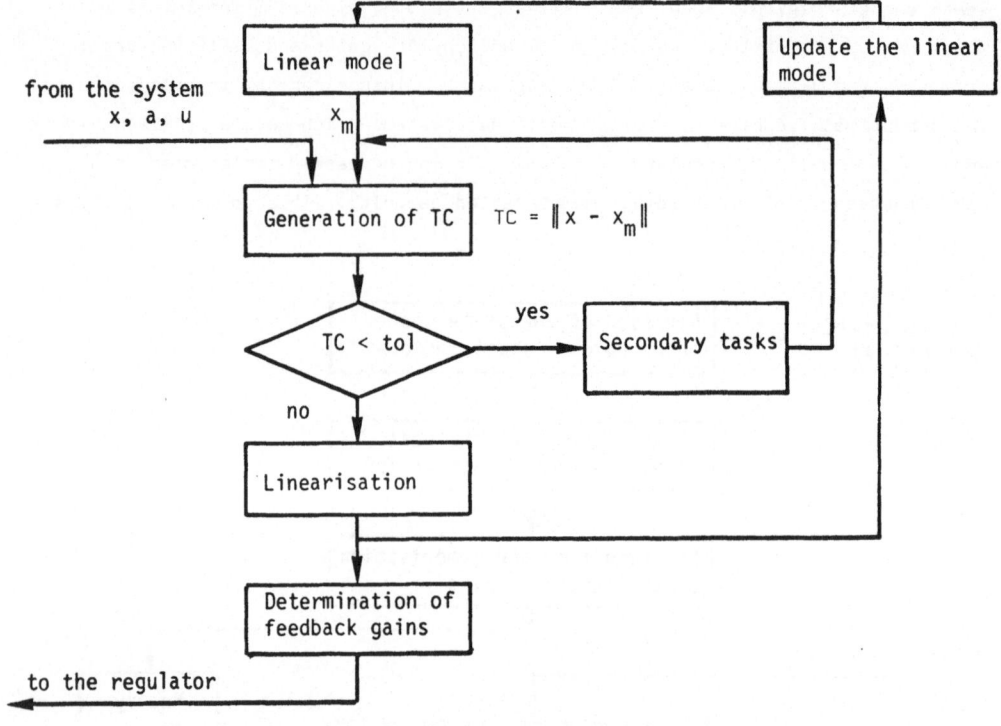

Fig. 3 : Algorithm 2 - On-line linearization

In both of the presented algorithms the supervising ordinator may perform secondary tasks as long as a correction signal is not detected.

5. EXAMPLE

The suboptimal control of a travelling overhead crane is chosen as an illustrative example. As the system initial and final conditions may change, the algorithm 2 is applied with the "receding horizon" approach for the determination of feedback control law. The performances of the suboptimal control algorithm are compared to performances achieved by feedback gains issued from the linearization around the final steady-state conditions.

Dynamical equations of the travelling overhead crane [7] have led to the following scaled state model :

$$\dot{x}_1 = - x_2$$

$$\dot{x}_2 = 0,981\,\delta\,\sin x_3 + u_1 + \delta\,\sin x_3\,u_2$$

$$\dot{x}_3 = x_4$$

$$\dot{x}_4 = - \frac{1}{\ell_c - 0,5\,x_5}\,(0,981\,\sin x_3 + 0,981\,\delta\,\sin x_3\,\cos x_3 + $$
$$+ 2\,x_4\,x_6 + \cos x_3\,u_1 + \delta\,\sin x_3\,\cos x_3\,u_2)$$

$$\dot{x}_5 = - 2\,x_6$$

$$\dot{x}_6 = 0,981(\cos x_3 - 1 - \delta\,\sin^2 x_3) + x_4^2\,(\ell_c - 0,5\,x_5) - $$
$$- \sin x_3\,u_1 - (1 + \delta\,\sin^2 x_3)\,u_2$$

$$y_1 = y_{1c} - x_1 + \sin x_3\,(\ell_c - 0,5\,x_5)$$

$$y_2 = - \cos x_3\,(\ell_c - 0,5\,x_5)$$

where ℓ_c : desired final cable length, and δ : mass ratio of the trolley and transported charge appear as system parameters.

The performance index used for the determination of the linear adaptative feedback is:

$$J = \frac{1}{2} \int_0^\Delta u^T R u\,dt \qquad R = \text{diag}\begin{bmatrix} 20 & 20 \end{bmatrix} \qquad \Delta = 2,4 + 3\,\|x\|$$

with the final state constraint $x(\Delta) = 0$.

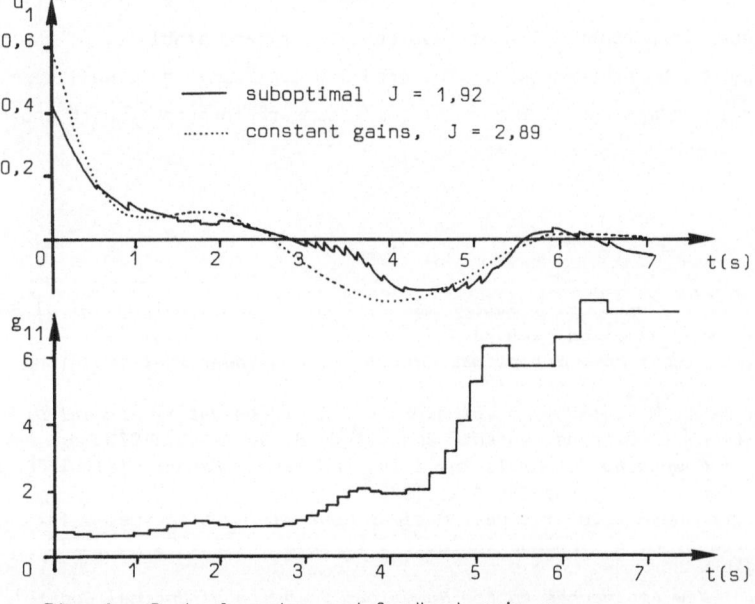

Fig. 4 : Control vector and feedback gain g_{11}

The control variations are compared on fig. 4. The behaviour of one of the feedback
gains is on the same figure. System trajectories are illustrated on fig. 5.

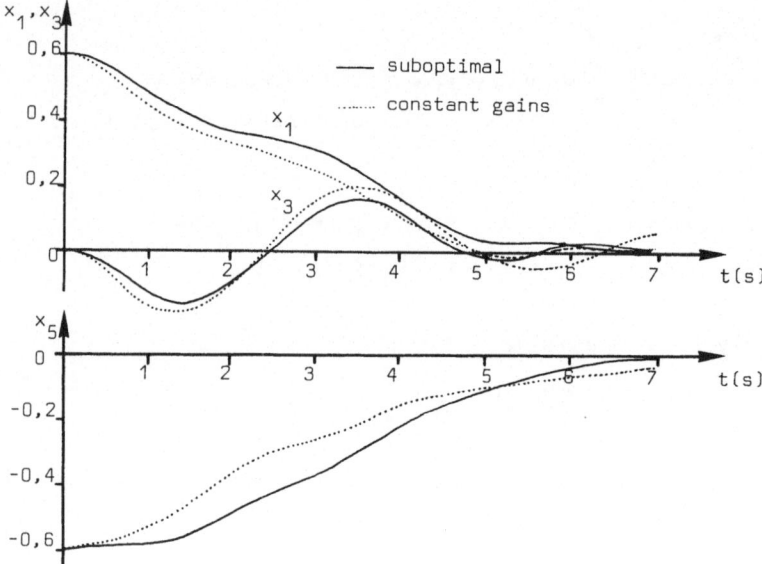

Fig. 5 : System trajectories

6. CONCLUSION

A fast and very flexible method for the suboptimal adaptative feedback control of
nonlinear systems is presented. Regulation and adaptation mechanisms are state dependent. Computation and adaptation of linear feedback gains rather than direct determination of an open-loop control vector assures security and stability of the system, even when computers have breakdowns. The proposed algorithms suit well with
decentralized control schemes and give results which compare favourably with those
obtained by optimal and conventional techniques.

REFERENCES

[1] Kriechbaum, G.K.L.; Noges, E. : *Suboptimal Control of Nonlinear Dynamical Systems via Linear Approximation*, Int. J. of Control, Vol. 13, No 6, 1961

[2] Weber, A.P.J.; Lapidus, L. : *Suboptimal Control of Nonlinear Systems*, AIChE
Journ., Vol 17, No 3, pp 641-659, 1971

[3] Davison, E.J.; Maki, M.C. : *The Numerical Solution of the Matrix Riccati Differential Equation*, IEEE Trans. on Aut. Cont., AC-18, pp 71-73, 1973

[4] Van Loan, C.F. : *Computing Integrals Involving the Matrix Exponential*, IEEE
Trans. on Aut. Cont., AC-23, pp 395-404, 1978

[5] Anderson, B.D.O.; Moore, J.B. : *Linear Optimal Control*, Prentice-Hall, 1971

[6] Thomas, Y.; Barraud, A. : *Commande Optimale à Horizon Fuyant*, Revue RAIRO, J-1,
pp 126-140, 1974

[7] Mårtensson, K. : *New Approaches to the Numerical Solution of Optimal Control
Problems*, Report 7206, Div. of Automatic Control, Lund Inst. of Technology, 1972

IDENTIFICATION STRATEGIES FOR TIME-DELAY SYSTEMS

J.E. Marshall
School of Mathematics, University of Bath,
England

1. Introduction

It is particularly important for the accurate control of time-delay systems that the delay elements, and the delay-free dynamic elements should be well modelled.

The chief principle on which time-delay control systems work is that of prediction, requiring that accurate models are available. Many time-delay system control schemes exhibit features in common with model-reference structures.(1,2)

Because of the close connection between model-reference and time-delay system control schemes it is natural that sensitivity methods should find use in the adaptive control of time-delay systems. It is important that mismatch between plant and model should be reduced to a minimum especially when time delays are 'large' in some sense.

We shall have in mind systems in which a time-delay, and the delay-free part of the plant (to be called the sub-plant) occur in cascade. When the sub-plant produces the input to the delay the phrase "delayed output" is used, and conversely "delayed control". We shall not consider the delayed-state case where delays appear within internal dynamic loops.

Following the usage of R.E. King (3) we speak of temporal sensitivity for partial derivatives w.r.t. delay, and parametric sensitivity for partial derivatives w.r.t. sub-plant parameters.

2. Time-delay system control structure

Many time-delay control structures have, or may be reduced to, the structure shown in Fig.1. $C(s)$ represents a controller, $G(s)$ a sub-plant, $T = \exp(-s\tau)$ the series delay, and the subscripted variables G_0, T_0 represent sub-plant and delay-models respectively.

It is often assumed, wrongly in the author's view, that $G = G_0$, $T = T_0$, especially in the application of Smith's principle. The transfer function of the system of Fig.1. is evidently

$$CGT/\left(1+CG_0+C(GT-G_0T_0)\right).$$

The term $GT-G_0T_0$, the mismatch term needs to be zero, or small for Smith's principle to apply. The smallness of this term implies accurate prediction and hence accurate control. In the absence of disturbance the signal $d(t)$ will be zero when system and model are matched. Disad-

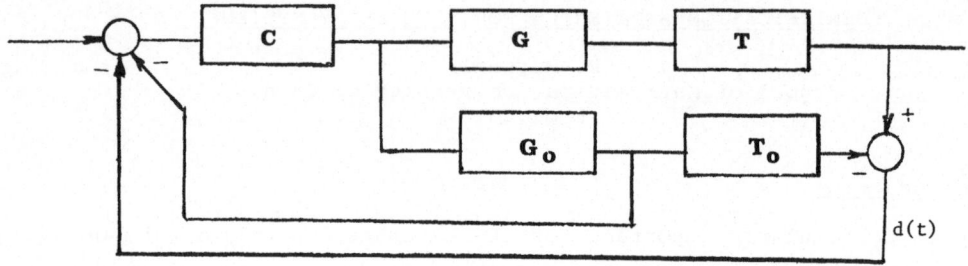

Fig.1.   Predictor Control Scheme

vantages, and wise extensions of such schemes, have been widely published.(4)

3. Application of sensitivity principles

It is well known that the techniques of Meissinger (5) and Osburn
(6), together with the use of sensitivity points techniques to generate
sensitivity coefficients may be used for identification in the delay-
free case.(7,8) It is helpful to recall briefly the essentials of such
schemes so that their application to systems with delay may be made
clear.

The structure of model reference schemes is such that error signals
are produced that are functions of the mismatch between plant and model
parameters. $e(t) = e(t, \Delta\underline{\alpha})$ where $\Delta\underline{\alpha}$ is a vector of parameter mismatch.
The elements of $\Delta\underline{\alpha}$ may be recovered by exploiting the convexity of an
integral $J = \int_{t_0}^{t_1} e^2(t, \Delta\underline{\alpha})dt$. This integral is not, of course, a performance criterion integral, but is sometimes mistaken for one. Provided
that $\Delta\underline{\alpha}$ is small, in a sense to be discussed, then it is possible to
show that

$$\Delta\underline{\alpha} = H^{-1} \cdot \left[\frac{\partial J}{\partial \alpha}\right].$$

Where H is the Hessian $\left[\frac{\partial^2 J}{\partial \alpha_i \partial \alpha_j}\right]$ in the region of the parameter plane
for which J is zero.

Indeed if $\Delta\underline{\alpha}_i$ is such that $e(t) = \Delta\underline{\alpha}_i \cdot e_i^*(t)$ and $e_i^*(t)$ is (sensibly)
independent of $\Delta\alpha_i$ then $\Delta\alpha$ may be recovered easily with H^{-1} a pre-
computed and possibly constant matrix. The sensitivity coefficients
$\frac{\partial J}{\partial \alpha_i}$ may be calculated via sensitivity relationships of the kind

$$\frac{\partial e}{\partial \alpha_i} = \mathcal{L}^{-1} \frac{\partial}{\partial \alpha_i} \cdot G(s, \underline{\alpha}) \cdot \overline{x} \cdot$$

where $G(s,\underline{\alpha})$ represents the transfer function relating $e(t)$ to $x(t)$. (8)

For delay-free systems the calculation of sensitivity coefficients in this way is straightforward.

The theory applies equally well to the case where one, say, of the α_i represents a delay τ, but problems now emerge in the realisation of the sensitivity coefficient.

4. Temporal sensitivity coefficient

Clearly $\frac{\partial}{\partial \tau}\left[G(s,\alpha)e^{-s\tau}\right] = -s \cdot \left[G(s,\alpha)e^{-s\tau}\right]$ so that (apart from sign) all temporal sensitivity realisations requires differentiation w.r.t. time. This difficulty is an important difference for realisation compared with delay-free identification.

As elsewhere, where pure differentiation appears a necessity it is often possible from knowledge of signal bandwidth to approximate to the differentiation. The expression "pseudo-sensitivity" coefficient is used where the sensitivity models are approximations to those derived theoretically. The successful use of pseudo-sensitivity models is demonstrated later.

5. Temporal identification

An adaptive scheme for a delay will be described. It is based in this example on the Smith prediction control scheme but the principles extend to other schemes.

The design problem reduces to the following steps, assuming a system of the type shown in Fig.1.

(i) Design the controller $C(s)$ assuming nominal values of delay and sub-plant parameters.

(ii) Using Sensitivity points methods find and hence realise the sensitivity model.

(iii) By simulation studies (or directly if possible) plot the sensitivity function $\frac{\partial e}{\partial \tau}$ for variation of delay about its nominal value. This should indicate the linear region of delay offsets – i.e. that range of offset for which $\frac{\partial e}{\partial \tau}$ is independent of offset. This implies that the second derivative is zero so that

$$e(t) = \Delta\tau \cdot e^*(t) \quad \text{giving}$$

$$J = \int_0^t e^2(t)dt = (\Delta\tau)^2 \cdot \int_0^t \left(e^*(t)\right)^2 dt$$

$$\text{and } \frac{\partial J}{\partial \tau} = (2\Delta\tau) \cdot \int_0^t e^*(t)^2 dt = \Delta\tau \cdot I(t), \quad \text{say}$$

and hence $\Delta\tau = \left(\frac{\partial J}{\partial \tau}\right)/I(t)$.

(iv) Recall the technique to obtain $\frac{\partial J}{\partial \alpha}$, due to Osburn (6), of using

$$\frac{\partial J}{\partial \alpha} = \frac{\partial}{\partial \alpha} \int_0^t e^2(t)\,dt = \int_0^t 2e(t)\frac{\partial e}{\partial \alpha}\cdot dt.$$

The obvious extension, letting α be a temporal parameter, gives

$$\frac{\partial J}{\partial \tau} = 2 \int_0^t e(t)\cdot\frac{\partial e}{\partial \tau}\cdot dt.$$

Both $e(t)$ and $\left(\frac{\partial e}{\partial \tau}\right)$ are generated so that the gradient $\frac{\partial J}{\partial \tau}$ is available *as a function of t*.

(v) $I(t) = 2 \int_0^t \left(e*(t)\right)^2 dt$ can be generated as a function of t, so that in principle $\Delta\tau$ which we assume constant can be found. One way in which this relation may be exploited is to calculate $\Delta\tau$ at time T by replacing the upper limit of the integral by T, and using $\Delta\tau = K\left(\frac{\partial J}{\partial \tau}\right)$, where $K = 1/I(T)$.

5.1 Modification of delay-model

Evidently in using $\Delta\tau$ to modify the control scheme it is possible only to modify (up-date) the delay-model. Modification of the plant delay itself (which presumably is already as small as possible) is undesirable or impossible.

The mechanism for up-dating model delay has features which have no analogy in the single parameter delay-free case.

The ease with which *fixed* delays may be produced using ADC, digital storage, and DAC has re-encouraged interest in time-delay control schemes of the predictor type. However, delay models capable of being up-dated introduce important questions of technique. The author has favoured the use of a hybrid computing technique for variable delays, using a store of fixed length with variable clock rate.(1,9) This method has the advantage over its obvious rival, a variable length store with fixed clock rate, in that it preserves stored information. Adding or removing elements of the store clearly involves loss of information when storage elements are 'removed', or the manufacture of information when the store length is increased. Of course modification is possible to minimise such effects.

A fixed-rate store which uses digital simulation via the differential equation for variable delay is an alternative which has been proposed by the author's colleagues, and others. That this technique may be used "on-line" has yet to be demonstrated.

5.2 An example of temporal adaptive control

Fig.2 shows the realisation of a temporal adaptive scheme based on these methods.

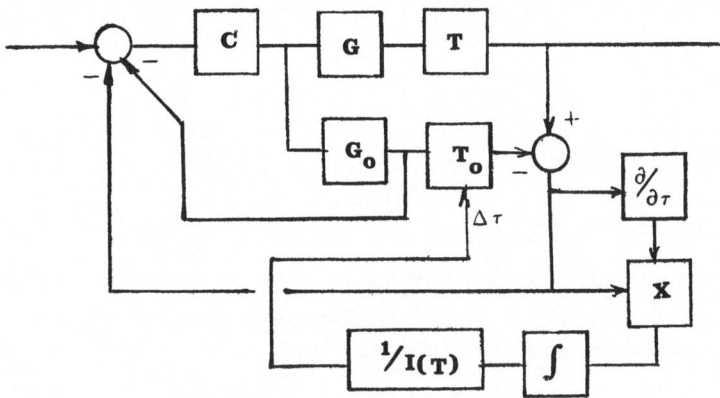

Fig.2. Temporal Adaptive Control

In this example it is assumed that
  (1) A Smith predictor scheme is appropriate
  (2) $x(t)$ is of known form, i.e. step-like
  (3) T is sufficiently large
  (4) $I(T)$ has been pre-computed
  (5) Error linear in offset
  (6) Noise-free system
Pre-simulation is needed to establish the validity of (3) and (5) in particular cases.

In this case where the correction to delay is to be made at time T there is no "loop gain" problem it is merely $K = \frac{1}{I(T)}$. Plant and model parameters are obvious from the diagram.

We discuss later the choice of loop gain when the model delay is updated continuously.

6. Mixed (hybrid) identification

Experience with the M.I.T. method for delay-free systems has shown that identification of a single parameter and its correction may be performed continuously without serious instability problems. However attempts to identify two or more parameters continuously lead to inter-action of loops and serious stability problems unless adaptive loop gains are chosen with particular care.

In this section will be discussed techniques for simultaneous identification of one temporal and one sub-plant parameter.

Here it is shown that interaction problems can be reduced, and single-shot identification is possible in the linear region. This linear region, to be found by pre-simulation, is not necessarily restricted to offsets small in a percentage sense. In one example it was found that time delay offsets of ±80% still gave linearity in the sense of

$$e(t) = \Delta\alpha \cdot e_{\alpha}^{*}(t) + \Delta\tau \ e_{\tau}^{*}(t)$$

being a good approximation.

6.1 Diagonalisation procedure 1

The case where $J = \int_{0}^{t} e^2(t)\,dt$ and the system is open-loop until time T (i.e. update at T) is very similar to the case treated in section 5. The loop gains are calculated as follows:

$$\text{As before define } J = \int_{0}^{t} e^2(t, \Delta\alpha, \Delta\tau) \cdot dt$$

It is necessary to find at time T the relationships between the gradients $\frac{\partial J}{\partial \alpha}$, $\frac{\partial J}{\partial \tau}$, which may be calculated, and integral which have functions of the sensitivity coefficients as their integrands by analogy with the single variable case given earlier.

The components of gradient of the surface J, w.r.t. $\Delta\alpha$ and $\Delta\tau$ are

$$\frac{\partial}{\partial (\Delta\alpha)} \int_{0}^{t} e^2(t)\,dt \ \text{ and } \ \frac{\partial}{\partial (\Delta\tau)} \int_{0}^{t} e^2(t)\,dt \ \text{ respectively.}$$

Neglecting higher terms (offsets in the linear region being supposed)

$$J = \int_{0}^{t} \left(\frac{\partial e}{\partial \alpha}\right)^2 (\Delta\alpha)^2 dt + \int_{0}^{t} \left(\frac{\partial e}{\partial \tau}\right)^2 (\Delta\tau)^2 dt + 2 \int_{0}^{t} \Delta\alpha\Delta\tau \left(\frac{\partial e}{\partial \tau}\right)\left(\frac{\partial e}{\partial \alpha}\right) dt.$$

Assuming that the offsets are in the linear region, and that differentiation under the integral is valid

$$\frac{\partial J}{\partial \alpha} = 2\Delta\alpha \int_{0}^{t} \left(\frac{\partial e}{\partial \alpha}\right)^2 \cdot dt \quad + \quad 2\Delta\tau \int_{0}^{t} \left(\frac{\partial e}{\partial \tau}\right)\left(\frac{\partial e}{\partial \alpha}\right) dt$$

$$\frac{\partial J}{\partial \tau} = 2\Delta\tau \int_{0}^{t} \left(\frac{\partial e}{\partial \tau}\right)\left(\frac{\partial e}{\partial \alpha}\right) dt \quad + \quad 2\Delta\tau \int_{0}^{t} \left(\frac{\partial e}{\partial \tau}\right)^2 \cdot dt$$

Further

$$\frac{\partial^2 J}{\partial \alpha^2} = 2 \int_{0}^{t} \left(\frac{\partial e}{\partial \alpha}\right)^2 \cdot dt \quad , \quad \frac{\partial^2 J}{\partial \tau^2} = 2 \int_{0}^{t} \left(\frac{\partial e}{\partial \tau}\right)^2 \cdot dt$$

$$\frac{\partial^2 J}{\partial \alpha \partial \tau} = \frac{\partial^2 J}{\partial \tau \partial \alpha} = 2 \int_{0}^{t} \left(\frac{\partial e}{\partial \tau}\right)\left(\frac{\partial e}{\partial \alpha}\right) dt$$

so that the integrals are simply related to the Hessian of J.

$$\begin{pmatrix} \dfrac{\partial J}{\partial \alpha} \\ \dfrac{\partial J}{\partial \tau} \end{pmatrix} = H \cdot \begin{pmatrix} \Delta \alpha \\ \Delta \tau \end{pmatrix}.$$

In the linear region the minimum is well defined so that H^{-1} will exist, indeed it may be expected that H will be diagonally dominant for sufficiently large T because its mean diagonal elements have positive integrands. The problem has reduced to finding the finite derivatives of J and hence $\Delta \alpha, \Delta \tau$ via

$$\begin{pmatrix} \Delta \alpha \\ \Delta \tau \end{pmatrix} = H^{-1} \cdot \begin{pmatrix} \dfrac{\partial J}{\partial \alpha} \\ \dfrac{\partial J}{\partial \tau} \end{pmatrix}.$$

From the fixed upper limit case pre-calculation of the elements of H and hence H^{-1} may be performed. $\dfrac{\partial J}{\partial \alpha}, \dfrac{\partial J}{\partial \tau}$ are found as before by using

$$\frac{\partial}{\partial \alpha} \int_{0}^{T} e^2(t)\,dt = 2 \int_{0}^{T} e(t) \cdot \frac{\partial e}{\partial \alpha} \cdot dt,$$

and the corresponding relation in τ. H^{-1} is the matrix of loop gains.

Fig.3 shows the realisation of such a scheme, i.e. where $J(T)$ is used and the parameters α, and τ are updated at $t = T$. The mechanism for

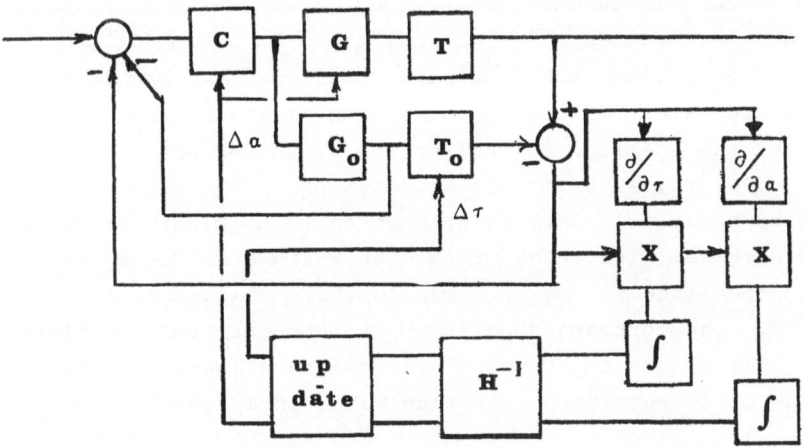

Fig.3. Mixed Temporal/Parametric Adaptive Control

updating the delay model is as before. Knowledge of $\Delta \alpha$ enables the plant parameter α to be corrected to nearer its nominal value or, where this is not possible, by modifying the model α_0 and 'correcting' the controller and H^{-1}. In the examples to be shown it is assumed that the former is possible, but this is a known impracticality.

6.2 Continuous up-dating of parameters

In the examples discussed so far we have discussed the open-loop case
where up-dating takes place once, at time t = T, and where diagonalisation (via H^{-1}) is straightforward both mathematically and practically
provided that $\Delta\alpha, \Delta\tau$ (not necessarily small) were such that the modelling
error $e(t)$ was linear in offset. It is surely desirable that up-dating
should take place as quickly as possible, and the possibility of continuous up-dating needs to be considered.

As we have remarked earlier continuous up-dating using the MIT scheme
for the scalar parametric case is possible finding the loop gain by
'trial' off-line. What is needed is a systematic procedure to determine
the loop gain from the continuous (scalar) temporal case and the continuous mixed case. In the following section an outline is given of the
procedure for the continuous case corresponding to the example in section 6.1.

6.3 Diagonalisation procedure 2

The only modification to Fig.3 is to indicate continuous up-dating
to modify the values of elements of the loop gain matrix, which strictly
should no longer be labelled H^{-1}. The new elements of the loop gain
matrix are found as follows via simulation studies. The method described
also indicates the extent of the linear region of the parameter plane
(α, τ). System and the adaptive controller are simulated and the following
values are found by trial, after all elements of "H^{-1}" are set to zero.
K_A, the value of parametric loop gain, when only the parameter channel
is allowed to operate. The range of $\Delta\alpha$ over which a unique K_A exists
corresponds to the 'linear' region. During these measurements τ is set
to equal τ_0. K_T, the corresponding value of loop gain for the temporal
loop when $\alpha = \alpha_0$ and only the temporal channel is allowed to adapt.
Again the range of $\Delta\tau$ resulting in unique value of K_T determines the
linear range in $\Delta\tau$.

Next, with both channels operative with loop gains K_A, K_T the cross-
terms K_T^A, K_A^T are found in turn by adjusting for minimum cross-coupling.
Viz: adjust K_T^A so that the zero offset gives rise to zero correction
from the temporal channel, and then K_A^T for zero offset of the parametric
channel. Finally, each of the elements K_A, K_T, K_T^A, K_A^T are divided by

$$1 - \frac{K_T^A \, K_A^T}{K_A \, K_T}$$ to give the new values for the elements of "H^{-1}". The reason

for this last step is that K_A, K_T were found with the other channel
inoperative, and unmodified they represent the 'wrong' partial differen-

tial coefficient.(9)

In finding the values of K_A experimentally it is helpful to display the offset in α after correction, as a function of time and adjust K_A until the final corrected offset is zero. There may well be overshoot before the parameter settles to its correct value.

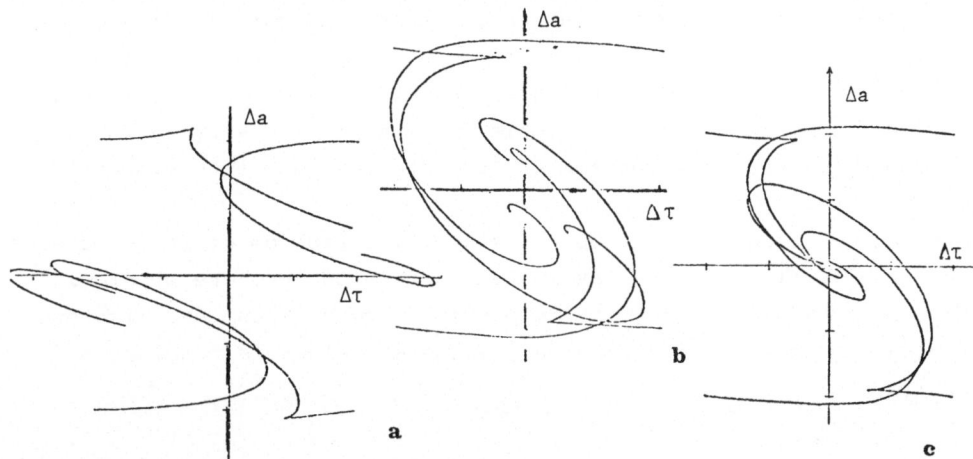

Fig.4. Parameter Plane Trajectories

Figs.4(a),(b),(c) show parameter-plane trajectories. In (a) cross terms in "H^{-1}" have been set to zero. In (b) the final stage of division by $1 - \dfrac{K_T K_A}{K_A K_T}$ is omitted. (c) indicates successful simultaneous adaptive adjustment of the offsets.

This procedure produces *constant* values of the "H^{-1}" parameters. It is undoubtedly possible by making elements of H time-dependent that the trajectories in the parameter offset plane could approach the origin optionally (not necessarily radially).

Clearly, use of H(t) rather than H(T), or diagonalisation procedure 2, would modify the trajectories but this requires continuous calculation of $H^{-1}(t)$. As the temporal sensitivity coefficients are delay dependent $H^{-1}(t)$ may not be well-defined for small values of t.

The author is aware of no research on "optimal" parameter plane trajectory studies, even for the delay-free case.

7. Adaptive control for more general inputs

The examples given have all been concerned with known step-like inputs. It is of course possible to calculate J(t) for any input but the corresponding surface may have a peculiar geometry in some cases. If the input is of known form but of unknown magnitude scale it is of course possible to calculate J(T) on line, and to perform the inversion.

But this does not extend simply the continuous case. Again it is possible to work iteratively for larger offsets if the input consists of block pulses of well-defined length, not necessarily periodic.

Conclusion

Some strategies for identification and adaptive control for systems with a single delay element have been given, and results demonstrated for a restricted class of input functions. Suggestions have been made of open research problems, for example the use of time varying loop gains, the optimisation of parameter trajectories, and extensions to general inputs.

No detailed discussion has been given of the problem of large offsets but in the control of time-delay systems for which accurate prediction is important, and where sensitivity problems are serious, even improvements close to the region of nominal parameter values is considered worthwhile.

References

(1) MARSHALL, J.E. (1979) *The Control of Time-Delay Systems*, Peter Peregrinus, I.E.E. Control Engineering Series.

(2) GARLAND, B. and MARSHALL, J.E. (1978) "On the Applicability of O.J.M. Smith's Principle", in GREGSON, M. *Recent Theoretical Developments in Control*, Academic Press.

(3) KING, R.E. (1967) "Sensitivity analysis of a class of Differential-Difference Systems", *Int. J. Control*, Vol.5, No.6, pp.583-588.

(4) HANG, C.C. (1979) "Modified Smith Predictors for the control of processes with dead-time", *Proc. I.S.A. Annual Conference*, Chicago, October 1979, pp.33-44.

(5) MEISSINGER, H.F. (1964) "Parameter Optimisation by an automatic open loop computing method" (Paper 69), *4th International Conference of A.I.C.A.*

(6) OSBURN, P.V. (1960) "Adaptive Control Systems: an analytical design method for MRAC systems", Doctoral Thesis, M.I.T.

(7) TOMOVIC, R. (1963) *Sensitivity Analysis of Dynamic Systems*, McGraw-Hill.

(8) GARLAND, B. and MARSHALL, J.E. (1975) "Application of the sensitivity points method to a linear predictor control system, *Int. J. Control*, Vol.21, No.4, pp.681-688.

(9) GARLAND, B. (1978) "Adaptive & Optimal Control of Time-Delay Systems", Ph.D. Thesis (Mathematics), University of Bath.

Acknowledgment

The author is pleased to acknowledge his gratitude to Dr. B. Garland and Mr. Arun Chotai for their collaboration in research on time-delay system control at the University of Bath.

ADAPTIVE CONTROL OF EXTREMUM SYSTEMS

Jan Sternby
Dept of Automatic Control
Lund Institute of Technology
Lund, Sweden

Abstract

Two model-based methods for extremum control are treated. The models consist of a dynamic linear part and a static nonlinearity which can be placed either at the input or at the output. It is shown that models with an input nonlinearity are easier to handle, but may give poor adaptive control laws. With a nonlinearity at the output, the optimal control law is dual, even if the system parameters are known. A way of rewriting this latter model is suggested to facilitate the use of parameter identification.

1. INTRODUCTION

Extremum control was a popular topic in automatic control some decades ago. A commonly considered problem was to find controller settings to minimize some nonlinear functional of the control error, such as the integral of the error squared. Nowadays that problem is rather solved using an adaptive scheme, and extremum control has not been discussed so much any longer.

There are, however, another class of problems, where the output should not be kept constant, but should rather be minimized or maximized. In these problems the nonlinearity is inherent in the problem and is not introduced by the designer. One example is the adjustment of the blade angles in a water turbine to achieve a maximum of produced power. This paper deals with the application of adaptive control laws to this type of extremum control problems.

In the past decades, computer technology has developed enormously. This is one reason why it might be rewarding to reconsider extremum control problems. It is now possible to implement rather complex control algorithms in low cost microcomputers, as has already been shown with adaptive control. It should then be possible to benefit from inserting more ideas from adaptive control and identification into the extremum control area. Moreover, with today's competition for market shares and increasing system complexity, even small gains may be very valuable.

A well-written survey of the older methods for extremum control is given in Blackman(1962). In recent years the idea of adaptive extremum control has been exploited by e.g. Keviczky and Haber(1974) and Bamberger and Isermann(1978).

In the present paper will be considered only methods based on the use of a system model. It will be shown that different models may differ drastically in their behaviour. For a simple example a comparison is also made between using a certainty equivalent control law and one which is dual in the sense of Feldbaum. It is noticable that this is a fairly realistic example where the dual control law is significantly better.

As already mentioned, extremum control systems have one major characteristic in common. In the absense of disturbances, the steady-state relation between input and output should be a function with an extremum. The object of control is to stay as close to this extremum as possible despite the influence from dynamics, noise or drifts. In order to use optimal control theory, this desire must be translated into a formal loss function. There are several ways of doing this. One possibility is to use a system model to estimate the slope. The control law can then be designed to keep the slope as close to zero as possible, e.g. with its variance as a measure.

However, it is not at all clear what is the best and most natural way of modeling such a nonlinear dynamic system. To be able to use system identification it is of course desirable to have a model which is linear in its unknown parameters. Any a priori knowledge about the process should then be utilised in the choice of regressors. In this way it may be possible to handle quite complicated, but partially known nonlinear systems.

In general cases it is however difficult to find model structures that are general enough, and still allow calculations to be done. One attempt is to separate the linear and nonlinear parts into two blocks in series. There are then two possibilities: the nonlinear part can be placed either before or after the linear part. This choice will have a large influence on the behaviour of the model as can be seen from the following example.

Example

Consider a first order linear system with white equation noise, and a nonlinearity in the form of a squaring device. Then with the nonlinearity at the input of the linear part the overall system is

$$y(t+1) = ay(t) + bu(t)^2 + e(t)$$

where $e(t)$ is a white noise process. Suppose a stationary solution exists ($|a| < 1$). Expected values then are

$$Ey = \frac{b \cdot Eu^2}{1 - a}$$

If the goal is to minimize Ey (and $b > 0$) the best performance is thus achieved by putting $u(t) = 0$! Furthermore, if $|a| > 1$ no stationary solution exists.

Now turn to the other case with an output nonlinearity described by

$$x(t+1) = ax(t) + bu(t) + e(t)$$

$$y(t) = x(t)^2$$

For $a = b = 1$ this is the problem considered by Jacobs and Langdon(1970). They show that because of the nonlinear measurement this is a dual control problem in the sense of Feldbaum. The conditional distribution of the state x is discrete, the possible values being $x = \pm|x|$. The

conditional mean of x can then be calculated. It is shown that it is
not optimal in the long run to have u(t) = -x̂(t). These results would
probably not change much if a = 1-ε< 1. Even if a is slightly greater
than one, a stationary solution still seems possible. □

There are thus significant differences between the two cases in spite
of their identical static response curves. In the first case with the
nonlinearity at the input, the optimal control is constant, and thus
contains no feedback. The solution to the second problem includes
feedback and therefore seems more attractive. It is however more
difficult to compute, because it has a dual nature even with known
parameters.

May be an output nonlinearity is in general more important than an
input nonlinearity for a good description of a nonlinear system. The
only possible effect of a known nonlinearity at the input is to
restrict the possible input values for the linear part. The nonlinear
control problem can then be transformed to linear control with
positive inputs. If the range of the nonlinearity is the whole of the
real axis, then a change of control variable will reduce the problem
into a linear one.

3. NONLINEARITY AT THE INPUT

With the nonlinearity at the input it is easy to set up a model which
is linear in the parameters, and thus directly lends itself to
parameter estimation and adaptive control. It was shown in the
previous section that the optimal control in the case of known
parameters is constant if the criterion is the mean output value. It
will now be shown that, as expected, adaptive control may give a poor
result in the corresponding case with unknown parameters.

In order to simplify notations and analysis only a special low order
case will be treated. The same type of problems that appear here will
however show up also with dynamical models of any order. The system
considered is of the Hammerstein type.

$$y(t) = k + b\cdot u(t-1) + c\cdot u(t-1)^2 + e(t) \tag{1}$$

The noise {e(t)} is supposed to be a sequence of independent random
variables with zero mean. Only the parameters k and b are supposed to
be unknown and have to be estimated, but an unknown c-parameter can
also be handled without any change of the results. This system should
be controlled so that its output is kept as small as possible. To
accomplish this, the criterion used is the steady state mean value of
the output. Admissible control laws may use all information available,
i.e. u(t) may depend on y(t), u(t-1) and all previous inputs and
outputs. If the parameters of the model (1) are all known the optimal
control law is

$$u(t) = -b/2c \tag{2}$$

The optimal control law is thus no feedback controller. For the model
(1) this controller minimizes the expected value of the output. But it
also minimizes the output of the next step. With a more general system
model where the output depends on the value of the input at several
sampling points, the best steady state and the best one-step
controllers will not coincide. For a further discussion on this point
see Keviczky and Haber(1974).

Adaptive Control and Estimation

When the parameters of the model (1) are unknown, the control law (2) has to be modified. One approach is to replace the parameters by their estimates to form an adaptive control law, i.e.

$$u(t) = -\hat{b}(t)/2\hat{c}(t) \qquad (3)$$

This control law is also one step ahead optimal for the chosen criterion. Since the process noise e in (1) is assumed to be white the process parameters can be estimated with an ordinary least squares estimator. It is also possible to use a stochastic approximation type algorithm. This variant is easier to analyze, and will be discussed in detail in the sequel. Let $\hat{x}(t)$ be the column vector of parameter estimates obtained at time t and

$$\theta(t) = [1 \quad u(t-1)] \qquad (4)$$

Then

$$\hat{x}(t) = \hat{x}(t-1) + \frac{C}{t}\cdot\theta(t)^T[y(t) - cu(t-1)^2 - \theta(t)\hat{x}(t-1)] \qquad (5)$$

Analysis

For the case of least squares estimation, the general Bayesian convergence results of Sternby(1977) show that the estimates will converge with probability one. But the limits will in general differ from the true values if the conditional variance does not tend to zero. This is what happens here.

The behaviour of the algorithm can also be analyzed using the technique derived by Ljung(1977). In order to apply his results a number of technical conditions must be fulfilled. Some of these conditions are difficult to check mathematically, but are relatively easy to accept intuitively. In this paper no strict proofs of non-consistency will be given. Instead the differential equations of Ljung will be used to show the expected paths of the parameter estimates. The results are confirmed by simulations of the original algorithms. Many identification procedures may be described in terms of the general recursive algorithm of Ljung(1977), which is in the time-invariant case with $\gamma(t)=1/t$

$$\hat{x}(t) = \hat{x}(t-1) + \frac{1}{t}\cdot Q(\hat{x}(t-1),y(t)) \qquad (6)$$

where the measurements y are generated as one component of the vector φ in

$$\varphi(t) = A(\hat{x}(t-1))\cdot\varphi(t-1) + B(\hat{x}(t-1))e(t) \qquad (7)$$

The noise vectors e($\cdot$) are supposed to be independent. According to Ljung(1977) (6) will asymptotically behave like the solution to

$$\dot{\underline{x}} = f(\underline{x}) = EQ(\underline{x},y) \qquad (8)$$

In calculating the expectation of (8) $\underline{x}$ is a fixed vector and the measurement y is generated from (7) with this fixed $\underline{x}$-value. In our case we have from (5)

$$f(\underline{x}) = E\ \theta(t)^T [(k - \underline{k}) + (b - \underline{b})u(t-1) + e(t)] \qquad (9)$$

where $\underline{x}$ is a column vector containing $\underline{k}$ and $\underline{b}$. The expectation shall be calculated for every fixed value of the vector $\underline{x}$. With u =$-\underline{b}/2c$ let

$$\varepsilon = (k - \underline{k}) + (b - \underline{b})\cdot u = (k - \underline{k}) - (b - \underline{b})\cdot\underline{b}/2c \qquad (10)$$

Then (8) is

$$\dot{\underline{k}} = \varepsilon \quad ; \quad \dot{\underline{b}} = \varepsilon\cdot u \qquad (11)$$

The parameters may converge to any point on the curve $\varepsilon=0$ for which (11) is stable. Convergence to the correct point may happen since it satisfies $\varepsilon=0$. There are, however, infinitely many other possible convergence points which form a parabola in the $\underline{b}$-$\underline{k}$-plane. The trajectories of (11) are easily found by dividing the two equations. They satisfy

$$\underline{b}(\tau) = \underline{b}(0)\cdot\exp\{[\underline{k}(0) - \underline{k}(\tau)]/2c\} \qquad (12)$$

Figs. 1 and 2 show parameter phase planes for the algorithm and its associated differential equations respectively with the true parameter values k=b=0.4, c=0.2 and the standard deviation of the noise $\sigma=0.03$. Note the parabola of stationary points (dashed in both figs.).

For small c-values part of the stationary points are unstable. This can be studied through a linearization around the stationary points, i.e. by looking at the derivative matrix of the right member of (11). The unstable points must satisfy

$$b - \sqrt{b^2 - 32c^2} < \underline{b} < b + \sqrt{b^2 - 32c^2} \qquad (13)$$

This will only happen if the square root exists. A phase plane for (11) in this case is shown in fig. 3.

The parameters are thus likely to converge to some point which gives a nonoptimal input value. The same thing will happen also in more general cases as shown in Sternby(1978a). Such cases may include least squares identification, dynamics in the model or the input modification of stochastic approximation type discussed by Keviczky and Haber(1974).

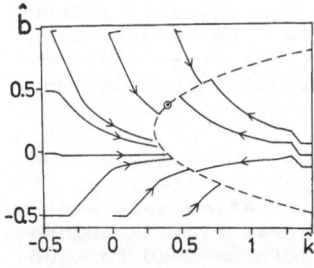

Figure 1-Phase plane
for algortithm

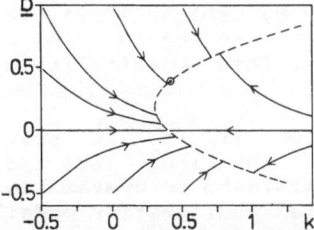

Figure 2-Phase plane
for diff. equation

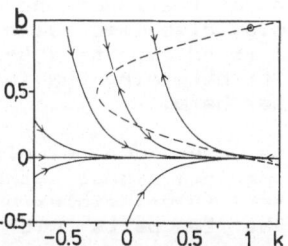

Figure 3-Unstable
stationary points.

Improved Control

The control law must be changed to get convergence to the correct
values. It is then important to find the reason for nonconsistency.
Here the problem is that the same factor ε appears in both equations
of (11). This happens because the input depends only on the estimates,
and is therefore constant in the calculation of $f(\underline{x})$ of (9). Both
components of the vector θ(t) are then constant. All attempts of
improvement must therefore aim at increasing the variation of the
input.

A straightforward way is to add a perturbation signal to the input.
This works, but convergence was very slow in a few test simulations
performed. Also, one more parameter, the perturbation signal
amplitude, has to be chosen in advance.

Another possibility could be to introduce a forgetting factor in the
identification procedure to prevent the estimates from converging too
fast. If the forgetting factor is allowed to tend to one, then the
estimates will finally converge. Simulations have indicated that this
method may give a nonsatisfactory behaviour. The estimates tend to
stay constant for a while and then suddenly jump to another constant
value.

The control law (3) is one-step ahead optimal for the chosen
criterion. But the dynamic programming could be pursued one more step
to give a two-step ahead optimal control. This would introduce a
tendency in the control law to actively reduce the parameter
uncertainty. In our case, however, this optimal control would still
depend only on the estimates, and the problem remains unsolved. But if
dynamics are included in the model (1) this might be a possible
method, since the input will then also be a function of the measured
output, and θ(t) is no longer a constant in the calculation of $f(\underline{x})$ of
(9).

4. NONLINEARITY AT THE OUTPUT

The example in the models section shows that a nonlinearity at the
output of a linear system is much more difficult to handle than one at
the input. Even if the system parameters are known, the optimal
control is e.g. of a dual nature. It tries actively to improve the
estimate of the input x to the nonlinearity at the price of worse
short term control.

In one of the oldest and most used methods of extremum control the
slope of the nonlinearity is estimated by applying a perturbation
signal at the input and observe its effect at the output. This method
was e.g. discussed already by Leblanc(1922). Taking the so estimated
slope as an output, a self-tuning regulator of suitable order could be
used to determine the input. This possibility will not be pursued any
further here.

Things are much simplified if it is possible to measure the
intermediate signal between the linear part and the nonlinearity. It
is then, in principle, possible to do system identification separately
for the two parts with the nonlinearity modelled as e.g. a second
order polynomial. A self-tuning regulator could again be used to keep
the output of the linear part with minimal variance around the
estimated position of the extremum of the nonlinearity. The results of

the previous section will then apply to the nonlinear part, the only difference being that its input is not determined directly, but through linear dynamics. Noise on this intermediate signal will then act as a perturbation signal and improve identifiability, but as stated in the previous section convergence may be slow.

When the intermediate signal is not measurable the problem is more difficult. Even with known parameters the optimal control law is dual. A simple example of this type will now be discussed to show the difficulties and suggest a solution method that can possibly be extended to the case with unknown parameters.

Example

Consider the system

$$y(t) = [x(t) - c]^2 + v(t) \tag{14}$$

$$x(t+1) = x(t) + u(t) + w(t+1) \tag{15}$$

where c is an unknown constant and v(·) and w(·) are zero mean disturbances. Only y(t) is measurable, and the object of control is to minimize the mean value of the output. First let c=0. This is no restriction here, since c can be subtracted from both members of (15). A reason for choosing this example is that the optimal control has been calculated by Jacobs and Langdon(1970) and is available for comparison.

One possibility is to apply certainty equivalence and let $u(t) = -\hat{x}(t)$. The problem is then to calculate a good estimate $\hat{x}(t)$. In the special case v(t)=0 the conditional distribution for x(t) can be tracked exactly as shown by Jacobs and Langdon(1970). Florentin(1964) treated the case v(t)≠0 by approximating the conditional distribution by the sum of two Gaussian distributions. Another possibility is to rewrite the system as follows to be able to use some identification method directly. If R is the known variance of w(t) and c=0, then inserting (15) into (14) gives

$$y(t) = y(t-1) + u(t-1)^2 + R + 2u(t-1)x(t-1) +$$

$$+ e(t) + (w(t)^2 - R) + 2w(t)[x(t-1) + u(t-1)] \tag{16}$$

where e(t) = v(t)-v(t-1).

Estimation

Assuming that {e(t)} and {w(t)} are independent zero mean sequences, the least squares method can be used to estimate x(t-1) from (16). The last row is then regarded as zero mean noise with zero autocorrelation. The first three terms are known at time t-1. The measurement equation (16) will thus give $\hat{x}(t-1|t)$ from $\hat{x}(t-1|t-1)$. Then (15) is used to get $\hat{x}(t|t)$ from $\hat{x}(t-1|t)$ as

$$\hat{x}(t|t) = \hat{x}(t-1|t) + u(t-1)$$

This is an approximation since w(t) is actually partly known at time t through the measurement y(t). Now denote $\hat{x}(t|t)$ by $\hat{x}(t)$. The approximate least squares estimation equations then are

$$\hat{x}(t) = \hat{x}(t-1) + u(t-1) + K(t)\varepsilon(t) \tag{17}$$

$$\varepsilon(t) = y(t) - y(t-1) - u(t-1)^2 - R - 2u(t-1)\hat{x}(t-1) \tag{18}$$

$$K(t) = 2P(t-1)u(t-1)/[\sigma(t-1)^2 + 4P(t-1)u(t-1)^2] \tag{19}$$

$$P(t) = \sigma(t-1)^2 P(t-1)/[\sigma(t-1)^2 + 4P(t-1)u(t-1)^2] + R \tag{20}$$

From (16) with $u(t) = -\hat{x}(t)$ a suitable value for the standard deviation $\sigma(t-1)$ of the measurement noise is

$$\sigma(t-1)^2 = \sigma^2 + 2R^2 + 4RP(t-1) \tag{21}$$

where σ^2 is the variance of e(t).

Control

As shown by simulations, the certainty equivalence controller $u(t) = -\hat{x}(t)$ does not work very well for this example. For the case R=0 this can be explained by the consistency results of Sternby(1977), which applied here say that for consistency, the sum of the inputs squared must diverge. With the certainty equivalence controller, (17) shows that $u(t) = -K(t)\varepsilon(t)$. But Corollary 3 of Sternby(1977) tells that K(t) is square summable, and the same thing is then true for u(t) since $\varepsilon(t)$ is mean square bounded. It is also obvious that the controller may be trapped at the value u(t)=0 as K(t+1) will then also be zero.

Some feature is needed in the controller to improve the estimation of x. This can be achieved by adding a perturbation signal to the input. The simulations show two disadvantages with that method. The perturbation amplitude must be chosen accurately by the user, and it is nevertheless not possible to get a performance close to the optimal.

Dual Control

Another method is to minimize the criterion two steps ahead as was suggested in Sternby(1978b) for linear systems with unknown parameters. Thus u(t) should be chosen to minimize $E[y(t+1) + y(t+2)|t]$. In the following derivation will be used the approximations

$$\hat{x}(t) = E[x(t)|t] \quad \text{and} \quad P(t) = Var[x(t)|t]$$

Then with $u(t+1) = -\hat{x}(t+1)$

$$\text{Min } E[y(t+2)|t+1] = y(t+1) + R - \hat{x}(t+1)^2 \tag{22}$$

The best two-step u(t) should therefore minimize

$$V[u(t)] = E[2y(t+1) + R - \hat{x}(t+1)^2|t] \tag{23}$$

With the use of (16)-(21) V can be written in the form

$$V(u) = (u - K)^2 + f(u) + \text{constant} \tag{24}$$

with $K = -\hat{x}(t)$ and

$$f(u(t)) = \sigma(t)^2 P(t)/[\sigma(t)^2 + 4P(t)u(t)^2] \tag{25}$$

Utilising the structure of f(u), V(u) can be approximately minimized. In the neighbourhood of u=0 a linear approximation of V"(u) is made from u=0 to the point where f"(u)=0. Outside this area V"(u) is approximated by a constant. In both areas the minimizing u is then found by equating the corresponding approximation of V'(u) to zero.

Simulation

The certainty equivalence control (CE), with and without an added perturbation signal, and the two-step control were tested by simulation. The system used was described by (14)-(15) with c=0, R=1 and v(t)=0 in order to allow a comparison with the optimal control derived by Jacobs and Langdon(1970). The amplitude of the perturbation signal was adjusted to best possible performance.

It was found that with no perturbation signal, the certainty equivalent control behaved much better with a constant $\sigma(t)$ than that given by (21). The reason is probably that with that control the uncertainty P is often large, so that $\sigma(t)$ of (21) would also be large and decrease the influence of the measurements on the estimation by keeping K(t) small. This control law was therefore tuned manually with respect to the constant $\sigma(t)$. The results are shown in fig. 4, where the mean values of the output together with estimated standard deviations over 20 runs of 500 steps are displayed.

The accuracy (standard deviation) in the mean value estimation is better than 0.1 and there is thus a significant difference in the behaviour of the three algorithms. It can be concluded that dual control is needed in this case. Fig. 5 shows the state x controlled by the twostep controller in a run of 400 steps with R=0.16 and σ=0.2. At t=200 c changes from c=0 to c=5. Because of the integrator, x will track the change in c.

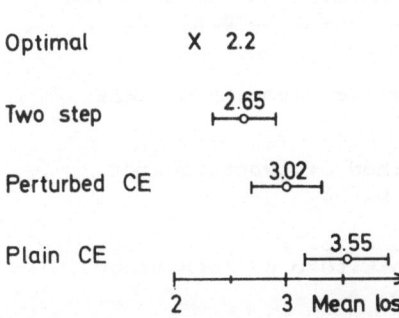

Figure 4-Control laws compared

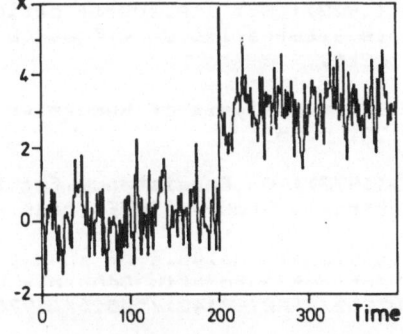

Figure 5-Tracking a moving optimum

5. CONCLUSION

Two possible models for extremum systems have been discussed, both consisting of a dynamic linear part and a static nonlinear part. It was shown that with the nonlinearity at the input, performance may be unsatisfactory when using a straightforward adaptive control law based on certainty equivalence.

With an output nonlinearity the calculations are more difficult even with known system parameters. For a special example the system equations were rewritten to allow the application of the least squares method for identification, and a dual control law could be computed. To make the method interesting, it should be extended to cases with unknown parameters and higher order dynamics. This has not yet been done, but it seems to be possible, maybe with slight extensions of existing identification procedures.

6. ACKNOWLEDGEMENT

This work was supported by the Swedish Board for Technical Development under contract No. 78-3763, which is gratefully acknowledged.

7. REFERENCES

Bamberger, W. and Isermann, R.(1978): Adaptive on-line steady-state optimization of slow dynamic processes. Automatica 14,223.

Blackman, P.F.(1962): Extremum-seeking regulators. In Westcott(Ed.): An exposition of adaptive control. Pergamon Press.

Florentin, J.J.(1964): An approximately optimal extremal regulator. J. Electronics Control 17,211.

Jacobs, O.L.R. and Langdon, S.M.(1970): An optimal extremal control system. Automatica 6,297.

Keviczky, L. and Haber, R.(1974): Adaptive dual extremum control by Hammerstein model. Proc. IFAC Conf. on Stochastic Control, Budapest.

Leblanc, M.(1922): Sur l'électrification des chemins de fer au moyen de courants alternatifs de fréquence élevée. Revue Générale de l'Electricité.

Ljung, L.(1977): Analysis of Recursive Stochastic Algorithms. IEEE Trans AC-22, p.551.

Sternby, J.(1977): On Consistency for the Method of Least Squares Using Martingale Theory. IEEE Trans AC-22, p.346.

Sternby, J.(1978a): Analysis of an Extremal Controller for Hammerstein Models. Dept. of Automatic Control, Lund Institute of Technology. CODEN: LUTFD2/(TFRT-7142)/1-015/(1978).

Sternby, J.(1978b): A Regulator for Time-varying Stochastic Systems. Proc. IFAC World Congress, Helsinki.

APPLICATIONS OF ADAPTIVE CONTROL SYSTEMS

P. C. Parks
Dept. of Mathematics and Ballistics
Royal Military College of Science
Shrivenham / Great Britain

W. Schaufelberger
Dept. of Electrical Engineering
Swiss Institute of Technology
(ETH) Zürich / Switzerland

Chr. Schmid
H. Unbehauen
Department of Electrical Engineering
Ruhr-University Bochum / Federal Republic of Germany

ABSTRACT

This paper reviews the application of adaptive control in three areas:
aircraft control systems, process control and electrical drives. In
the aeronautical field, where pioneering work on adaptive control was
carried out, relatively few applications have been achieved on account
of competition of rival techniques such as air data scheduling and also due to deficiencies in basic adaptive control theory. From the
field of applications in the cement industry, the steel and metallurgical industry, the chemical industry, the paper industry, power
plants, and some miscellaneous areas of process industry, the most
interesting developments in adaptive control during the last five
years are discussed, although only relatively few applications have
been reported. It is pointed out that the reason for this situation is
due to the complicated dynamic behaviour of these industrial processes
as well as the (healthy) conservative management of many industrial
processes. The field of electrical drives seems to be one of the most
promising areas for the application of adaptive control. The dynamics
of such systems are well understood and limitations of theory are less
restrictive.

In general it is recommended that more experimental work should be
done in connection with proven theoretical methods. These methods,
however, must provide a systematic design procedure, which can be understood and also implemented by industrial engineers.

1. INTRODUCTION AND PROBLEMS

A detailed technical review of adaptive control covering the entire
range of applications and methods is clearly beyond the aims of this
paper. The paper deals with applications of adaptive control in aeronautics, in process industries and for electrical drives. The review
does not in any way claim to be complete, but it tries to discuss the

most interesting developments especially those published during the
last five years. There are a number of survey papers on adaptive control - partly in book form - dealing with general aspects [1-4] as
well as surveys of more specific fields of adaptive systems [5-9]. However, only very few papers are directly concerned with practical applications [10]. With the rapid appearance of very cheap microprocessors during the last five years new possibilities for the industrial
application of adaptive systems have been opened up [11]. Therefore,
adaptive control has become very attractive again during the last few
years. In addition significant progress has been made recently in the
theory of adaptive systems, including a drawing together of general
ideas and a better grasp of details such as global stability properties.

Given the limits of this paper, it would seem most useful to concentrate attention on the following broad questions:

1) Why is adaptive control necessary in industry?
2) Which adaptive design methods fulfil the requirements for application?
3) What is the actual status of adaptive control in the various areas?
4) What conclusions can be drawn from the actual situation for the
   further development of adaptive control?

Later sections try to give short answers to these questions in the
three fields under review.

During the past 30 years it has often been predicted that adaptive
control would become a common sophistication. This control technique
claims to compensate for changes in process parameters attaining or
holding a defined, usually optimal, set of system conditions given a
certain initial indeterminateness or varying working conditions of a
time-varying system. This can normally be attained by adaptation of
parameters, system structures or signals using the information received during the adaptive process. High costs of development, implementation and operation of such control systems must be justified eco-
nomically if the control performance is affected significantly by such
changes or if the adaptive control system serves as a substitute for
the time-consuming process of plant experiments, such as extensive
measurements, modelling, evaluation, controller design, and controller
tuning.

The number of proposed adaptive control algorithms is very large relative to the number of actual applications. Successful applications are mostly based on particular solutions, in which variations of plant parameters are known and used directly to adapt a controller improving performance significantly. From the present state of application, however, it can be concluded that progress in the industrial application of adaptive control systems appears only if some methods become standard and thus generally applicable. From the domain of theory we have an arsenal of methods, but there is a lack of experience of the efficiency of the proposed methods as well as of the existing possibilities of converting methods into practice. The innovation process of introducing adaptive control as a standard of technique is time-consuming, despite the acceleration of new hardware techniques, so more emphasis should be put on the clarification of essential prerequisites, such as practicability, limits of applicability, assumptions for design and influence of real restrictions. From the practical point of view this requirement can be rephrased by the following questions, which should be answered before deciding on the implementation of adaptive control.

- What are the *variations* in process dynamics and are they significant?
- What degree of *improvement* can be expected using adaptive methods and are the costs justified?
- Are *alternative* solutions a better means to solve the problem?

From the methodological point of view additional requirements must be fulfilled by a suitable adaptive control and design method:

- guaranteed stability or a maximal stability region,
- small amount of a priori knowledge about the process,
- robustness in the case of disturbances,
- general applicability of different processes,
- high speed of the adaptive algorithm to compensate for parameter variations,
- a small amount of computer time and capacity.

A lot of adaptive structures and methods violate these requirements. Only a small set of such structures are really suitable. The correct selection is a critical task.

2. A SHORT SURVEY OF ADAPTIVE STRUCTURES AND DESIGN METHODS

General applicability of adaptive methods would have to become a reality if these systems are to be designed in a systematic way. This applies both to theory and practice. It would allow one to generalise results and would make applications simpler and more commonplace. A first step in the *systematic design* is the characterizing of the control system features, such as the basic structure and the principle of design. The mode and principle of action are basic features, which can be described in a basic structure, thus a rough description of the structure is the first step in the design of an adaptive control system.

Basic Structures

For the design of adaptive control systems today three main basic control system structures are relevant. The first structure, the *parallel reference model structure* (Fig. 1), allows the identification of plant dynamics and the adaptation of the controller parameters to achieve a given model behaviour for the closed loop. The second structure, *the*

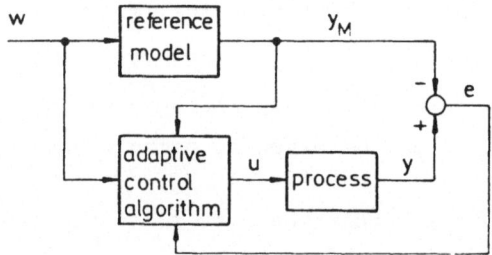

Fig. 1. Model reference adaptive control systems with a parallel reference model

closed loop structure, is based on a classical control loop (Fig. 2). Changes of parameters or other disturbances are detected by an identification block. Depending on the chosen criterion (adaptive set point)

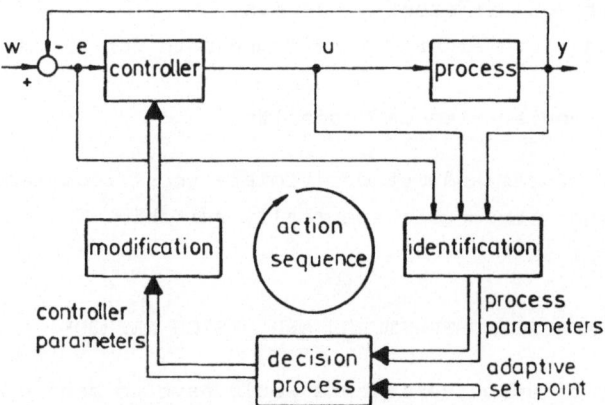

Fig. 2. Adaptive control system structure with closed loop adaptation

the controller parameters are obtained by a decision process and the controller is corrected accordingly. In this case, the adaptive system is realized by a second closed-loop control system, since the effect of controller modification is fed back to the decision process by the basic control loop and the identification process. Thus an adaptive error forces the adaptation process. In the third basic structure, *the open loop structure*, the decision process is reduced to a fixed mapping of the process parameters to the controller parameters. The original decision process is already realized in the design phase of the a-daptive control system. This type of structure is widespread and in vogue today, since it allows one to tune a wide range of controllers using a manifold of popular on-line process identification methods. The assumption for the faultless operation of systems with this structure is a good knowledge of the actual process dynamics. Thus the identifi-cation block has to satisfy high requirements of precision and speed. Only powerful identification methods can be applied in this situation and the requirements for good control performance are met only in the

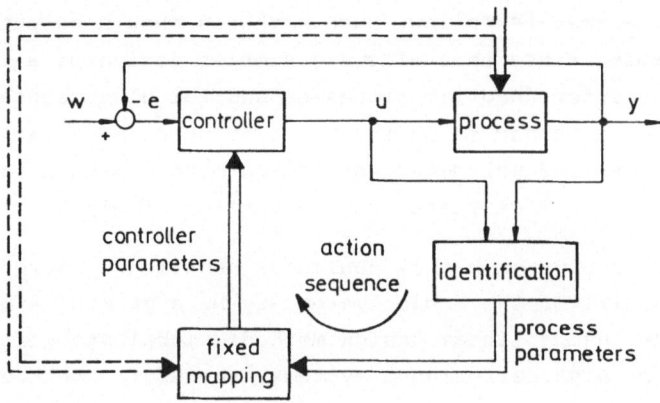

Fig. 3. Adaptive control system structure with open-loop adaptation

case of robust and reliable on-line identification algorithms. All sys-tems designed using the so-called certainty equivalence principle [9] show this structure, since the uncertainties of estimated process pa-rameters are not taken into account in the controller design procedure.

Another form of open loop adaptive control is also indicated in Fig. 3. This is the so-called "gain scheduling" technique where external meas-urements are used to adjust this controller parameters. The resulting system is adaptive but not self-adaptive.

Principles of Design

The advances in design of adaptive systems are oriented according to
their principles. These principles fix the theoretical concept of design, which is divided into two phases. In the first phase, the main
problem is to find out the adaptive law for the controller parameters
after some pre-experimental work. The result is an adaptive control
algorithm. Finally in the second phase the free parameters are chosen
and the realization of the algorithm is taken into account. Four different design principles must be considered [12].

In the *continuous* design principle all signals are treated as continuous signals in the first phase. Then in the second phase the designed
system is prepared for digital signal processing by discretization. In
the *discrete* principle the process with its continuous input and output signals is treated as a sampled data system. This allows one to
apply the theory of discrete systems. The result of the first phase is
a system of nonlinear difference equations describing the entire control structure. Thereafter in the second phase numerical specification
of the free design parameters follow. Much emphasis should be put on
the discrete principle, since this allows a simpler design of a larger
class of systems, e.g. for deadtime processes and for disturbances inside the process and disturbances in the measured process variables.
In addition various on-line parameter estimation methods which have
been developed during the last years, can be applied simply.

Adaptive control systems show a highly nonlinear behaviour. Therefore,
the guaranteed stability of the whole system may be a general aim of
the design procedure. Using linear design methods global stability is
never guaranteed. For high performance systems, therefore, methods
from nonlinear theory must be used. In the *stability* design principle
the global stability of the whole system is the basis of the design
procedure. Therefore, in the first phase the aim is to find an algorithm which guarantees stability in the whole parameter and signal
space. For that purpose the direct method of Lyapunov or the hyperstability theory of Popov is frequently used. Such theories applied to
the adaptive control problem usually guarantee global stability. Only
quite recently has the question of global asymptotic stability been
resolved for certain self adaptive control systems with "sufficiently
exciting" input signals [26].

For the *stochastic* design principle in the first phase the influence
of unknown stochastic disturbances on the control system is already

considered. This principle has to be applied in the case of large disturbances. An adaptation to the process itself as well as the kind of
disturbances can be performed.

3. ADAPTIVE CONTROLS IN AIRCRAFT

Introduction

The first research contracts for self-adaptive aircraft controls were
awarded in the USA some 25 years ago and the first symposium on self-
adaptive flight control [13] was held at the Wright Air Development
Center, Ohio, in January 1959. It is surprising therefore to find today that the aeronautical engineering industry has produced so few in-
service self-adaptive systems after a quarter of a century of effort.

The basic reason for this has been summarised succinctly by Stein [14]:
"Our control engineering predecessors lost the competitive battle to
the sensor builders". The difficulties lie partly in the aircraft itself and partly in deficiencies in the available adaptive control theory. It is first necessary to look quickly at the aircraft control problem.

Rigid aircraft dynamics

Coordinate axes are defined as fixed in the rigid aircraft as shown in
Fig. 4. The moving axes relationships of dynamics are then used to obtain the complete non-linear equations which, using British notation,
take the form

$$M_o(\dot{u} - rv + qw) = M_o g1 + X \qquad (1)$$

$$M_o(\dot{w} - q[U + u] + pv) = M_o gn + Z \qquad (2)$$

$$B\dot{q} - (C - A)rp - G(p^2 - r^2) = M \qquad (3)$$

$$- \dot{\ell} = - rm + qn \qquad (4)$$

$$M_o(\dot{v} - pw + r[U + u]) = M_o gm + Y \qquad (5)$$

$$A\dot{p} - G\dot{r} - (B - C)gr - Gpq = L \qquad (6)$$

$$C\dot{r} - G\dot{p} - (A - B)pq + Gqr = N \qquad (7)$$

$$\dot{m} = pn - r1 \qquad (8)$$

$$\dot{n} = q1 - pm \qquad (9)$$

(1, m, n) being the direction cosines of the downward vertical.

For certain motions such as rapid rolling these equations have to be
considered in full, but for many applications linearised versions of
these equations are acceptable. The equations then split into two sets,
one for symmetrical pitch-plane or "longitudinal" motions, the other

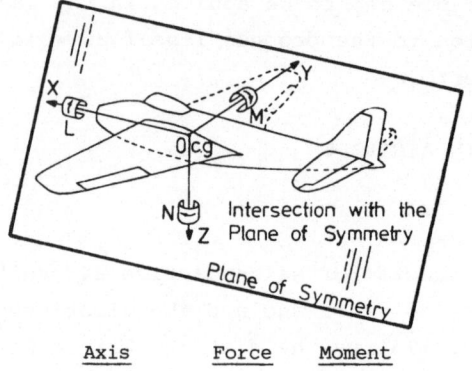

Axis	Force	Moment
OX Longitudinal	X	Rolling L
OY Transverse	Y	Pitching M
OZ Normal	Z	Yawing N

Positive in Direction of Arrows

Fig. 4. Aircraft axes and notation

for "lateral" yawing and rolling motions:

$$M_o \dot{u} = -M_o g\theta + X_u + X_w w + X_q q$$

$$M_o(\dot{w} - qU) = Z_u u + Z_w w + Z_q q$$

$$B\dot{q} = M_u u + M_w w + M_q q$$

$$\dot{\theta} = q$$

$\left.\vphantom{\begin{array}{c}1\\1\\1\\1\end{array}}\right\}$ "Longitudinal" dynamics   (10)

and

$$M_o(\dot{v} + rU) = Y_v v + Y_p p + Y_r r + M_o g\phi$$

$$A\dot{p} - G\dot{r} = L_v v + L_p p + L_r r$$

$$C\dot{r} - G\dot{p} = N_v v + N_p p + N_r r$$

$$\dot{\phi} = p$$

$\left.\vphantom{\begin{array}{c}1\\1\\1\\1\end{array}}\right\}$ "Lateral" dynamics   (11)

where A, B, C, F, G, H are moments and products of inertia ($F = H = 0$ on account of symmetry), M_o is the mass of the aeroplane and the X, Y, Z, L, M, N's are force and moment aerodynamic 'stability derivatives'. Here ℓ, m, n appearing in (1) – (9) are given by

$$\ell = -\theta, \quad m = \phi, \quad n = 1 .$$

These equations lack feedback control. For designing a longitudinal autopilot it is customary to consider the pitch-plane equations neglecting u and the so-called "phugoid oscillations" having approximate frequency $\sqrt{2}$ g/U rad/s, and to add in an elevator angle δ giving two equations

$$M_o(\dot{w} - qU) = Z_w w + Z_q q + Z_\delta \delta$$

$$B\dot{q} = M_w w + M_q q + M_\delta \delta \tag{12}$$

For a feedback control system in which the pitch-plane angular velocity q, or the normal acceleration n = $\dot{w}$ - qU is measured and fed back to control δ we need to obtain a transfer function relating normal acceleration n = $\dot{w}$ - qU to δ. This, written in the Laplace transform notation, is

$$\frac{\bar{n}(s)}{\bar{\delta}(s)} = \frac{U Z_w M_\delta}{BM_o s^2 - BZ_w s - UM_o M_w} \tag{13}$$

(where we have neglected Z_δ and M_q). The denominator of (13) yields the characteristic equation

$$s^2 - \frac{Z_w}{M_o} s - \frac{UM_w}{B} = 0 \tag{14}$$

giving the "short period" or "weathercock" oscillations of frequency $\sqrt{-UM_w/B}$ rad/s. For a conventional aircraft the lift and moment coefficients Z_w and M_w are negative quantities. They each depend linearly on both air density ρ, and forward velocity U explicitly, and implicitly on Mach number. Neglecting the Mach number effects we can rewrite (14) in standard form

$$s^2 + 2\zeta\omega_o s + \omega_o^2 = 0 \tag{15}$$

in which the undamped natural frequency, ω_o, varies as $\sqrt{\rho} U$ and ζ, the damping ratio, as $\sqrt{\rho}$. Variations of ω_o and ζ are shown for a modest subsonic operational flight regime in Fig. 5. These variations would seem at first sight to make a strong case for adaptive flight control.

However pilots are themselves very adaptable and do not mind changes of a factor 10 in ω_o and a factor of 4 in ζ (Rynaski [15])! What pilots do *not* like are time lags greater than 100 milliseconds and non-minimum phase controls in which the response to a step input at first moves in the wrong direction [15]. Moreover many of the coefficients appearing in the equations of motion, such as (12) above, depend in a simple, almost linear fashion on the dynamic pressure $\frac{1}{2}\rho U^2$. Thus simple adjustments can often be made by this one pressure measurement leading to simple and quite acceptable adaptive (but not *self*-adaptive) control systems. This technique, which may include measurements of ρ as well as $\frac{1}{2}\rho U^2$, is known as "air data scheduling" and is at the heart of Stein's comment [14] noted above.

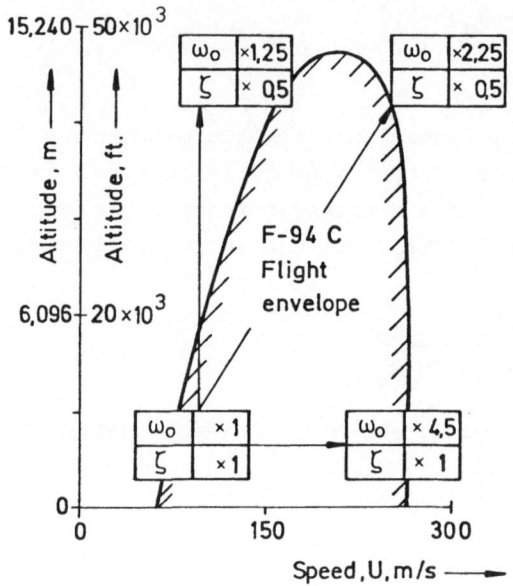

Fig. 5. Flight envelope and variations of ω_o and ζ for F-94 C aircraft

Flexible aircraft dynamics

The pitch-plane model developed above in (13) neglects not only the low frequency phugoid motion, but also higher frequency structural vibration frequencies; thus the aircraft is in fact a "distributed parameter system". Even without feedback loops these elastic vibration or "modes" can couple together to give rise to "flutter" instabilities. A classical bending-torsion flutter situation is shown in Fig. 6: here the aerodynamic lift force does positive work during a cycle thus causing a growing oscillation.

The phase difference which makes this possible can be produced naturally at a sufficiently high forward speed U, but can also result from phase-lags produced in a feedback control loop of an autopilot leading to "servo-induced flutter", a particularly troublesome phenomenon in missile design.

The uncertainties associated with structural resonances which can cause flutter instabilities are strikingly illustrated in Fig. 7 taken from [14]. Here a frequency response function which is well defined for low frequencies can have almost any amplitude and phase at higher frequencies. The frequency response of a typical lightly damped metal structure is plotted as an Argand diagram in Fig. 8, illustrating the rapid amplitude and phase changes that are possible.

Any adaptive control theory which is applied to aircraft must be able

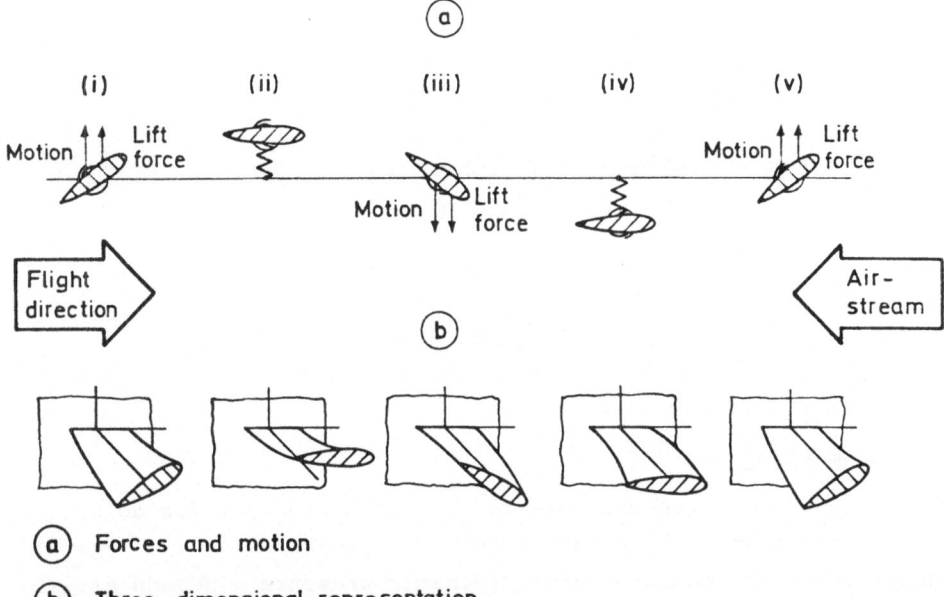

(a) Forces and motion

(b) Three-dimensional representation

Fig. 6. Classical bending-torsion flutter instability

either to deal with these structural resonances, or to neglect them, in a mathematically rigorous way. In fact adaptive control for active flutter suppression has been considered in recent years [16] as the coefficients in the equations of motions for flutter calculations depend in a more complex way on ρ, U and Mach number than do the coefficients of equation (13) above, for example.

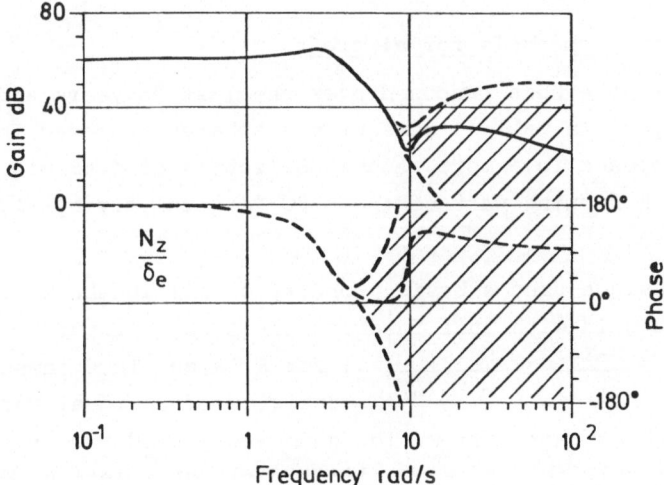

Fig. 7. Uncertainty of pitch-plane transfer function at high frequencies due to structural resonances

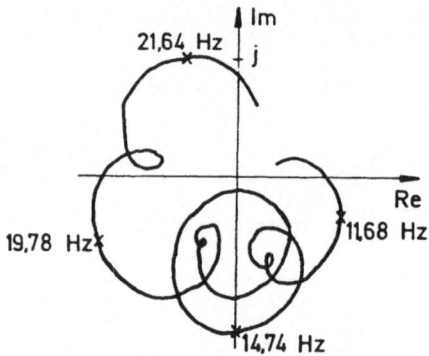

Fig. 8. Rapid variation of frequency response function due to structural resonances

Air data scheduling

As we have seen, adaptive controls for aircraft and missiles do not
have to be *self*-adaptive: in fact many successful schemes are "open-
loop adaptive" using measurements of dynamic pressure $\frac{1}{2}\rho U^2$ and air
density ρ. Sensors are required for these measurements and, in order to
achieve high standards of reliability, may have to be triplicated or
even quadruplicated. An important advantage is that the adjustments do
not depend on the existence of suitable inputs as is the case for self-
adaptive control. However as requirements grow for high performance
over ever larger flight envelopes there is a feeling that air data
scheduling may not be able to cope so successfully with future autopilot design requirements. Thus the competition between air data scheduling and self-adaptive control design continues.

Proposed self-adaptive controls for aircraft

A number of schemes have been proposed over the last 25 years and it is
instructive to compare in Table 1 some of the schemes proposed in 1959
[13] with those proposed in a more recent collection of IEEE of papers
[17] describing work forming part of the F-8C "digital fly-by-wire"
programme in the USA.

Most of the 1959 designs used a basic property of all feedback systems
that a high forward gain with unit feedback gives small input/output
error since $\frac{\text{output}}{\text{input}} = \frac{\theta_o(s)}{\theta_i(s)} = \frac{KG(s)}{1 + KG(s)} \approx 1$ for K large. Thus these a-
daptive autopilots consisted of an ideal model in
series with such a high gain system. The gain was pushed almost to instability usually resulting in small a limit cycle oscillation. The
M.I.T. scheme however had a parallel model and was the prototype of the
highly important model reference concept.

1959 Self-adaptive Flight Control Symposium [13]

Nature of system	Contractor
High gain, with limit cycle oscillation	Minneapolis-Honeywell
Model reference, with sensitivety functions	M.I.T.
Impulse response identification and gain control	Sperry, Aeronutronics
Reciprocal transfer function	Convair

1977 IEEE papers [17]

Nature of system	Contractor
Maximum likelihood estimation and gain control	Honeywell
Multiple model adaptive control	M.I.T.
Digital adaptive control using Riccati equation	Rensselaer
Moving window identification	Langley Research Center

Table I. Comparison of adaptive control schemes proposed in 1959 and 1977

The 1977 designs use various identification algorithms which were virtually unknown in 1959 and so are of the type of system shown in Figs. 2 and 3. In the discussion of this papers in [17] the proposals from academic institutions were particularly severely criticised by Rynaski as being impracticable, see also [15].

The status of the theory of self-adaptive control

While the lack of success in application of self-adaptive controls to aircraft is due partly to the factors outlined above, aircraft control specialists have noted also some deficiencies of self-adaptive control theory itself [14]. We shall conclude this section by examining this problem.

An early deficiency of proposed self-adaptive controls was the question of their stability. Indeed some of the more promising schemes such as the 1959 M.I.T. scheme shown in Table I could be shown to be unstable even with quite simple inputs [18, 19]. This led to the important concept of designing adaptive controls from a stability point of view, for example Parks [18] from which developed, with increasing sophistication, work by Monopoli [20], Johnson [21], Narendra [22, 23], Landau [24] and their associates.

However even with the systems designed from a stability point of view doubts remained, for example, about parameter convergence. In [18] Parks had proposed a simple Liapunov redesign of a model-reference control for adjusting the gain of a first order plant (see Fig. 9).

If e is the model-plant error signal and x the gain error, $K - K_v$, then the Lyapunov function

$$V = e^2 + \frac{1}{BT} x^2$$

for which

$$\frac{dV}{dt} = -\frac{2e^2}{T}$$

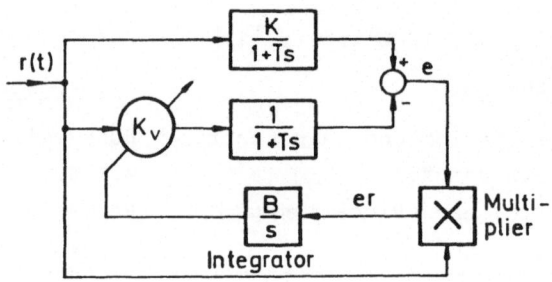

Fig. 9. Simple model-reference adaptive control system

was used to design the plant gain adjustment. It is stated in [18] that "this scheme is guaranteed to be asymptotically stable, provided of course that r(t) ≠ 0". This statement implies that provided r(t) ≠ 0 then e → 0 and x → 0 as t → ∞. Unfortunately this is not true and a likely outcome is, that e → 0 but x → constant ≠ 0 as t → ∞. For example consider the input r(t) = exp(-λt) then one can find power series solutions in exp(-2λt) to represent e and x. One of the two power series solutions for x has a non-zero constant term, and so in general x → constant as t → ∞, although e → 0.

This difficulty has only quite recently been tackled by Morgan and Narendra [26] where the concept of a "uniformly exciting" signal is introduced. Such signals must not decay to zero in the way r(t) does in the example above. It looks therefore as if transient pilot inputs occurring in specific manoeuvres may be insufficient for complete parameter adjustment.

Another problem which has been mentioned for example in [27], but has received insufficient attention, is the question of noise contamination. At its worst noise can cause parameter adjustments to become unbounded or exhibit a constant [27], off-set or bias, or at least to cause such a deterioration in performance that the supposed rapid response qualities of the self adaptive schemes based on stability theory are lost [28].

Self-adaptive control theory has undergone considerable development in recent years and a certain convergence of ideas is apparent – for example, the paper of Ljung and Landau [29] which discusses the similarities of "self-tuning regulators" and "model-reference adaptive control systems" which were initially developed from quite different points of view. Similarly the Lyapunov and hyperstability design techniques are shown to be very similar and to have similar difficulties [30]. However even these recent papers are limited to very specific plant transfer functions G(s), specifically (see [14]):

$$G(s) = k\frac{q(s)}{p(s)}$$

with (i) k having known sign, (ii) p(s) a Hurwitz polynomial, (iii) known bound on the degree of p(s), (iv) known relative degree m between q(s) and p(s), implying $G(s) \to k/s^m$ as $|s| \to \infty$.

Thus the latest theory falls short of the requirements of an aircraft feedback system at higher frequencies illustrated starkly by Figs. 7 and 8.

The question of flexibility of the aircraft structure has been mentioned above. Self-adaptive control of distributed parameter systems is a new topic that, so far, has hardly been studied at all. A theory of feedback control of distributed parameter systems is emerging using functional analysis concepts [31, 32]. Certain design principles may be of use here, such as the rule "measure where you excite" [33] – this avoids instabilities caused by the time delay due to wave transmission in a flexible structure. It is this which is an essential cause of "servo-induced flutter". New principles of control and observation "spill over" are emerging [34, 35], and a fully fledged theory of self-adaptive control of distributed parameter systems may be expected before too long and will be of considerable interest to the aircraft control engineer.

Even with such a new sophisticated theory the aircraft autopilot designer will still encounter old problems such as servo rate and power limitations, to say nothing of aerodynamic limitations such as stalling characteristics of surfaces and the basic dynamic non-linearities appearing in equations (1) – (9).

It is not surprising therefore that the many sophisticated self adaptive controls have first appeared in applications areas other than the field of aeronautics, as the following parts of this paper show.

4. ADAPTIVE SYSTEMS IN PROCESS CONTROL

4.1. Adaptive Control System in the Cement Industry [36-45]

The process for cement production includes the following three subsystems:

a) *Raw material blending:* The raw materials mainly consisting of four oxides CaO, SiO_2, Al_2O_3, Fe_2O_3 of different composition are contained in feeder tanks, from which they are fed into the raw mill by a conveyor belt. The rawmeal output is stored in the silo.

b) *Rotary kiln:* The homogenized silo content is fed to the kiln. As a result of the sinter process a very hard clinker is produced.

c) *Cement grinding mill:* The closed loop cement ball grinding mill works very similarly to the raw mill. The input material flow consists of the clinker usually mixed with gypsum and slag. The fresh material flow, together with the reflux, is fed into the ball grinding mill. The mill output is brought by an elevator to the separator where the final product (cement) is separated from the reflux (breeze).

Different adaptive control schemes have been applied to raw material blending and to cement grinding mills (Table II), which will be discussed briefly.

Adaptive Control of Raw Material Blending

As the main goal in cement production is to produce cement of a given quality at minimum cost, two control tasks should therefore be fulfilled

- control of the maximum quantity of rawmeal,
- control of the chemical composition of rawmeal.

Both control loops are coupled. However when a good quantity control system is applied, e.g. by using an autonomous adaptive extremum controller, then the quality control can be investigated separately. For the raw mill the control of chemical composition is especially important. The quality characteristics (Ca, Si, Al-moduli) must be kept on given values with minimum variance. Because of variable physical characteristics of the raw materials a self-tuning algorithm for adaptive control of the quality was developed [36-38]. Applying this MIMO self-tuning control strategy the minimization of the variance of the output signals (oxide moduli) from varying reference values is assured.

Adaptive Control of Cement Grinding Mills

The basic operation is very similar to that of the raw mill. The static characteristic of the mill has an extremum in the final product quantity for the fresh input material feed. The purpose of an optimal control may be to seek the optimal quantity under varying process conditions and to ensure the optimal values of the mill and final product output.

In addition to the extremum control concept several models for the dynamic properties of the closed circuit mill have been proposed during the last few years. Based on different models, various strategies for adaptive control systems have been developed.

In [40] a self-tuning extremum controller was used to generate the control law for the mill. The Blain value of the cement as a measure of quality is adjusted to a given reference value with minimum variance by self-tuning control.

The adaptive PID breeze flow control is another concept [41], where the reflux is conrolled by a PID-controller. Since the separator setting is a nonlinear function, the plant dynamics depend strongly on the mill output. So the separator setting is directly used to adapt the PID-controller in order to keep the control action of the breeze flow rate constant.

Sensitivity models are used in [42] to adapt two cascaded controllers. The inner control loop has a PI-controller for the mill loading (measured by a microphone) by acting on the fresh input material feed while the outer control loop has a PID-controller for the power consumption of the elevator system. All five controller parameters had been made adaptive so as to minimize the integral criteria of the two error signals. To do this a direct parameter adaptation was applied, using a sensitivity model based on the gradient method. This sensitivity model consists of a linear model of the plant and the controller. Therefore the non-linearity of the separator had to be linearized for the sensitivity model. The proportional coefficients of the gradient algorithm are normalized and automatically adapted to the behaviour of the mill.

4.2. Adaptive Control in the Steel and Metallurgical Industries [46-60]

Many processes in the steel and metallurgical industries are nonlinear and nonstationary. Therefore, the application of classical control methods has not led to the desired results. More than ten years ago the first proposals were made for improving these processes by application

of adaptive control systems. However, the processes in the steel and
metallurgical industries are characterized by a large variety of very
different properties. Very often the specific target variables cannot
be directly measured, or at least not without considerable time delay.
In some parts of these multivariable processes both continuous and discontinuous nonlinear processes with small reaction times are coupled.
Furthermore the disturbances are usually distributed through the entire
complex process. On the other hand there are also very slow processes
(e.g. blast furnaces) where no continuous measurements of the target
variables are possible.

In metallurgical plants the self-tuning regulator has been given serious consideration. This technique has been applied with dramatic success, e.g. to ore crushing plants [56]. The main advantage of this
technique is that it identifies a process from operating data without
the need for lengthy experiments.

The main success of adaptive control has been in various applications
to steel mills. Therefore, in the following these applications will be
discussed briefly.

Applications in the steel industry

Metal rolling mill systems appear to represent the most successful
applications in the iron and steel industry of sophisticated process
control systems. Adaptive control modes are successfully used on a
broad basis. While the later discussion uses a hot strip mill as a
specific example, it relates also to plate mills and tandem cold reduction mills.

The control modes include feedback, feedforward, predictive and adaptive control [46]. Feedback control is used for speed, roll position
and automatic gauge control. Feedforward control is applied in several
of the process model calculations, including roll force and temperature. These process models are extremely important for the operation of
the plant. The order in which model programs run, iteration and adaption of the model permit accurate performance with relatively straightforward algorithms.

Predictive control with bar-to-bar feedback is applied for head end
thickness and temperature using the above mentioned process models.
Predictive control is also used for in-bar control of finishing and
coiling temperatures because of significant transport lags between

points of control and points of measurement.

Thermal and dimensional conditions in the mill change with the pace and duration of rolling. Rolling characteristics may vary from bar to bar even for the same stated steel grade code. The predictive control must adapt to this changeable environment for accurate control of the physical properties of the steel strip. The control system must be and is adaptive, implying three criteria:

1) Definition of optimum operating conditions, e.g. product dimensions and temperatures.

2) Comparision of actual with desired values of controlled variables, e.g. roll force and temperature.

3) Adjustment of system parameters for correction of the actual performance towards that desired. The system makes automatic on-line adjustments to the process model to represent more accurately the parameters related to the product chemistry, mill condition, and other environmental factors.

Having determined the reduction distribution along the mill stands, the reference values for conventional control of roll gap, thickness after stand, strip speed, roll force and tension are calculated for each stand by means of mathematical process models. These models consist of roll force, roll gap, rolling torque, forward slip and material hardness models and are constantly updated by means of *adaptation constants* during rolling in order to achieve a more optimal set-up calculation for the mill. For the model adaptation regional adaptation is preferred because it is almost impossible to employ reasonable techniques to a complex equation that represents the global range of a process. Thus the actually measured values are inserted in the mathematical partial models and new preliminary adaptation constants calculated [49]. By comparing the old adaptation constants and these preliminary adaptation constants the old adaptation constants are adapted. Using the newly calculated adaptation constants new rolling parameters are calculated and issued to the mill. Constant recalculation and resetting results in diminishing differences between the calculated and actual mill parameters. Once a predetermined number of adaptations has been made a regression analysis of the successively calculated adaptation constants is made to adapt the material constants.

4.3. Adaptive Control in the Chemical Industry [61-68]

Applications of adaptive control schemes in chemical plants are very rare. Nevertheless some work is in progress and aimed at the design and implementation of simple and robust adaptive controllers to replace conventional ones. Investigations are being carried out on some pilot plants like distillation columns, evaporators or reactors according to Table II. Robust algorithms capable of self-tuning or adaptation of parameters are desirable, particularly if large excursions in the plant operation occur. Model reference adaptive control (MRAC) systems are used to adjust the inputs and the parameters of a plant so that its outputs track those of a reference model [61, 62]. This is usually based on the lowest model order that adequately represents the process dynamics. MRAC systems are insensitive to inherent modelling errors, they do not require the time-consuming calculations for process identification and optimization and offer a feasible approach for tuning multivariable control systems. Difficulties in controlling chemical processes are often associated with variable time constants and time delays and process nonlinearities. Any digital algorithm used therefore must be capable not only of achieving indefinite steady state regulation but must be able to deal satisfactorily with any sudden large disturbance in the set point or measured plant output. This and load disturbance rejection are problems which can be solved by self-tuning regulators [63, 64].

As a *representative* application of adaptive control let us consider the following example of a distillation column.

Adaptive control of distillation columns

Distillation columns can be regarded as typical chemical plants to demonstrate the feasibility of adaptive control techniques. A schematic and very simplified diagram of a (binary) distillation column is shown in Fig. 10. The mixture of two components, fed into the column, is split into a light component (top product) and into heavy components (bottom product). The principal control inputs are the internal reflux rate M_R and the heating power, which is proportional to the steam flow rate M_S. These inputs control the concentration c_T and the exit flow rate M_T of the top product.

The main control objective is to hold the top product concentration c_T and flow rate M_T on constant values despite feed flow rate disturbances ΔM_F. From time to time it is necessary to change the throughput of the column to meet changing demands. In order to minimize the disturbances to the column conditions resulting from these necessary changes, a supervisory control scheme (not indicated in Fig. 10) ap-

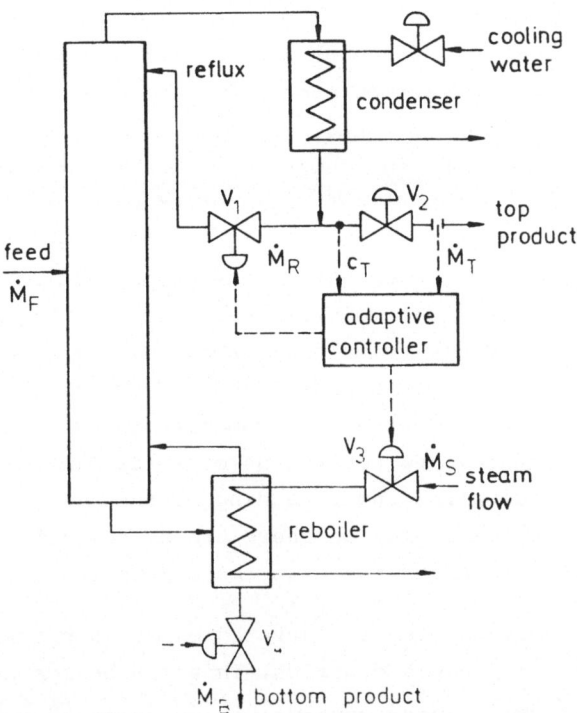

feed $\dot{M}_F$

V_1 $\dot{M}_R$ c_T $\dot{M}_T$

V_2

cooling water

condenser

top product

adaptive controller

V_3 $\dot{M}_S$ steam flow

reboiler

V_4

$\dot{M}_B$ bottom product

Fig. 10. Schematic diagram of a distillation column and the control scheme

plies the changes in set point to the feed flow controllers. Since the column behaviour depends on the feed rates, feedforward control of c_T is an accepted control scheme, but it requires measurement of disturbances, which may not be feasible or not always possible. As the operating conditions change, the controller parameters should be adapted since the column behaviour is nonlinear.

Gain and time constants of the plant are strongly dependent upon the feed rate M_F and also upon the direction and magnitude of ΔM_F. In addition, stochastic fluctuations influence the process variables. So a self-tuning controller [63, 64] or a dual control strategy [65, 66] can be an adequate component in such interactive plants.

4.4. Adaptive Control in the Paper Industry [69-72]

Paper machines have been controlled digitally for many years. Digital control of base weight and moisture content is an example where adaptive control techniques are applied. The control objectives are to maintain the best possible uniform quality, under the production rate chosen and with the given grade of paper. Further, waste is to be mini-

mized. The specific aims are to keep the base weight and moisture as
close as possible to a chosen setpoint and the main moisture content
as high as possible; the deviations should have minimum statistical
variance. Design of an advanced control system based on modern control
theory certainly requires more knowledge about the paper process than
is necessary in most cases for direct application of some conventional
control schemes. The object of control is to operate the thick stock
flow valve and also the steam pressure in the drying section in such a
way that the base weight and the moisture deviation of the manufactured
paper from a predetermined set point are minimized. For the purpose of
control it is of major interest to understand the dynamics which describe how fluctuations in concentration propagate through the system.
The process involved, mixing and separation of water fibre mixture, are
very complicated and a detailed understanding of the mixing and separation phenomena is not available. So there are no possibilities to
determine the characteristics of the disturbances in the system.

In this case minimum variance control strategies are appropriate, which
in addition adapt to the process dynamics. This can be successfully
performed by means of self-tuning controllers as shown in Table II.

4.5. Adaptive Control in Power Plants [73-79]

Although adaptive control is particularly well suited to many applications in nuclear and fossil-fuelled power plants as well as in hydroelectric power plants, its application up to now has been limited by
several technical and human obstacles. Power plants are highly complex,
nonlinear, time-varying, constrained systems. As accurate mathematical
models are seldom available, control behaviour is also often determined by operator actions. On the other hand many control problems in
plant operation can often be solved by hardware design changes or ad-
ditions. Unfortunately there is also very often much human resistance
to adaptive control systems on power plants, because they are usually
hard to analyse.

The problems which seem to be very appropriate for adaptive control
[73] can be divided into the following time-scaled types:

a) *Slowest level:* Changes in plant performance; constant ongoing main-
   tenance; changes in nuclear core reactivity or changes in fossil
   fuel quality (especially coal).

b) *Faster level:* Turbine exhaust pressure variations or imperfect air-
   fuel ratio with consequent inefficiency and increased emissions

caused by daily weather changes; level changes in hydroelectric
plants.

c) *Fast level:* Power changes cause the plant dynamics to vary considerably because of inherent nonlinearities. Also the contingency control, where the plant must respond to a component failure or to external disturbances could benefit from adaptive control.

Controls had been successfully applied for starting-up and shutting-
down of steam power-plants as well as for fast load changes in steam
and hydroelectric power plants in order to satisfy the constraints imposed, by using the principles of variable-structure systems. Their
essential property is the insensitivity of performance to variations
in the parameters of the controlled section [74 to 76]. In the domain
of adaptive systems these controls represent structural adaptive systems. Other investigations are related to studies for adaptive load-
frequency control [5, 6], which however have not yet been realized up
to now. Another very interesting study deals with the application of
a model reference multivariable adaptive control to the steam generator of a fast breeder nuclear reactor [79].

4.6. Adaptive Control of Miscellaneous Areas of Process Industry [80-86]

Also in other areas of process control there is an interest on application of adaptive systems. One remarkable application is the MIMO-
adaptive control of a refrigerant compressor test plant with 3 inputs
and 3 outputs [81]. Since in refrigerant compressor performance tests
the static characteristics must be examined in a wide range as accurately and quickly as possible, a very fast control system is necessary, which shows stable and optimal behaviour for the whole range of
operating conditions. Thus the control objective is to change the process from one set of operating conditions to another in an optimal
manner. This was unsuccessfully accomplished using single-variable PID-
control with fixed parameters. Because of the interaction of the process variables and the unknown nonlinear behaviour, PID-control shows
low performance and instabilities. In an adaptive solution a decomposed
discrete state space model is used to estimate both states and parameters by Kalman filters. Based on the estimated result the optimal control is determined using a quadratic performance index. In spite of a
high computing time (compared with PID-control) a satisfactory control
behaviour could be obtained over a wide range of operating conditions.

Another interesting example of the successful application of adaptive control has been pilot experiments on a heat-exchanger [86]. This system is characterized by a strongly load-dependent time-variant dynamical behaviour. Three different adaptive control systems have been realized and compared, as can be seen from Table II. As a result of this study it has been shown that the method using a parallel model of the closed loop combined with hyperstability theory gave an excellent adaptive control action and, at the same time, needed the smallest effort for realization.

These two applications and others (see Table II), demonstrate the practicability of adaptive control methods which are based on modern control theory.

5. ADAPTIVE CONTROL OF ELECTROMECHANICAL AND RELATED SYSTEMS

The field of electromechanical systems may be characterized as follows: There are fairly good and accurate models available. The time constants are rather short (milliseconds to seconds), requiring a high sampling rate. Measurements can easily be obtained for most physical variables. The systems may range from small single units to large networks. The small units can usually be modelled by low order models.

Different causes may be responsible for a need for adaptive control. Unknown disturbances or load conditions may be responsible or linear control theory may be used to control a nonlinear system in a wide range of operating conditions. In the latter case there are two solutions which are both frequently observed in electromechanical systems control. The first one is adaptive control, the second one is scheduled adaptation. The term "scheduled adaptation" is used for systems, where the controller adjustment is only based on the actual state of the system and no identification is used [101].

Many successful applications of adaptive control techniques have become known during the last years in this field. The major trends will be discussed for some important areas in the sequel.

Generators [77, 87 to 89]

Load and structure may change in wide ranges in a power system. Model reference adaptive controllers and self-tuners have been used to improve stability and transient behaviour of generators. Field test results and industrial experience are encouraging.

Drives [90 to 96]

Parameter changes due to reduced field current, discontinuous current

from SCR's and varying load conditions lead to control problems in this
area. Self-tuning control systems and model reference adaptive controllers have been used. Use of scheduled adaptation is the normal state of the
art [93]. The goals of the control systems are usually fast settling
time and no overshoot.

Position systems [97 to 100]

Nonlinear coupling between the axis of a telescope and elevators with
changing load are typical control problems in this area. Accuracy of
pointing or positioning without overshoot are goals that can be better
reached with adaptive control in some cases.

Autopilots for ships [28, 101]

Several successful implementations of adaptive autopilots have been reported. A fixed controller is not well suited if speed, depth of water,
wind and other conditions change. Accurate course keeping or reduced
fuel consumption may be control goals. A design procedure for a controller with scheduled and self-tuned adaptation to many operating conditions is described in [101].

Machine tools [102 to 109]

This is an interesting field. Many articles on very elaborate designs
of self-tuning and self-optimizing systems were published in the late
sixties and early seventies. These designs did not live up to the ex-
pectations that had been placed on them mainly due to measurement problems and due to the difficulty to define a goal for adaptivity. They
were replaced by much simpler scheduled adaptive systems or limiting
controllers in the sequel [102, 103]. The goal of the systems is smooth
operation without overshoot and maximum throughput, even if parameters
change due to load, workpiece and tool condition. Often, some of the
state variables are in addition severly limited.

Adaptive systems have also been successfully used to control robots
and manipulators.

In summary, different kinds of adaptive controllers have been used to
control many electromechanical systems. Simple solutions based on scheduled adaptation are often possible in this field because many electromechanical systems are very transparent and well understood.

The controllers must in many applications been kept fairly simple because of the high sampling rate that is required if implementation is
done on a microprocessor [110, 111].

field	references	plant	behaviour	output variables	time response (time constants)	manipulated variables	method	adapted variables	status
CEMENT INDUSTRY	[36] to [37]	cement raw material belanding	a) stochastic varying amounts of the constituent oxides b) static nonlinear characteristic with extremum	a) control of chemical composition of rawmeal (with minimum variance) b) control of maximum quantity of rawmeal (gradient evaluation)	time constants in minutes	weight feeder rates	modified selftuning minimum variance regulator (MIMO-system)	set point of the weight feeder (moduli values)	pilot experiments 1976/78
	[39] [40]	cement grinding mills	a) stochastic variations in the quality of cement b) static nonlinear characteristic with extremum	a) control of fineness (Blain value) of the final product (cement) b) control of maximum quantity of final product	time constants in minutes	a) speed of separator spreading plate b) mill input feed rate	a) selftuning minimum-variance control of fineness b) selftuning extremum regulator of quantity	parameters of selftuning regulator	pilot experiments 1976/78
	[41]			breeze flow control	$T_1=T_2=3-6$min $T_t=4$min	fresh input material feed	parameter adaptation by feed forward control	controller parameters	in operation 1977/78
	[42]			a) mill loading b) power consumption	$T_1=3.3$min $T_2=1.1$min $T_t=2.5$min	a) fresh input material b) set point of the cascaded inner loop	parameter adaptation by sensitivity models using a gradient method	controller parameters	pilot experiments 1976/78
STEEL INDUSTRY	[47] [48]	plate mill	-statistical properties -nonlinear dynamics	-rolling torque -roll separating force -power -final thickness -final shape -final width (final pass schedule)	very fast (seconds)	-screw down -load roll gauge -speed	step by step adaptation of model equations for optimal reference values (and optimal pass schedule)	adapted model equations	in operation 1973
	[47]	hot strip mill		-roll gap -thickness -final temperature -roll force -strip tension					in operation 1973/75
	[49]	tandem cold reduction mill		-roll gap -thickness -strip speed -roll force -strip tension					in operation 1975
	[50]	hot strip coiling	maintaining coil temperature constant is very difficult	coiling temperature	very fast	spray banks	matching model equations for controlling the error from strip to strip	model parameters	in operation 1975
	[51]		varying radius of coil	speed control	very fast	armature voltage	feed-forward parameter adaptation	controller parameters	in operation 1979
	[52]	LD-converter	incomplete knowledge of dynamic behaviour of the converter and content of input	-decarburization rate -oxygen content in the slag	long dead times (~20sec)	-oxygen flow rate (stop point of blowing) -lance height above the bath surface	predictive algorithm using a model and Wiener filter ("learning algorithm")	model parameters	in operation
	[53]			-carbon content -metal temperature			self-adaptive model technique (self-tuning predictor)		in operation 1975
	[54]	verter		-decarburization -carbon content of the steel			predictive model adaptation using the technique of hyperstable model reference systems		in operation 1975

Table II. Practical applications of adaptive systems

field	references	PROCESS					ADAPTATION		
		plant	behaviour	output variables	time response (time constants)	manipulated variables	method	adapted variables	status
METALLURGICAL INDUSTRY	[55]	blast furnace	time-varying, nonlinear characteristics of reduction process	quality and quantity of hot iron	very slow (hours)	-blast temperature -blast humidity	self-adaptive model technique (self-tuning predictor)	model-parameters	proposed for application 1978
	[56]	ore crusher	stochastic variation of input material quality	maximum quantity of crushed ore	long dead time (minutes)	mill input feed rate	self-tuning minimum variance regulator	parameters of the self-tuning regulator	in operation 1974
	[57]	sinter-plant	time-varying nonlinear behaviour, large measurement noise and input disturbances	permeability of sinter mixture (sintering rate)	PT_1T_t-process (T_t=20s, T_1=30s)	water addition (water valve position)	fuzzy-set theory	levels of input and output fuzzy sets	in operation 1976
	[58]	bulk flotation in nickel recovery	-nonlinear behaviour -statistical properties depending on ore minerals	-pH-value (varying between 3.5 and 5) -metal value of concentrates	slow	reagent feed	model adaptation to the process using regression analysis for prediction	model parameters	in operation 1975/76
							optimum seaking using evolutionary operation principle	adaptation of pH-value	in operation 1975/76
	[59]	wet grinding mill in Ni-ore-concentrator plant	-nonlinear behaviour -statistical changes in grindability	grinding noise (as a measure for mill power)	slow dead time	-feed rate of fine ore -lumps -water	tuning of parameters of a normal PI-controller with changing process conditions	controller parameters	in operation 1975
	[60]	titanium dioxide kiln	statistical properties	composition of process outlet	very slow (hours)	fuel rate	self-tuning regulator	parameters of the regulator	in operation 1977
CHEMICAL INDUSTRY	[61]	double-effect evaporator	a) deterministic disturbances -feed flow rate -feed concentration -feed enthalpy b) dynamics depending on feed rates	a) first effect hold up b) second effect hold up c) second effect concentration	1/2 - 1 hour	a) steam flow b) first effect bottoms c) second effect bottoms	model reference adaptive control Lyapunov-design	state-feedback controller with feedforward	pilot experiments 1972/73
	[62]	packed bed tubular reactor	nonlinear behaviour, high sensitivity to load changes (wall temperature)	extent of a) n-butane b) propane c) hydrogen	~10min	feedrate of a) n-butane b) hydrogen	model reference adaptive control Lyapunov-design	state-feedback controller	pilot experiments 1977
	[63]	distillation column	a) deterministic disturbances of feed flow rate b) dynamics depending on direction and magnitude of setpoint and feed flow rate changes	top product composition	T_t=1min	reflux flow-rate	self-tuning regulation (compared with PI-controller)	parameters of the self-tuning regulator	pilot experiments 1977
	[65] [66]			top product a) concentration b) flow rate	slow	a) reflux rate b) heating power	dual control algorithm (cautions control algorithm)	parameters of the algorithm	pilot experiments 1977
	[67]	cracking furnace	a) drift b) limited rate of cooling or heating	temperature	~100sec	fuel flow	variable control period	control period	pilot experiments 1977/78
	[68]	neutralization of pH-value	nonlinear titration curve	pH-value	not significant	acid/bases-flow	PD-controller with deadband characteristic	width and gain of deadband	in operation 1973
PAPER INDUSTRY	[69]	paper machine	stochastic disturbances with unknown characteristics	a) moisture b) basis weight	large known time delays	a) steam pressure b) thick stack flow	MIMO-self-tuning regulator	parameter of the regulator	in operation 1972
	[70] [71]			moisture		steam pressure	self-tuning regulator with feed-forward		in operation 1973
	[72]	continuous digester	a) nonlinear time varying b) stochastic disturbances	Kappa-number (quality of the pulp)	slow	steam pressure	self-tuning regulator		in operation 1973

Table II. (continued)

field	references	plant	behaviour	output variables	time response (time constants)	manipulated variables	method	adapted variables	status
P O W E R P L A N T S	[74]	water-gate system of a hydroelectric power station	varying gain and time constants T because of jack load variations	water-gate position	1sec < T 4sec	motor supply voltage	structural adaptive systems with restricted control action	variable structure regulator	in operation 1974
	[75] [76]	interconnected heating power stations	varying steam generation and consumption	steam flow rate	slow	steam pressure	structural adaptive systems with constraints	variable structure regulator	in operation 1975
	[77]	interconnected power plant in power system	stochastic fluctuations in power consumption	frequency (load)	slow	power command	self-tuning PID-regulator	controller parameters	simulation studies 1979
	[79]	steam generator of a fast breeder nuclear reactor	nonlinear load-dependent behaviour	-steam flow rate -steam pressure -sodium temperature	settling time ∿50 to 200 sec	-sodium steam flow rate -steam valve aperture -water flow rate	reference model approach using parallel and series-parallel schemes in connexion with a self-tuning controller	controller parameters	simulation studies 1979
M I S C E L L A N E O U S A R E A S O F P R O C E S S I N D U S T R Y	[80]	glasshouse heating systems	disturbances a) wind b) temperature c) radiation	glasshouse temperature	∿1/2h	heating pipe temperature	least-squares-like gradient identification of process gain factor and adaptation of a PI-controller	gain of a master PI-controller	pilot experiments 1977
	[81]	refrigerant compressor test	nonlinear and unknown characteristics	a) evaporating pressure b) condensing pressure c) temperature at compressor suction inlet	∿20min	a) expansion b) flow rate of cooling water c) electrical power of the heater	nonlinear state and parameter estimation using Kalman-filter technique. Optimum control using quadratic performance index	all parameters of feedback law	experiments 1977
	[82]	air heater	nonlinear behaviour (air flow dependant)	air-temperature	slow	mixing valve position	recursive least squares parameters estimation	parameters of a deadbeat-controller	pilot experiments 1978
	[83]	condenser cooling processes	nonlinear static characteristics	effective power	fast	r.p.m. of condenser pump	extremum control using Hammerstein model	r.p.m. of condenser pump	pilot experiments 1976
	[84]	epitaxy reactor	time-variant dynamics	a) thickness b) sensitivity of epitaxy layers	slow	a) dropping gas flow b) duration c) reaction	parameter identification by Kalman filtering	inverse model parameters	in operation 1974
	[85]	vaporization reactor	unknown process dynamics	rate of vaporization	slow	electrical vaporization heating	recursive least squares parameter estimation	parameters of a deadbeat controller	in operation 1976
	[86]	heat exchanger	load dependent time-variant dynamics	outlet temperature	2-4 min	heat flow rate	comparison of 3 methods: a) adaptation of a PI-controller using test signals b) adaptation of compensating controller by direct identification c) parallel model technique using hyperstability theory	controller parameters	pilot experiments 1976
ELECTROMECHANICAL SYSTEMS	[87]	synchronous generator	steady state stability problem	voltage speed active power	fast	excitation	series-parallel MRAS	controller parameters	industrial use 1979
	[88] [89]	synchronous generator	stability problem	voltage power operating conditions	fast	excitation	self-tuning	controller gains	field test 1979
	[110]	DC-generator	turbine speed is changing	voltage	fast	field voltage	self-tuning	controller gains	laboratory implementation 1976
	[90]	DC-drive	varying load	speed	fast	SCR controller	MRAS	controller parameters	implementation 1974
	[91]	DC-drive	varying load	speed	fast	input to current control loop	self-tuning	parameters of PI-controller	experiments 1978

Table II. (continued)

field	references	PROCESS					ADAPTATION		
		plant	behaviour	output variables	time response (time constants)	manipulated variables	method	adapted variables	status
ELECTROMECHANICAL SYSTEMS	[92] [93]	DC-drive	nonlinear	speed	fast	SCR controller input	scheduled adaptation	structure and parameters of controller	industrial applications 1974
	[94]	test bench automatic gears	varying gain	speed	fast	input voltage	self-tuning	gain of controller	industrial use 1972
	[95]	DC-drive	varying load	speed	fast	SCR controller input	hyperstable adaptive model & control	coefficients of state variable feedback	experiments 1973
	[96]	synchroneous motor	nonlinear	speed	fast	SCR controller for DC-AC converter	scheduled adaptation	coefficients of PID controller	experiments 1977
	[97]	optical tracking telescope	variations in inertia, friction	speed position	fast	power amplifier input	MRAS Lyapunov design	controller parameters of tachometer loop	field test 1973
	[98]	radio telescope	variations in inertia, friction	speed position	fast	power amplifier input	MRAS augmented error	parameters of tachometer control loop	field test 1979
	[99] [100]	elevator	load variations	position	fast	armature current of DC-motor	self-tuning	coefficients of state variable feedback	laboratory implementation 1974
	[101]	autopilot for tanker	variations in parameters	speed-heading	slow	rudder command	self-tuning	controller parameters	experiments 1977
	[28]	autopilot for ship	variations in parameters	heading	slow	rudeer command	MRAS	rate feedback gain	experiments 1973
	[104] [105]	cutting machine tool lathe	changes in cutting force	speed	fast	input to drive for feed	self-tuning	controller gain	implementation 1973
	[106]	metal cutting process	variations in cutting force & depth	cutting force and speed	fast	SCR controller input	scheduled PI-controller with constaint	controller parameters	implementation 1973
	[107]	4-colour printing-machine	variations in inertia, elasticity	position difference of figures	fast	length of sheet between printing stations	scheduled PD-controller	controller parameters	industrial use 1977
	[108]	5 axis industrial robot	nonlinear oscillatory	speed position	fast	input to speed control subsystem	scheduled state variable feedback	feedback coefficients	experiments 1979
	[109]	compliant manipulator	nonlinear oscillatory	position	fast	power amplifier input	model parameter identification selftuning	model and controller parameters	experiments 1979

Table II. (continued)

6. DISCUSSION OF THE GENERAL SITUATION

Although use of adaptive control techniques is quite widespread in the three fields under review, the number of significant applications is really quite small.

In the aeronautical field we note the use of rival techniques and fundamental shortcomings of theory that have restricted the use of adaptive autopilots up to the present day, although the very idea of adaptive control probably originated in the aircraft industry.

During the last 20 years a large number of reports was published on the application of identification and estimation techniques in process industries. However, only very few reports deal with the adaptive control schemes of the type considered here and especially with the continuous use of such control systems for more than a short pilot-scale experimental period.

Here the reason for this situation cannot only be explained by the relatively complicated dynamic behaviour of these various processes under external and internal disturbances, drifts, parameter changes etc. Often the main reason is blamed on conservative management, who will not hazard their very expensive plants by applying modern control methods which are unproven and still "in leading reins". In addition among the production and instrumentation staff there is an absence of sufficiently trained personnel who understand these control strategies.

Although there has recently been some experimental work on the implementation of modern adaptive techniques in nearly all areas of process industry plants, in laboratories as well as in some factories there is still no great interest in accepting these techniques. Despite the advantages of more sophisticated algorithms they will not be adopted unless it can be clearly demonstrated that they are far superior to classical control and they do not introduce other problems into an interactive process industry plant. Optimum long term steady state regulation, good set point following properties, and the need to ensure that any control action applied and the effect it produces on the plant does not in any way endanger plant safety, are the main general conflicting requirements.

Perhaps the most successful field of application of adaptive controls so far is the field of electrical drives. Here the system equations and dynamics are well understood and the parasitical vibration modes are of higher frequency and are more easily eliminated in the design process than is the case for aircraft control.

However the application of adaptive control in most fields is still to this date in a phase of discovery. It is necessary to identify problems in which adaptive control theory can offer real advantages for control. The hitherto known applications should not conceal the fact that often the financial expense did not yield the desired commercial benefit, since a presumably inadequate knowledge of the processes limited the success.

From this brief discussion of the actual situation of adaptive control it becomes obvious that both research engineers as well as industrial engineers must cooperate closely in the further developments in this field. The research engineers should try in the future not only to develop other new adaptive methods, but to investigate common features of various methods that already exist. This would also allow the transfer of knowledge and results to other methods and applications. As only a few such trials, have been conducted, e.g. [12], much remains to be done in the future.

The research engineer should provide for practical application only those methods which fulfil the following conditions:

- possibility of a systematic design of a guaranteed stable and efficient adaptive controller using only little a priori knowledge;

- the expense of realization by mini or micro computers must be low, especially with regard to the number of numerical operations during one sampling period and the memory capacity;

- general applicability of the method to a variety of processes.

Furthermore for practical application, rules for a systematic design must be provided. It would be desirable for the industrial engineer to have program packages for a complete interactive design procedure which can be understood easily. As only few possibilities are available up to now [112], much effort has still to be made.

On the other side the industrial engineer must provide more possibilities for applications of adaptive methods. However, for this he must fulfil the following conditions:
- a good possibility of describing dynamical processes properly;
- a better education in modern control and thus more confidence in adaptive control systems;

- intensive cooperation with the research engineer during the design
  and realization phase.

In spite of all this, more experimental work should be done crystalli-
zing a group of generally applicable adaptive controllers from theory
and thus arousing a large scope and interaction for application.

7. CONCLUDING REMARKS

While we can agree that adaptive control methods are in use, reports
on applications are thin and performance data from such systems even
thinner. Until these voids in the technical field are filled, industrial engineers are likely to continue to use time-proved classical methods understood and valued from proved performance. Therefore, more
interaction between industry and research institutes is necessary. Industry should initiate orientation to provide examples of real world
process control systems and problems. The academics should avoid struc-
turing hypothetical problems to justify minor extensions to basic theory, and instead study real world problems with the intention of devel-
oping the extensions needed to apply basic theory to these problems.
What is needed is interactive communication on a regular basis, and
not merely transmittal of papers. However, engineers in industry must
also learn to describe clearly their control problem in a way which may
be accepted by research engineers.

REFERENCES

[1] Unbehauen, H. and Schmid, Chr.: Status and industrial application
     of adaptive control systems. Automatic Control Theory and Appli-
     cation 3(1975), No. 1, pp. 1/12.

[2] Asher, R.B., Andrisani, D. and Dorato, P.: Bibliography on adap-
     tive control systems. Proceed. IEEE 64(1976), No. 8, pp. 1226/40.

[3] Zypkin, J.: Adaption und Lernen in kybernetischen Systemen. Olden-
     bourg Verlag, München 1970.

[4] Zypkin, J.: Adaptive system theory today and tomorrow. Proceed.
     IFAC-Symposium Ischia (1973), pp. 47/67.

[5] Åström, K.: Self-tuning regulators - design principles and appli-
     cation. Proceed. of the Workshop on Application of Adaptive Con-
     trol, Yale University 1979.

[6] Unbehauen, H. and Schmid, Chr.: Application of adaptive systems
     in process control. Proceed. of the Workshop on Applications of
     Adaptive Control. Yale University 1979.

[7] Lindorff, D. and Carroll, R.: Survey of adaptive control using
     Lyapunov design. Int. J. Control 18(1973), pp. 897/914.

[8] Lüders, G. and Narendra, K.: Lyapunov functions for quadratic dif-
     ferential equations with applications to adaptive control. IEEE
     Trans. Aut. Control 17(1972), pp. 798/801.

[9] Wittenmark, B.: Stochastic adaptive control methods - a survey. Int. J. Control 21(1975), No. 5, pp. 705/730.

[10] Landau, J.D. and Unbehauen, H.: The development of adaptive systems in Germany and France. ASME Trans. J. of Dynamic Systems, Measurement and Control 96(1974), pp. 405/413.

[11] Huguenin, F.: Entwicklung von Regelalgorithmen für Mikrorechner. Elektroniker (1978), No. 11, pp. 11/20.

[12] Schmid, Chr.: Ein Beitrag zur Realisierung adaptiver Regelungssysteme. Ph. D. Dissertation, Ruhr-Universität Bochum 1979.

[13] Gregory, P.C. (Ed.): Proceedings of the self-adaptive flight control systems symposium. Wright Air Development Center 1959.

[14] Stein, G.: Adaptive flight control - a pragmatic view. Proceed. of the Workshop on Applications of Adaptive Control, Yale University 1979.

[15] Rynaski, E.G.: Adaptive control application to aircraft. Proceed. of the Workshop on Applications of Adaptive Control, Yale University 1979.

[16] Noll, T.E. and Huttsell, L.J.: Wing store active flutter suppression - correlation of analysis and wind tunnel data. AIAA Journ. of Aircraft 16(1979), pp. 491/497.

[17] IEEE: Special mini-issue on the F-8C program. IEEE Trans. AC-22 (1977), pp. 752/806.

[18] Parks, P.C.: Lyapunov redesign of model-reference adaptive control systems. IEEE Trans. AC-11(1966), pp.362/367.

[19] James, D.J.G.: Stability of a model reference control system. AIAA Journ. 9(1971), pp. 950/952.

[20] Monopoli, R.V.: Model reference adaptive control with augmenthed error signal. IEEE Trans. AC-19(1974), pp. 474/484.

[21] Johnson, C.R.: Input matching, error augmentation, self-tuning and output error identification - algorithmic similarities in discrete adaptive model following. Report Virginia Polytechnic Institute 1979.

[22] Narendra, K.S. and Valavani, L.S.: Direct and indirect adaptive control. Automatica 15(1979), pp. 653/664.

[23] Narendra, K.S. and Valavani, L.S.: Stable adaptive controller design - direct control. IEEE Trans. AC-23(1978), pp. 570/583.

[24] Landau, Y.D.: Adaptive control - the model reference approach. Marcel Dekker 1979.

[25] Cesari, L.: Asymptotic behaviour and stability problems in ordinary differential equations. Springer Verlag 1959.

[26] Morgan, A.P. and Narendra, K.S.: On the uniform asymptotic stability of certain linear time varying differential equations with unbounded coefficients. S and IS Report 7807, Yale University 1978.

[27] Hang, C.C. and Parks, P.C.: Comparative studies of model-reference adaptive control systems. IEEE Trans. AC-18(1973), pp. 419/428.

[28] van Amerongen, J. and Udink ten Cate, A.J.: Model reference adaptive autopilots for ships, Automatica 11(1975), pp. 441/449.

[29] Ljung, L. and Landau, Y.D.: Model reference adaptive control systems and self-tuning regulators - some connections. Proc. IFAC Congress, Helsinki (1978), pp. 1973/1979, Pergamon Press 1978.

[30] Narendra, K.S. and Valavani, L.S.: A comparison of Lyapunov and Hyperstability approaches to adaptive control. S and IS report 7804, Yale University 1978.

[31] Lions, J.L.: Optimal control of systems governed by partial differential equations. Springer-Verlag 1970.

[32] Curtain, R.F. and Pritchard, A.J.: Functional analysis in modern applied mathematics. Academic Press 1977.

[33] Parks, P.C. and Pritchard, A.J.: On the construction and use of Lyapunov functionals. Proc. IFAC Congress Warsaw (1969),paper 20.5.

[34] Balas, M.J.: Modal control of certain flexible dynamic systems. SIAM J. Control and Optimisation 16(1978), pp. 450/461.

[35] Meirovitch, L. and Öz, H.: An assessment of methods of the control of large space structures. Proc. JACC, Denver 1979.

[36] Keviczky, L., Kolostori, J. and Hilger, M.: On simultaneous optimal control of a raw material blending and ball grinding mill. Proceed. IFAC-Symposium Johannesburg (1976), pp. 143/158.

[37] Keviczky, L., Hetthessy, J., Hilger, M. and Kolostori, J.: Self-tuning adaptive control of cement raw material blending. Automatica 14(1978), pp. 525/532.

[38] Hilger, M., Keviczky, L., Kolostori, J. and Szijj, F.: Moderne Regelungsmethoden und die bei ihrer Anwendung erworbenen Erfahrungen. Preprints Fachtagung "Regelungstechnik in Zementwerken", Bielefeld 1978.

[39] Csaki, K., Keviczky, L., Hetthessy, J., Hilger, M. and Kolostori, J.: Simultaneous adaptive control of chemical composition, fineness and maximum quantity of ground materials at a closed circuit ball mill. Proceed. IFAC-Congress Helsinki (1978), pp. 453/460.

[40] Keviczky, L. and Vajk, I.: A self-tuning extremum regulator. Proceed. 12th Asilomar Conference on Circuits, Systems and Computers 1978, pp. 69/72.

[41] Heinrich, K. and Kinske, H.: Durchsatzoptimierung von Sichterumlaufmühlen. Preprints Fachtagung "Regelungstechnik in Zementwerken", Bielefeld 1978.

[42] Kunze, E. und Salaba, M.: Praktische Erprobung eines adaptiven Regelungsverfahrens an einer Zementmahlanlage. Preprints Fachtagung "Regelungstechnik in Zementwerken", Bielefeld 1978.

[43] Schulz, R.: Prozeßregelung einer Kugelmühle. Preprints Fachtagung "Regelungstechnik in Zementwerken", Bielefeld 1978.

[44] Schulze, H.: Erfahrungen mit einem Systemprogramm für Adaption und Regelung in Zementwerken. IFAC/IFIP-Symposium Zürich (1974), pp. 436/450.

[45] Zhivogladov, V. and Mirkin, B.: Adaptation and direct digital control of the cement industry technological process with application of distributed check. IFAC/IFIP-Symposium Zürich (1974), pp. 451/62.

[46] Müller, W.E.: Automation in metal processing. Proceed. IFAC-Symposium Johannesburg (1976), pp. 15/30.

[47] Seyfried, H.W. and Stöle, D.: Application of adaptive control in rolling mill area, especially for plate mills. Proceed. IFAC-Congress Boston 1975, paper 46.1.

[48] Stelzer, R., Heideprim, J. and Griese, F.W.: Adaptivalgorithmen für die Prozeßmodelle einer rechnergesteuerten Grobblechstraße. Stahl und Eisen 93(1973), No. 23, pp. 1100/1105.

[49] Alberts, R. and Joubert, A.: Automation of Iscor's five stand tandem cold mill. Proceed. IFAC-Symposium Johannesburg (1976), pp. 597/605.

[50] Ono, M., Kurokawa, T., Hirao, F., Takeda, E. and Nakano, K.: The computer control system of hot strip coiling temperature. Proceed. IFAC-Congress Helsinki (1978), pp. 159/166.

[51] Dittmar, E.: Mikrocomputer-Einsatz in der Automatisierung. Vogel-Verlag, Würzburg 1979.

[52] Voll, H., Danby, P. and Ramelot, D.: Improvement in oxygen furnace controllability, Prepr. IFAC/IFIP-Symposium Zürich (1974), pp. 99/111.

[53] Yasui, T.: Dynamic control of LD converter with digital computer. Proc. IFAC/IFIP-Symposium Zürich (1974), pp. 112/121.

[54] Landau, I.D. and Muller, L.: A new method for carbon control in basic oxygen furnace. Proceed. IFAC-Symposium Johannesburg (1976), pp. 257/261.

[55] Wellstead, P.E., Mantagner, G. and Gale, S.: A self-adaptive model for the prediction of blast furnace hot metal production and quality. Control System Centre Report UMIST 1978.

[56] Borisson, U. and Syding, R.: Self-tuning control of an ore crusher. IFAC-Symposium on Stochastic Control Budapest (1974), pp. 491/99.

[57] Carter, G. and Rutherford, D.: A heuristic adaptive controller for a sinter plant. IFAC-Symposium Johannesburg (1976), pp. 315/324.

[58] Mattila, O. and Pentillä, H.: Experience of computer automation at Kotalahti concentrator. Proceed. IFAC-Symposium Johannesburg (1976), pp. 461/473.

[59] Uronen, P., Tarvainen, M. and Aurasmaa, H.: Audiometric control system of wet semiautogenous and autogenous mills. Proceed. IFAC-Symposium Johannesburg (1976), pp. 215/233.

[60] Dumont, G.A. and Bélanger, P.R.: Self-tuning control of a titanium dioxide kiln. Trans. IEEE AC-23 (1978), No. 4, pp. 532/538.

[61] Oliver, W.K., Seborg, D.E. and Fisher, D.G.: Model reference adaptive control based on Lyapunov's direct method, part I and II. Chem. Eng. Commun (1974), pp. 125/140.

[62] Tremblay, J.P. and Wright, J.D.: Multivariable model reference adaptive control of a pilot scale packed bed tubular reactor. Proceed. IFAC/IFIP-Symposium Den Haag (1977), pp. 513/525.

[63] Sastry, V.A., Seborg, D.E. and Wood, R.K.: Self-tuning regulator applied to a binary distillation column. Automatica 13 (1977), No. 4, pp. 417/424.

[64] Morris, A.J., Fenton, T.P. and Nazer, Y.: Application of self-tuning regulators to the control of chemical processes. Proceed. IFAC/IFIP-Symposium Den Haag (1977), pp. 447/455.

[65] Badr, O., Rey, D. and Ladet, P.: Adaptive dual control for a multivariable distillation process. Proceed. IFAC/IFIP-Symposium Den Haag (1977), pp. 141/147.

[66] Badr, O., Rey, D. and Ladet, P.: On the dual-adaptive control and its practical applications. Automatica 15 (1979), pp. 91/96.

[67] D'Hulster, F.M. and van Cauwenberghe, A.R.: An adaptive controller for slow processes with variable control periods. Proceed. IFAC-Congress Helsinki (1978), pp. 461/467.

[68] Shinsky, F.G.: Adaptive pH controller monitors nonlinear process.

Control Engineering (1974), No. 2, pp. 59/59.

[69] Cegrell, T. and Hedquist, T.: Successful adaptive control of paper-machines. Proceed. IFAC/IFIP-Symposium Den Haag (1973), pp. 485/491.

[70] Borisson, U.: Self-tuning regulators - Industrial application and multivariable theory. Report 7513 Lund Institute of Technology, Lund 1975.

[71] Aström, K.J., Ljung, L., Borisson, U. and Wittenmark, B.: Theory and applications of adaptive regulators based on recursive parameter estimation. Proceed. IFAC-Congress Boston (1975), paper 50.1.

[72] Cegrell, T. and Hedquist, T.: A new approach to continuous digester control. Proceed. IFAC/IFIP-Symposium Zürich (1974), pp. 300/311.

[73] Mehra, R. and Eterno, J.: Adaptive power plant control: Problems and prospects. Proceed. of the Workshop on Applications of Adaptive Control. Yale University 1979, pp. 69/78.

[74] Erschler, J., Roubellat, F. and Vernhes, J.: Automation of hydroelectric power station using variable-structure control systems. Automatica 10(1974), pp. 31/36.

[75] Glattfelder, A.: Ein neues Entwurfskonzept für Weitbereichsregelungen. VGB Kraftwerkstechnik 55(1975), No. 7, pp. 428/431.

[76] Glattfelder, A. and Gross, L.: Weitbereichsregelung eines Wärmeverbundnetzes. Brennst.-Wärme-Kraft 29(1977), No. 1, pp. 27/33.

[77] Haber, R., Hetthessy, F., Keviczky, L. and Kovacs, K.: Some investigations on classical and adaptive load-frequency control. Proceed. of the Workshop on Applications of Adaptive Control. Yale University 1979, pp. 88/94.

[78] Davison, E. and Tripathi, N.: Decentralized tuning regulators: An application to solve the load and frequency control problem for a large power system. To appear in: Large Scale Systems: Theory and Application (1980).

[79] Irving, E. and Dang van Mien, H.: Discrete-time model reference multivariable adaptive control: Applications to electrical power plants. Proceed. 18th IEEE Conference on Decision and Control, Fort Lauderdale 1979.

[80] Udink ten Cate, A.J. and van de Vooren, J.: Digital adaptive control of a glasshouse heating system. IFAC/IFIP-Symposium Den Haag (1977), paper M2-2.

[81] Fujii, S., Fujimoto, H., Shibata, A., Shimada, S. and Hajiri, H.: Application of an adaptive control in a refrigerant compressor test. IFAC/IFIP-Symposium Den Haag (1977), paper M2-5.

[82] Kurz, H., Isermann, R. and Schumann, R.: Development, comparison and application of various parameter-adaptive digital control algorithms. Proceed. IFAC-Congress Helsinki (1978), pp. 443/452.

[83] Bamberger, W.: Adaptive on-line-Optimierung des statischen Verhaltens dynamisch träger Prozesse mit dem Programmpaket "OKIOPT". VDI-Bericht 276(1977), pp. 131/140.

[84] Rabut, C.: Self adaptive numerical control of an epitaxy reactor using Kalman-Bucy Filtering. IFAC-Congress Boston 1975, paper 52.4.

[85] Kapfer, W.: On-line Identifikation einer Aufdampfanlage mittels Prozeßrechner. VDI-Bericht 276(1977), pp. 101/107.

[86] Unbehauen, H. and Schmid, Chr.: Adaptive Regelung eines Wärmetauschers mit dem Prozeßrechner. Proceed. 21. Internat. Wiss. Koll. TH Ilmenau 1976, pp. 109/117.

[87] Irving, E., Barret, J.P., Charcossey, C. and Monville, J.P.: Improving Power Network Stability and Unit Stress with Adaptive Generator Control. Automatica 15(1979), pp. 31/46.

[88] Bonanomi, P., Güth, G., Blaser, F. and Glavitsch, H.: Concept of a practical adaptive regulator for excitation control. IEEE Summer Power Meeting 1979, paper A-79-753-2.

[89] Craven, R. and Glavitsch, H.: Large disturbance performance of an adapted linear excitation control system. IEEE Summer Power Meeting 1979, paper A-79-452-4.

[90] Courtiol, B. and Landau, I.D.: High Speed Adaptation System for Controlled Electrical Drives. Automatica 11(1975), pp. 119/127.

[91] Hribar, R. and Maron, M.: Gleichstromprüfantrieb mit adaptivem Drehzahlregler. Diplomarbeit AIE 7338, ETH-Zürich 1979.

[92] Buxbaum, A.: Einsatz von adaptiven Reglern bei geregelten Stromrichter-Stellgliedern. Deutsch-Französischer Aussprachetag Industrielle Anwendung adaptiver Systeme, Freiburg 1973, pp. 99/113.

[93] Buxbaum, A. and Schierau, K.: Berechnung von Regelkreisen der Antriebstechnik. AEG-Telefunken-Handbücher, Band 16, Elitera 1974.

[94] Speth, W.: Adaptivregelkreise in der Antriebstechnik. Interkama 1971, pp. 240/249.

[95] Sinner, E.: Adaptiver Zustandsregler für industrielle Strecken. Deutsch-Französischer Aussprachetag Industrielle Anwendung adaptiver Systeme, Freiburg 1973, pp. 115/136.

[96] Leimgruber, J.: Untersuchung des stationären und dynamischen Verhaltens drehzahlgeregelter Stromrichter-Synchronmotoren unter Berücksichtigung verschiedener Regelverfahren. Ph. D. Diss. ETH 6053, Zürich 1977.

[97] Gilbart, J.W. and Winston, G.C.: Adaptive Compensation for an Optical Tracking Telescope. Automatica 10(1974), pp. 125/131.

[98] Haque, S.I. and Monopoli, R.V.: Discrete Adaptive Control of a Radio Telescope. Proceed. of the Workshop on Applications of Adaptive Control, Yale University 1979, pp. 170/181.

[99] Maletinsky, V. and Schaufelberger, W.: Suboptimum Adaptive Control. IFAC Zürich(1974), pp. 129/143, Springer 1974, Control Theory 93.

[100] Maletinsky, V.: On-line parameter-estimation of continuous processes. IFAC Congress Boston (1975), paper 11.6.

[101] Källström, C.G., Aström, K.J., Thorell, W.E., Eriksson, J. and Sten, L.: Adaptive Autopilots for Tankers. Automatica 15(1979), pp. 241/254.

[102] Pressmann, R.S. and Williams, J.E.: Numerical Control and Computer Aided Manufacturing. Wiley 1977.

[103] Koren, V. and Ben-Uri, J.: Numerical Control of Machine Tools. Khanna 1978.

[104] Götz, F.R.: Regelsystem mit Modellrückkopplung für variable Streckenverstärkung - Anwendung bei Grenzregelungen an spanenden Werkzeugmaschinen. Ph. D. Diss. Stuttgart (1977), ISW 18, Springer 1977.

[105] Stute, G. and Götz, F.R.: Anwendung adaptiver Systeme bei spanenden Werkzeugmaschinen. Deutsch-Französischer Aussprachetag Industrielle Anwendung adaptiver Systeme, Freiburg 1973, pp. 263/278.

[106] Pfeifer, T. and Gieseke, E.: Entwicklung von Adaptive-Control-Einrichtungen für die spanende Bearbeitung. Deutsch-Französicher

Aussprachetag Industrielle Anwendung adaptiver Systeme, Freiburg 1973, pp. 279/291.

[107] Meyer, S., Rubruck, M. and Tröndle, H.P.: Adaptiver Abtastregler für das Längs-, Seiten- und Schnittregister von Tiefdruck-Rotationsmaschinen. Siemens-Zeitschrift 51(1977), pp. 394/398.

[108] Hesselbach, J.: Optimierung linearer Lageregelkreise (Kapitel: Winkelregelung der Drehachse eines Industrieroboters) in die Lageregelung an Werkzeugmaschinen, ISW Stuttgart (1979), pp. 263/269.

[109] Liégeois, A., Dombre, E. and Borrel, P.: Learning and Control for a Compliant Computer-Controlled Manipulator. IEEE Conference on Decision & Control (1979), pp. 1024/1027.

[110] Glattfelder, A.H., Huguenin, F. and Schaufelberger, W.: Microcomputer Based Self-Tuning and Self-Selecting Controllers. Automatica 1(1980), No. 1, pp. 1/8.

[111] Clarke, D.W. and Gawthrop, P.J.: Implementation and application of microprocessor - based self-tuners. Proceed. IFAC-Symposium Darmstadt (1979), pp. 197/208.

[112] Schmid, Chr.: CAD of adaptive systems. Preprints IFAC-Symposium on computer aided design of control systems, Zürich 1979.

MODEL REFERENCE ADAPTIVE CONTROL APPLIED TO
STEERING OF SHIPS

J. van Amerongen

Control Engineering Laboratory

Electrical Engineering Department

Delft University of Technology

This paper describes the application of Model Reference Adaptive Control Systems (MRAS) to steering of ships. The main goal is to improve the steering performance and to facilitate the adjustment of the controller. Therefore MRAS is applied to direct adaptation of the controller parameters as well as to parameter identification and state estimation with optimum noise reduction. Solutions are given how to deal with non-linearities in the system's dynamics. Results of experiments on board ships are given.

1. Introduction

The technique of model reference adaptive control systems (MRAS) has received a lot of attention during the last decade. For linear, noise-free processes of known order a solid theoretical base is available which enables the design of adaptive controllers. In practice however, these nice properties are not always present. In that case the controller design is less straightforward and sometimes ad-hoc solutions have to be sought.

In this paper the application of MRAS will be illustrated with the design of an adaptive autopilot for steering of ships. The reasons to apply adaptive control to this system will be shortly summarized here. A more extensive description is given by Van Amerongen and Van Nauta Lemke (1978, 1979). Conventional autopilots for ships mainly consist of a PID-controller, extended with a dead band, which should remove "high-frequency" rudder motions in bad weather conditions, and with a rudder limit, to prevent large rudder angles when they are not wanted. This is illustrated in Figure 1. Although several explanatory terms are used to explain the P, I and D-controller actions to the user, it is common practice that the autopilot is not very well adjusted. This is not only due to lack of knowledge of control theory, but also because the controller adjustment depends on the desired performance and on the ship's dynamics, which are

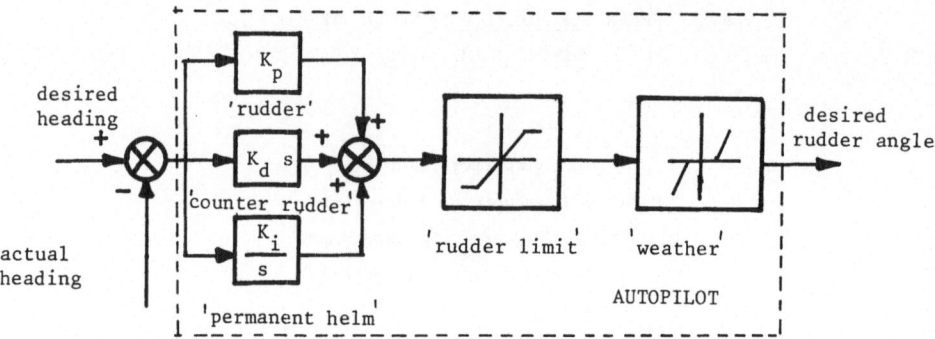

Fig. 1.   Conventional autopilot

influenced by the forward speed, load condition, waterdepth etc. The latter do not have
a clear and simple relationship with the controller settings. The desired performance
may differ between accurate course changing in narrow waters and maximum steering economy at the open sea.

By applying adaptive control, the conventional autopilot settings can be adjusted
automatically. The settings which remain for the user are:

-   selection of the desired rate of turn (during course changing)
-   selection between maximum course keeping accuracy or maximum rudder economy
    (during course keeping) .

The first setting defines in fact the slope of a step response (Figure 2).

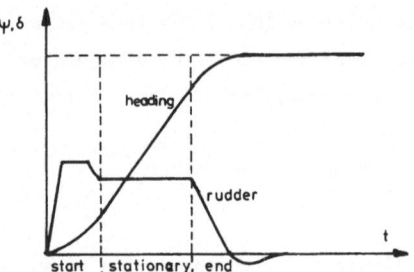

Fig. 2

course changing manoeuvre

There is no need to vary the shape of the begin and end of the response; for instance
there must be no overshoot.

The choice between accuracy and economy is related with the increased resistance due
to steering. This can be expressed in a quadratic criterion:

$$J = \frac{1}{T} \int_0^T (\epsilon^2 + \lambda \delta^2) \, dt \qquad (1)$$

    where ϵ denotes the heading error

             δ denotes the rudder angle

    and λ is a weighting factor.

Besides optimization of the controller with respect to the criterion (1), the heading
signal ψ, and it's derivative, the rate of turn signal $\dot{\psi}$, must be filtered. Especially
on small ships the amplitudes and frequencies of the disturbances caused by the waves

are so high that correction by rudder motions is impossible.

Based on these requirements the problem of designing an adaptive controller can be split into two parts.

1. Course changing controller

The desired performance is defined as a stepresponse, which can be expressed with the aid of a reference model. The controller should be adjusted in order to let the ship's response follow the reference model as close as possible. This problem can effectively be solved by applying MRAS.

2. Course keeping controller

The controller parameters should be adjusted in order to minimize criterion (1). This requires knowledge of the parameters of a mathematical model of the ship by means of an on-line identification algorithm. The identification problem and the filter problem, mentioned before, can be solved simultaneously by applying MRAS to parameter identification and state estimation.

In Section 2 the theoretical base for the design of the two adaptive controllers for an idealized system will be given. Section 3 discusses the modifications of the basic algorithms which are required to make the theory applicable to ship's steering. Section 4 gives results of full scale experiments on board ships.

2. Design of the adaptive controller

2.1 Direct adaptation of the controller parameters

During course changing the desired performance can be defined with the aid of a reference model. Application of MRAS yields adjustment laws for direct adaptation of the controller without explicit identification.

In order to derive the adjustment laws a mathematical model of the process has to be selected. Although the ship's steering dynamics may be well described by a non-linear third order model, it has turned out that it is advantageous to use the most simple linear description (Van Amerongen et. al. 1975). The model proposed by Nomoto, is well suited:

$$\frac{\psi}{\delta} = \frac{K}{s(s\tau + 1)} \tag{2}$$

where ψ is the heading signal

δ is the rudder angle

and s is the Laplace operator.

This proces will be controlled by an autopilot which is basically a PID controller, but the integrating action is designed from an adaptive point of view (Figure 3).

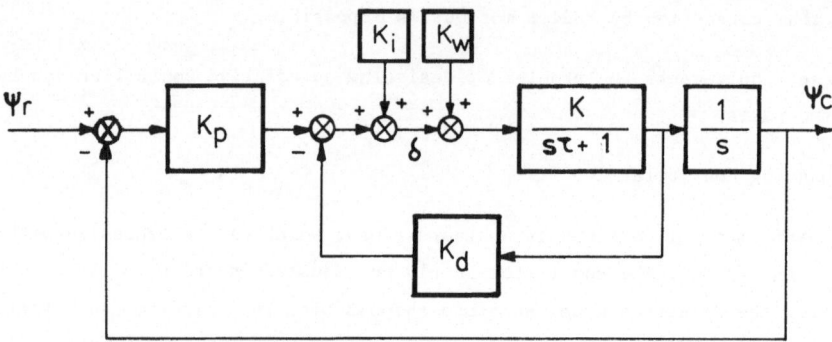

Fig. 3.   Course control system

Constant, or slowly varying disturbances, K_w, can be compensated by an equivalent rudder angle of opposite sign, K_i.

This control system is described by the following equations:

$$\dot{x}_p = A_p \, x_p + B_p \, u \tag{3}$$

with

$$x_p^T = (\psi_c , \dot{\psi}) \tag{4}$$

$$u^T = (\psi_r , 1) \tag{5}$$

and

$$A_p = \begin{bmatrix} 0 & 1 \\ \dfrac{-K_p K_s}{\tau_s} & \dfrac{1+K_d K_s}{\tau_s} \end{bmatrix} \quad , \quad B_p = \begin{bmatrix} 0 & 0 \\ \dfrac{K_p K_s}{\tau_s} & (K_w+K_i)\dfrac{K_s}{\tau_s} \end{bmatrix} \tag{6}$$

The reference model is described by:

$$\dot{x}_m = A_m \, x_m + B_m \, u \tag{7}$$

with

$$A_m = \begin{bmatrix} 0 & 1 \\ -\omega_n^2 & -2z\omega_n \end{bmatrix} \quad , \quad B_m = \begin{bmatrix} 0 & 0 \\ \omega_n^2 & 0 \end{bmatrix} \tag{8}$$

Stable adjustment laws can be found by using Liapunov's second method, or by applying the hyperstability concept (Landau, 1979).
This yields:

$$\frac{dK_p}{dt} = \beta \{ (p_{12} \, e + p_{22} \, \dot{e}) \, \varepsilon \} \tag{9}$$

$$\frac{dK_d}{dt} = - \alpha \{ (p_{12} \, e + p_{22} \, \dot{e}) \, \dot{\psi} \} \tag{10}$$

$$\frac{dK_i}{dt} = \gamma \{ (p_{12} \, e + p_{22} \, \dot{e}) \, 1 \} \tag{11}$$

where $\underline{e}^T = (e , \dot{e}) = \underline{x}_m - \underline{x}_p$ $\qquad(12)$

and P_{12} and P_{22} elements of the P-matrix which can be found by solving

$$A_m^T P + P A_m = -Q \qquad(13)$$

after choosing an arbitrarily positive definite matrix Q; α, β and γ are positive adaptive gains.

2.2 Parameter and state estimation

During course keeping it is essential that the process parameters are known, in order to be able to optimize criterion (1). In this situation, where only small rudder angles are used, the model (2) gives a very good description of the ship's steering dynamics. When the ship's parameters are known, the controller gains are found by straightforward application of optimum linear control theory, which yields:

$$K_p = 1 / \sqrt{\lambda} \qquad(14)$$

$$K_d = 1 / K \{ \sqrt{(1 + 2 K\tau / \sqrt{\lambda})} - 1 \} \qquad(15)$$

$$K_i = \gamma \int_0^t \varepsilon \, d\tau \qquad(16)$$

The parameters K and τ can be obtained with the aid of MRAS, used in an identification structure. An adjustable model is placed parallel with the transfer between δ and $\dot{\psi}$.

The proces is now described by the equation:

$$\dot{x}_p = -a \, x_p + K'u + K_w \qquad(17)$$

where $x_p = \dot{\psi}$

$u = \delta$

$K' = K / \tau$

$a = 1 / \tau$

and the adjustable model is described by

$$\dot{x}_m = - a_m \, x_m + K'_m u + K_{im} \qquad(18)$$

In a similar way as before the simple adjustment laws are found:

$$\frac{dK'_m}{dt} = -\beta \, e \, u \qquad(19)$$

$$\frac{da_m}{dt} = \alpha \, e \, x_m \qquad(20)$$

$$\frac{dK_{im}}{dt} = \gamma \, e \qquad(21)$$

where $e = x_m - x_p$ and α, β and γ are positive adaptive gains.

It has been shown (Landau, 1979) that this system gives unbiased parameter estimates, also when the process states are corrupted with noise.

Besides, x_m is a noise free estimate of x_p. This is a nice property because it has been mentioned already that filtering of noisy measurements is important in order to get a good course keeping performance. Compared with ordinary low-pass filters this type of filter has the advantage that it introduces hardly any phase lag.

3. Modifications required for practical application

The basic adaptive algorithms have been derived supposing that the process and reference model are both linear and of the same order. In the ship's steering system these assumptions are clearly not valid. First of all, the transfer between δ and $\dot{\psi}$ is non-linear and of higher order. Experiments have shown however that also in this situation the adaptive system remains working well when the simple adjustment laws (9) - (11) are used.

The varying controller gains compensate for non-linear process dynamics and neglected time constants.

More serious are the saturation-type non-linearities: the rate of turn limiter, which has to be added to the reference model in order to get the desired response, and the non-linearities in the steering machine: the rudder limiter and the maximum rudder speed. Without taking special precautions these non-linearities will seriously detoriate the performance of the adaptive system.

These problems can be circumvented when besides the parallel reference model also a series model is introduced. This modifies the process input in such a way that the maximum rate of turn (in the reference model) and the maximum rudder angle (in the process) are never exceeded.

From eqn. (7) it follows that

$$- \dot{x}_{m,max} < \omega_n^2 (u - x_m) < \dot{x}_{m,max} \tag{22}$$

where $\dot{x}_{m,max}$ is the maximum value of the rate of turn selected by the user. Equation (22) can be rewritten into

$$- \frac{\dot{x}_{m,max}}{\omega_n^2} + x_m < u < \frac{\dot{x}_{m,max}}{\omega_n^2} + x_m \tag{23}$$

The input-signal u, limited according to eqn. (23), can be seen as a modified input-signal u'. Using u' instead of u as an input for both the process and the reference model, allows the rate of turn limiter to be removed from the parallel model. The signal u' is computed with the aid of a series model.

In a slightly different way also the rudder limiter can be taken out of the control loop. Because the relation between the rudder angle and the input signal depends on the variable gains of the controller it is not possible to use a fixed limiter. It is however, possible to compute a factor f according to the formula:

$$f = \frac{\delta_{max}}{\delta_r} < 1 \tag{24}$$

which determines the ratio between the desired rudder angle δ_r and the maximum

rudder angle δ_{max}. When this factor f is introduced again into the series model
the influence of the rudder limiter is also taken off the control loop. This leads to
the structure of Fig. 4.

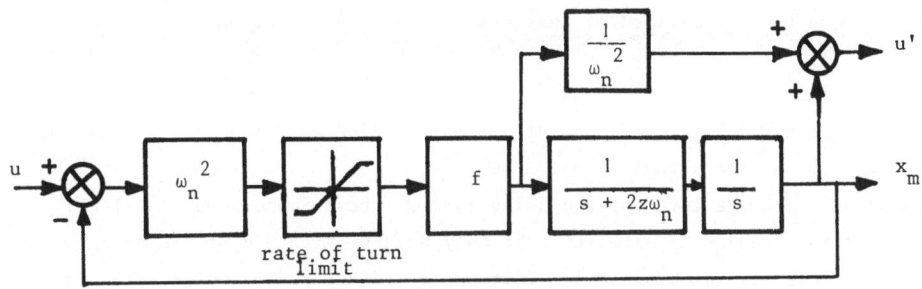

Fig. 4   Series model to modify the input

Due to these manipulations the non-linearities are in fact removed from the control
loops. So far the proof of stability of the adaptive system remains thus in tact.

The influences of the limited rudder speed cannot be compensated for in such an elegant
way. It can however, be taken into account when the difference between the desired
rudder angle δ_r and the actual rudder angle δ_c is measured on-line. A variable
"time-constant" τ_δ is introduced:

$$\tau_\delta = \frac{|\delta_r - \delta_c|}{\delta_{max}} \tag{25}$$

Adding the transfer function

$$H_{\tau_\delta} = \frac{1}{s\,\tau_\delta + 1} \tag{26}$$

into both the series and the parallel model, prevents instability of the adaptive
system, which otherwise could be expected for large values of τ_δ. Because this
(variable) time constant was not taken into account when deriving the adaptive laws
(9) - (11), a theoretical proof of stability cannot easily be given anymore.

Another problem which is met when MRAS is applied in practice, is the influence of
noise which causes drift of the controller parameters. Because the signals in the
case of ship's steering may be very noisy this problem cannot be neglected.
In the designed autopilot a combination of measures is taken against the influence
of noise.
- The parameter adjustment by means of direct adaptation is switched off when the
system is not sufficiently excited. (that is during course keeping).
- The process states are approximated by the model states which are much less noisy.
This is theoretically not correct but it appears to work well in the designed system.

When MRAS is used for identification some similar problems are met, although these are less complicated because the adjustable model is now placed parallel to a small part of the system only.

The parameter estimation works satisfactorily without any further precaution, although the performance may be improved by measures such as decreasing adaptive gains (Landau, 1979).

The state estimation however, can be improved. When there is much noise in the system (bad weather conditions) the output of the adjustable model is a reasonable filtered version of the actual signal.

In situations where the level of the noise is low, there is no need to filter, however. But the prediction will still be only an approximation of the actual signal. A better estimate can be obtained by adding a second adjustable model, especially meant for state estimation which is updated with the difference between the predictions and the actual signals. The parameters of this second model are similar to those used in the first adjustable model.

Because the aim of the filtering is to remove the high-frequency components of the disturbances, while low-frequency components should not be damped, the amount of filtering can be determined by the ratio

$$K = \frac{\sigma^2_{lf}}{\sigma^2_{lf} + \sigma^2_{hf}} \quad , \tag{27}$$

where σ^2_{lf} is the mean value of e_{lf}^2, where e_{lf} is the low-frequency component of the error between the predicted signal and the actual signal; in a similar way σ^2_{hf} is defined for the high frequency component of the error.

When the system is implemented digitally the most simple way to adjust the predicted value of the rate-of-turn signal, $\hat{\dot{\psi}}$ (k+1/1) is according to the formula:

$$\hat{\dot{\psi}} \ (k+1/k+1) = \hat{\dot{\psi}}(k+1/k) + K_\sigma \ (\ \dot{\psi} \ (k+1) - \hat{\dot{\psi}} \ (k+1/k) \ ) \tag{28}$$

A similar strategy can be followed to get optimum estimates of the heading signal. This second adjustable model has relations with Kalman-filtering, but it should be noted that the filter gains K_σ and $K_{\sigma\psi}$ are determined in a way which differs from Kalman-filtering.

4. Results and conclusions

The algorithms defined before have been implemented in a small digital computer, the PDP 11/03 DECLAB- system. Because the sampling interval can be chosen sufficiently small, the continuous time equations given before, can easily be approximated by equivalent discrete equations. On the other hand, the flexibility of the digital

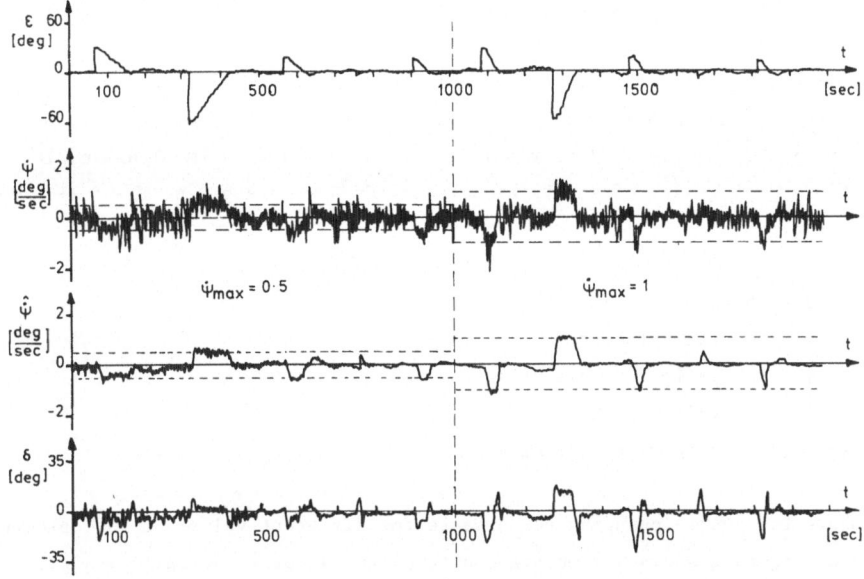

Fig. 5. Course changing performance of the adaptive autopilot

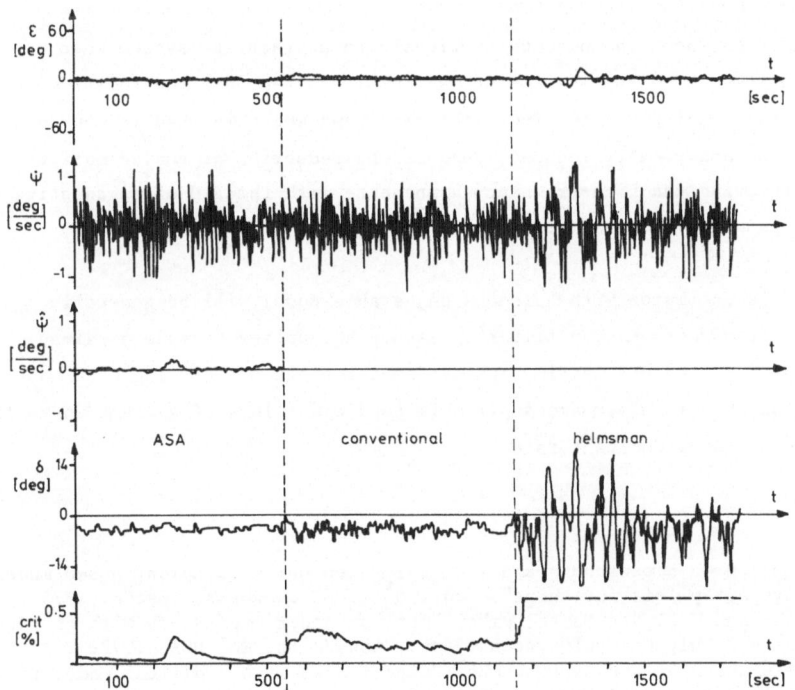

Fig. 6. Comparison between an adaptive autopilot, a conventional
autopilot and manual steering, during course keeping.

computer can be used to add some refinements, such as decreasing adaptive gains, etc.

After extensive tests with the aid of an analogue simulation set-up at the laboratory, full scale experiments on board various ships have been carried out and they are still going on. In Fig. 5 and Fig. 6 some results from the tests with HMs Tydeman, the oceanographic survey vessel of the Royal Netherlands Navy are given. The following signals are plotted:

ε , the course error
$\overset{\wedge}{\dot\psi}$, the rate of turn
$\dot\psi$, the estimated rate of turn
δ , the rudder angle
crit , the criterion (1), with $\lambda = 10$.

Figure 5 shows the course changing performance for different values of the selected rate of turn. Figure 6 shows a comparison between the adaptive autopilot (ASA), a conventional autopilot and the helmsman. At the moment the experiments are mainly directed towards the question which savings in fuel consumption can be obtained by the improved steering performance of the adaptive autopilot.

The theory of MRAS has been shown to be applicable to a practical system when a few measures are taken. The application of MRAS has resulted in a controller with a reduced number, now meaningful, settings. The dead band which is used in conventional autopilots could be drastically reduced, because the reduction of rudder motions has been solved by removing the high-frequency components with the aid of an adaptive state estimator. This is principally a more correct solution.

The suggested state estimator with a second adjustable model will be generally applicable to systems where Kalman-filtering cannot be applied because a Kalman-filter only suppresses the observation noise. The solutions for dealing with non-linearities, given in this paper are applicable to a wide class of systems where the saturation effects play a dominant role.

5. Literature

- J. van Amerongen, H.C. Nieuwenhuis and A.J. Udink ten Cate, "Gradient based model reference adaptive autopilots for ships", Proc. 6th IFAC Congress, Boston, 1975.
- J. van Amerongen and H.R. van Nauta Lemke, "Optimum steeringof ships with an adaptive autopilot", Proc. 5th Ship Control Systems Symp., Annapolis, Md, USA, 1978.
- J. van Amerongen and H.R. van Nauta Lemke, "Experiences with a digital model reference adaptive autopilot", Proc. Int. Symp. on Ship Operation Aut. , Tokyo ,1979.
- I.D. Landau, "Adaptive control - the model reference approach", Marcel Dekker, Inc., 1979.
- K. Nomoto, T. Taguchi, K. Honda, and S. Hirano, "On the steering qualities of ships", International Shipbuilding Progress, Vol.4, 1957.

MODEL REFERENCE ADAPTIVE SATELLITE ATTITUDE CONTROL

P.P.J. van den Bosch
Control Engineering Laboratory
Electrical Engineering Department
Delft University of Technology

W. Jongkind
Fokker
Space Division
Schiphol-Oost

An adaptive attitude control system of a satellite will be described that enables a fast reorientation of a satellite by means of one single slew around the so-called Euler axis. Additional measures have been taken to deal with saturation effects, associated with large angle slews. An improvement of the adaptive controller is realized by the introduction of the model updating concept. Results of simulations are included.

1. Introduction

Astronomical satellites are used to observe several celestial phenomena in space. However, some of these phenomena occur at irregular intervals and have a relative short life time. Therefore, after detecting such phenomena, it is highly desirable to reorientate the satellite to such a phenomenon as quickly as possible to increase the available observation time. This reorientation of the satellite is realized with the aid of the attitude control system.
Many attitude control systems for satellites are now a day based on a sequence of rotations about each of the three principal axis of the satellite separately. In that case the highly non-linear gyroscopic coupling between the axes disappears. However, the time to realize a reorientation or slew increases with a factor two to three, compared with a simultaneous slew around the three principal axes.

In this paper an attitude control system for satellites will be discussed that enables a rapid reorientation in space, by means of one single slew around the so-called Euler axis, because such a slew offers well-defined and smooth responses. The satellite is considered to be a rigid body, controlled by means of three orthogonally placed reaction wheels. A model reference adaptive controller has been used to realize the desired control objectives. However, due to the non-linearities in the equations that describe the satellite, it has not been possible to prove stability for any attitude. Still, extensive simulations indicate that the adaptive laws, derived and based on the linearized model of the satellite, are able to control the satellite. Satisfactory control behaviour is maintained, even in the presence of disturbances and (large) parameter variations in the satellite or in the controller. Some additional measures have to be taken to include saturation effects (maximum wheel speed, maximum control torque) into the adaptive controller. In all derivations and figures, only the formulas and structure of the X-axis will be given. The relations for the Y-axis and Z-axis are analogous.

2. Modelling

2.1 Dynamics of the satellite

A body fixed reference frame is taken with axes coincident with the satellite principal axes of inertia and the axes of the reaction wheels. If the external torques are neglected, the total angular momentum H of the satellite is constant with respect to an internally fixed reference frame and determined by the inertia matrix I of the satellite (with diagonal elements I_x, I_y and I_z), the satellite angular velocity vector Ω (with components Ω_x, Ω_y and Ω_z along body fixed reference axes), the moment of inertia of each reaction wheel J and the angular wheel velocity vector ω with respect to the satellite (with components ω_x, ω_y and ω_z).

$$H = I.\Omega + J.\omega \tag{1}$$

The time derivative of this equation offers the dynamic relations of the satellite and wheels:

$$I.\dot{\Omega} + J.\dot{\omega} + \Omega.(I.\Omega + J.\omega) = 0 \tag{2}$$

So the angular velocity of the X-axis of the satellite is defined by (if no external disturbances added):

$$\dot{\Omega}_x = \{(I_z.\Omega_z + J.\omega_z).\Omega_y - (I_y.\Omega_y + J.\omega_y).\Omega_z + J.\dot{\omega}_x\}/I_x \tag{3}$$

Expression 3 clearly indicates the gyroscopic coupling in the satellite, due to Ω_y, Ω_z, ω_y and ω_z, and the control action, due to the acceleration and deceleration of the reaction wheel.

2.2 Kinematics of the satellite

The attitude of the satellite is specified by means of a quarternion. The use of a quaternion in describing the orientation of a rigid body allows all possible attitudes. The problem of gimbal lock, encountered when using the more commenly understood Euler angles, is avoided. Moreover, they are very well suited to be calculated with the aid of an on board computer since only products exists in the formula and no goniometric parameters.

$$q_0^2 + q_x^2 + q_y^2 + q_z^2 = 1 \tag{5}$$

$$
\begin{pmatrix} \dot{q}_o \\ \dot{q}_x \\ \dot{q}_y \\ \dot{q}_z \end{pmatrix} = \frac{1}{2} \begin{pmatrix} 0 & -\Omega_x & -\Omega_x & \Omega_z \\ \Omega_x & 0 & \Omega_z & -\Omega_y \\ \Omega_y & -\Omega_z & 0 & \Omega_x \\ \Omega_z & \Omega_y & -\Omega_x & 0 \end{pmatrix} \begin{pmatrix} q_o \\ q_x \\ q_y \\ q_z \end{pmatrix} \tag{4}
$$

So, four parameters q_o, q_x, q_y and q_z, coupled by means of equation 5, describe uniquely the orientation of the satellite in space.

2.3. Reaction wheel

If the rotor of the reaction wheel is a permanent magnet or suchlike then the torque
T_{cx}, on the satellite due to the reaction wheel, is directly proportional to the stator
current, so $T_{cx} = g \cdot u_x$. In the model of the reaction wheel two non-linearities should
be added. The first one is the constraint that the wheel can deliver a maximum torque
T_{cx}, so $|T_{cx}| \leq T_{cx,max}$. The second one is the absolute angular velocity, the wheel speed
ω_x relative to the satellite is not allowed to exceed a maximum value, so $|\omega_x| \leq \omega_{x,max}$.
In Fig. 1 the reaction wheel, the dynamics and the kinematics (of the X-axis) of the
satellite are illustrated, together with the control torque T_{cx} ($= g \cdot u_x$), the disturbance torque T_{dx} and the torque due to the gyroscopic coupling T_{gx}.

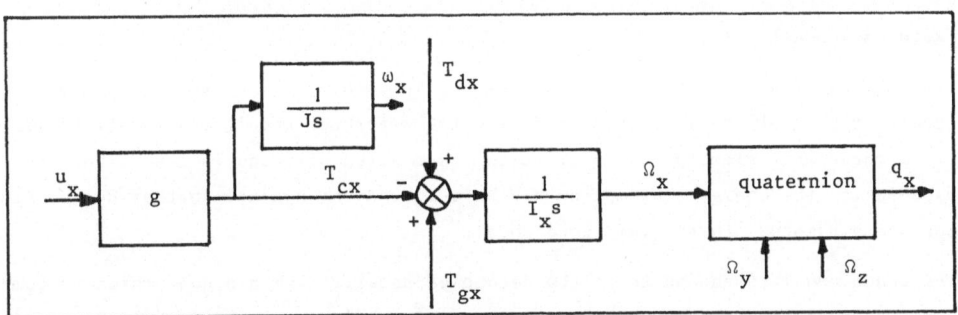

Fig. 1. Satellite (X-axis) with reaction wheel, gyroscopic coupling T_{gx} and
disturbance torque T_{dx}.

3. Controller Design

In the preceding part a model has been derived of a satellite controlled by means of
reaction wheels. This model describes the satellite as a non-linear multivariable system
with much interaction in the dynamics (gyroscopic coupling) and in the kinematics (quaternion). No general design technique is available to design an appropriate controller for
this nonlinear multivariable system. Therefore, an attempt can be made to decrease the
non-linear interaction in the satellite. Consequently, three scalar, linear systems will
result that can be controlled more easily. Due to the constant relation, $T_{cx} = g \cdot u_x$,

T_{gx} can be counteracted. So its influence on the X-axis will disappear. With this decoupling scheme, the dynamics of the satellite may be considered as three scalar systems. The three axes are only coupled by means of the non-linear quaternion equation 4.

Form a theoretical point of view, this decoupling scheme offers many advantages. Control laws can be derived and implemented for each axis separately. However, simulation results indicate the the proposed controller is able to counteract for the gyroscopic coupling, by which the decoupling scheme is no longer necessary. So, in this section we assume that the dynamics are decoupled, although the proposed decoupling scheme need not be implemented.

The purpose of this paper is to propose a controller scheme to realize a fast, three axis slew around the Euler axis even in the presence of parameter variations inside the satellite and external disturbances. A classical controller cannot meet all these requirements. Therefore an adaptive controller will be proposed and implemented. Among the adaptive controller schemes a model reference adaptive controller (Van Amerongen 1980, Landau 1979) will turn out to be extremely useful in controlling a satellite. For the rigid body satellite has a fixed, well defined structure, whose states are easily accessible, that can be measured very accurately. Moreover, many design requirements can be put into the reference model.

3.1. Reference model

The realization of a correct three axis slew is implemented in the reference model and because the satellite is forced to follow the reference model, the satellite will perform a three axis slew too and will reach its desired attitude in space. The speed of response and the energy consumption can be weighted one against another by choosing appropriate parameters in the reference model.

The reference model is chosen to be the decoupled satellite with a quaternion to compute its attitude out of the angular velocities. According to Mortensen (1968), feedback signals u_m for the reference model are derived from the corresponding quaternion q_m and the angular velocity Ω_m for each axis separately:

$$u_{mx} = K_{mpx} \cdot \frac{q_{mx}}{q_{mo}^3} + K_{mvx} \cdot \Omega_{mx} \qquad (6)$$

This type of control with feedback of Ω_m with K_{mv} and with feedback of q_m with K_{mp} can proven to be asymptotic stable for all attitudes Mortensen (1968). Moreover, this control strategy guarantees no steady state error in the model response. The reference model will always reach the desired attitude in space by a single three axis slew. With additional measures, to be taken when the input or the wheel velocities become saturated, this slew is a slew around the Euler axis. Before a new slew will be made, a new reference frame is calculated, such that the required orientation will be the origin of this new reference frame. So, at the end of each slew $q_{mx} = q_{my} = q_{mz} = \emptyset$ and $q_{mo} = 1$. In general, q_{mo} will be large compared with q_{mx}, q_{my} and q_{mz}. So, the quaternion equation 4 can be linearized and decoupled:

$$\dot{q}_{mx} = \tfrac{1}{2} \, \Omega_{mx} \tag{7}$$

This approximation, by which the model of one axis becomes linear and decoupled, allows the gains K_{mp} and K_{mv} to be calculated analytically. In choosing the natural
frequency ω_n and the relative demping z as design parameters, the feedback gains can
be calculated:

$$K_{mpx} = 2 \, \omega_n^2 \, I_{mx}/g_m \qquad\qquad K_{mvx} = 2 \, z \, \omega_n \, I_{mx}/g_m \tag{8}$$

A deviation of this control scheme for the reference model is necessary to maintain
a slew around the Euler axis, when the input or the velocity of the reaction wheels
reach their limits. Following a slew around the Euler axis, the angular velocities Ω_m
have a fixed, mutual relation.

So, when one of the inputs becomes saturated, the other two inputs have to be set to
an a priori calculated value. This value depends on the initial quaternion and the
inertia matrix I_m.

When a wheel reaches its maximum velocity, it can no longer apply a control torque.
Therefore, all control torques are set to zero, by which the reference model continues
the coast around the Euler axis.

By taking these two additional measures, the reference model will never leave its
slew around the Euler axis. In order to enable the satellite to follow the reference
model, the wheels of the satellite may not reach their maximum velocity, by which
control actions can become impossible. Therefore, it is necessary to set all control
torques of the reference model to zero before the wheels of the reference model reach
their maximum velocity. In our implementation we have reduced this value to 90% of
the maximum wheel velocity of the satellite. In Fig. 2 the reference model is illustrated:

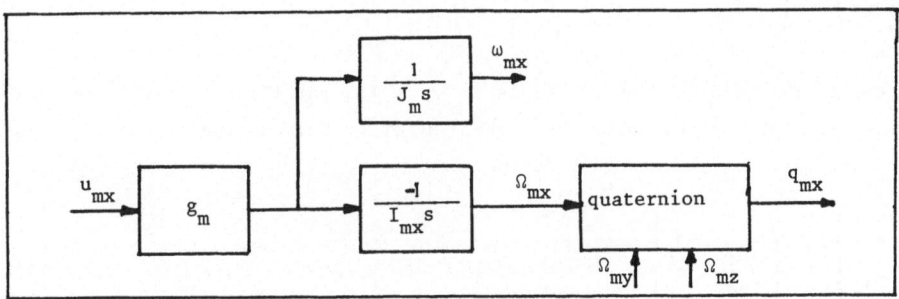

Fig. 2. Reference model (X-axis) with quaternion.

3.2. Adaptive Controller

In the preceding part a reference model has been derived that satisfies the control
requirements, like realizing a three-axes slew around the Euler axis and a nice dynamical behaviour. In this section an adaptive law will be derived to force the satellite
to follow this reference model as close as possible. So model following is the ultimate
goal of the adaptive controller, in spite of many differences between satellite and

reference model. These differences are mainly caused by differences in structure between satellite and reference model (gyroscopic coupling in the satellite, switches in the reference model), by differences in corresponding parameters (I and I_m, J and J_m, g and g_m) and also attributable to external and internal disturbances on the satellite.

Due to its success in other applications (Amerongen 1980, Landau, 1979) we have chosen the adaptive laws based on the Lyapunov stability theory as derived by Winsor and Roy. An improved integral action has been implemented and the "model updating" concept (Ten Hacken, 1976) has been applied to improve the convergence of the parameter adjustment and consequently the speed of response and energy consumption of the three axes slew.

It is assumed that the satellite is decoupled and linearized, by counteracting the gyroscopic coupling and by the approximation of the quaternion equation 4 with equation 7. Then, both the satellite and the reference model can be considered as three linear, independent systems. Using state feedback by means of K_{px} and K_{vx} and using an input v_x with gain K_{ix} to counteract for the disturbance torque T_{dx}, the linear model of the X-axis of the satellite becomes:

$$\begin{pmatrix} \dot{q}_x \\ \dot{\Omega}_x \end{pmatrix} = \begin{pmatrix} 0 & \frac{1}{2} \\ -\frac{K_{px}g}{I_x} & -\frac{K_{vx}g}{I_x} \end{pmatrix} \cdot \begin{pmatrix} q_x \\ \Omega_x \end{pmatrix} + \begin{pmatrix} 0 \\ \frac{K_{ix}g}{I_x} \end{pmatrix} \cdot v_x \tag{9}$$

Similarly, the model of the X-axis of the reference model becomes (with $K'_{mpx} = K_{mpx}/q^3_{m0}$):

$$\begin{pmatrix} \dot{q}_{mx} \\ \dot{\Omega}_{mx} \end{pmatrix} = \begin{pmatrix} 0 & \frac{1}{2} \\ -\frac{K'_{mpx}g_m}{I_{mx}} & -\frac{K_{mvx}g_m}{I_{mx}} \end{pmatrix} \cdot \begin{pmatrix} q_{mx} \\ \Omega_{mx} \end{pmatrix} \tag{10}$$

The purpose of the adaptive controller is to force the satellite to follow the reference model as close as possible. Therefore, the error between the states of the satellite and the states of the model is defined:

$$\begin{pmatrix} e_{1x} \\ e_{2x} \end{pmatrix} = \begin{pmatrix} q_{mx} - q_x \\ \Omega_{mx} - \Omega_x \end{pmatrix} \tag{11}$$

Sufficient conditions can be derived for the adjustment of the gains K_{px}, K_{vx} and K_{ix}, in order to decrease the error $(e_{1x}, e_{2x})^T$. Using the Lyapunov function as proposed by Winsor and Roy, the following adjustment laws are derived (Jongkind, 1977):

$$\begin{aligned} \dot{K}_{px} &= -\alpha_1 (P_{21}e_{1x} + P_{22}e_{2x}) \cdot q_x \\ \dot{K}_{vx} &= -\alpha_2 (P_{21}e_{1x} + P_{22}e_{2x}) \cdot \Omega_x \\ \dot{K}_{ix} &= \alpha_3 (P_{21}e_{1x} + P_{22}e_{2x}) \end{aligned} \tag{12}$$

The input v_x is not yet defined. There are no restrictions posed on it. Therefore, v_x may be chosen arbitrarily. In this case v_x is chosen to be constant and equal to one. In order to maintain stability the parameters α_1, α_2 and α_3 have to be positive and the elements p_{ij} of the matrix P have to satisfy the Lyapunov equation 13.

$$A_m^T P + PA_m + Q = 0 \tag{13}$$

with A_m the system matrix of the reference model and Q an arbitrarily symmetrical, positive definite matrix. The proposed adaptive controller is illustrated in Fig. 3.

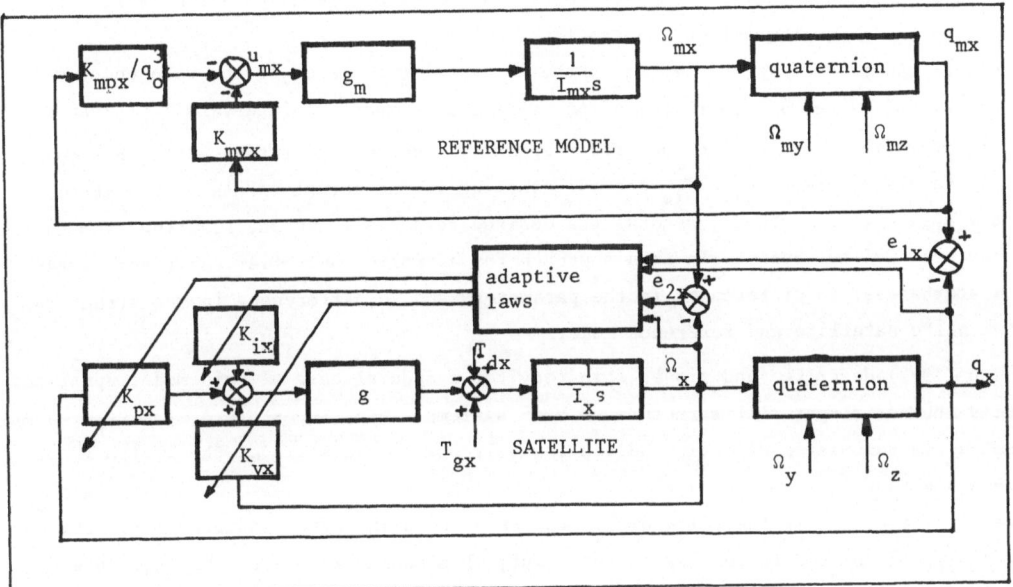

Fig. 3. Model Reference Adaptive Attitude Control (X-axis).

Another improvement can be implemented, both to speed up the responses and to decrease the energy consumption, namely in introducing the model updating concept.

3.3. Model Updating

Adaptive schemes, like the adaptive law derived in the preceding section, intent to minimize an error between the states of a reference model and the states of a system in modifying the parameters of a controller. Each time a parameter set is selected that minimizes an error between a model and a system.

Three main reasons may cause this error:

- differences in structure between model and system;
- differences in the states (due to disturbances or differences in initial conditions)
- differences in the parameter values.

Without knowing which of these reasons causes the error, the adaptive law only adjust the parameters. Even in the case the structure and the parameters of the model and of the system are equal and only the states differ, (e.g. caused by disturbances acting on the system), the parameters will be changed. After some time the error decreases and the parameters will return to their old values. This adjusting procedure, and therefore the responses of the system, can be improved by using the model updating concept.

The idea behind this model updating is to reduce the influence of old differences in the states in the error function, which error function is an input for the adaptive controller. Consequently, the actual differences between the parameters and the actual differences between the structure of a referencemodel and a system become more important. Especially these two differences have to be eliminates by the adaptive controller in adjusting the parameters of the controller.

This goal can be achieved by replacing the values of the states of the references model by the actual values of the states of the system at fixed time intervals T_{ic}. Then the reference model does not offer desired responses, independent of the system, but desired state trajectories, each T_{ic} seconds starting in the actual values of the states of the systeem. So, each T_{ic} seconds the desired response is known, starting from the actual state of the system. In this approach the adaptive controller can react faster to disturbances, to differences in the parameters and to differences in the structure between the satellite and reference model.

Landau (1979) describes and proves stability for a special case of this model updating, namely the serial-parallel structure of MRAS systems. Then, the state of the reference model x_m is calculated directly and continuously out of the state of the system x, so
$x_m = A_m x + B_m u$.
However, the choice of the value of T_{ic} can be used to introduce additional freedom in the design of the controller. As it turns out, this choice influences both the energy consumption and the response time, as illustrated in Fig. 4.

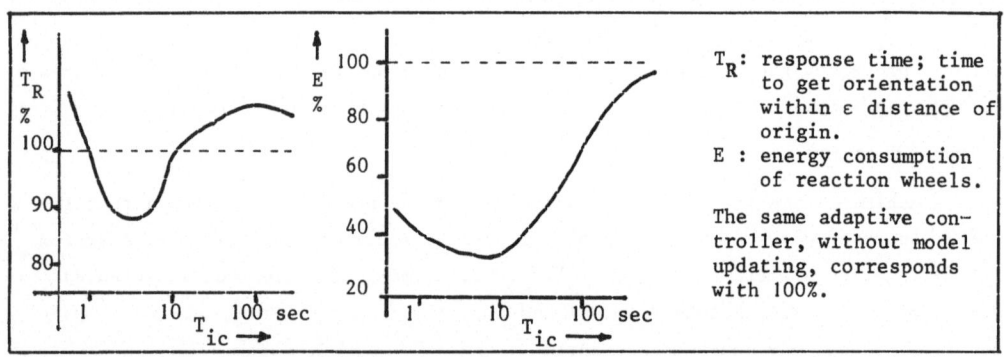

Fig.4. Response time T_R and energy consumption E as function of the model updating interval T_{ic}.

The functions T_R and E in Fig. 4 illustrate a trend. The actual value of T_R and E as a function of T_{ic}, measured with the attitude control system of the satellite, do not offer such a smooth function. This effect will be illustrated . The following results are observed in applying a small and a large angle slew:

slew	T_{ic}	0.5	1	4	5	6	8	9	10	∞	sec
small	T_R	293	137	84	82	80	78	98	73	100	%
	E	37	26	22	22	25	47	54	27	100	%
large	T_{ic}	1	4	5	6	8	9	10	25	∞	sec
	T_R	99	93	94	96	94	95	96	96	100	%
	E	40	37	33	38	45	57	44	37	100	%

In general, the optimal value of T_{ic}, to be used in the satellite described in the next section, turns out to be in the range of $\pm$ 2 to $\pm$ 6 seconds. Compared with an adaptive controller without model updating, the responses become faster (5 à 20 %) and more quiet, resulting in a saving on energy of up to 80%. Small values of T_{ic} introduce small damped oscillations. Large values of T_{ic} require much control effort to follow the reference model. Optimal values of T_{ic} reduce the energy consumption because unnecessary control effort is avoided by adapting the desired state trajectories to the actual states of the system.

It should be pointed out that the adaptive law (12) offers stability for a satellite and a reference model described by means of the equations (9) and (10). These equations are good approximations for small slews and at the end of large slews. However, large slews make the inputs and/or the wheel velocities saturated. Then equation (9) and (10) are no longer valid. Also the influence of model updating (a continuous-discreet non-linear problem) can not yet derived theoretically. Therefore, extensive simulations have to be made to study the behaviour of the proposed adaptive controller.

4. Simulation Results

Extensive simulations have been made to study the performance of the proposed attitude control system for large angle slews. The calculations are performed with a digital computer using double precision arithmetic (17 significant digits). The fourth order Runge Kutta integration method with an integration step of 0.5 seconds offers enough accuracy. For example, the maximum error in the quaternion equality (5) is 10^{-11}.

In Fig.5 a set of typical responses are shown (T_d added). Due to model updating the slew is no longer an exact slew around the Euler-axis. Still, the energy consumption is only 1/3 compared with an exact Euler-axis slew, obtained with the same controller only now without model updating. In using a reference model without additional measures to follow the Euler-axis, increases the energy consumption considerably (285 % with and 730% without model updating). The response time changes in that case also (-6% resp. +6%) compared with the proposed control scheme of this paper.

Many sensitivity studies have been made. No modification (of up to 50%) of the parameters of the satellite or of the controller affects the ability to force the satellite to follow the reference model and to realize a stable three axes slew. However, $T_{c,max}$ and ω_{max} have to be large enough to follow the reference model and to counteract the disturbance torque T_d and the gyroscopic coupling T_d.

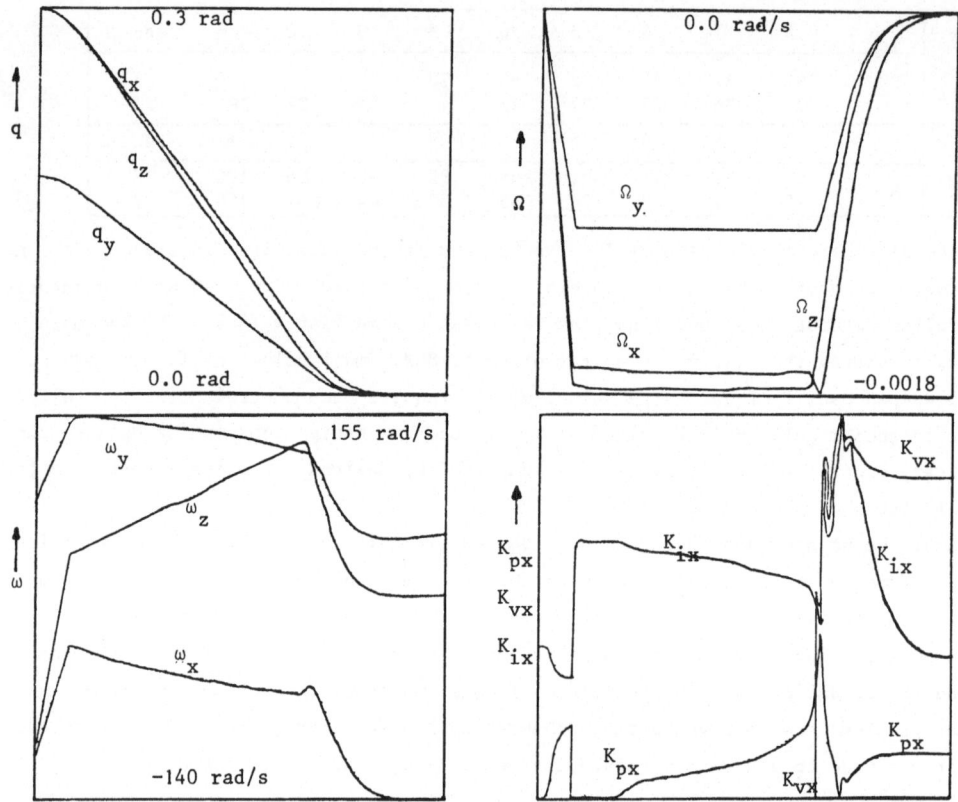

Fig.5. Responses of a large angle slew. The following parameters are used:
$I_x, I_y, I_z = 4000, 5000, 6000$ kgm^2; $I_{mx}, I_{my}, I_{mz} = 4336, 5420, 6504$ kgm^2; $J = J_m = 0.08$ kgm^2; $g = g_m = 0.01$ Nm/A; $\omega_{max} = 170$ rad/s; $T_{c,max} = 0.27$ Nm; $\alpha_1 = 10^4, \alpha_2 = 10^5, \alpha_3 = 100; p_{21} = 21, p_{22} = 626$; $\omega_n = 0.043, z = 0.92; T_{ic} = 5.; T_{max} = 500.$ s; $q(0) = (.3, .17, .3)^T; \omega(0) = (-100, 90, -90)^T$.

Conclusions

One single three-axes slew can be made with the proposed adaptive controller, even in the presence of disturbances and parameter variations. A slew around the Euler-axis offers reasonable fast and smooth responses with modest energy requirements. The model updating concept turns out to be very valuable for the adaptive attitude control system especially from the point of view of energy consumption.

References

Amerongen, J. van (1980). Model Reference Adaptive Control Applied to Steering of Ships. Proceedings International Symposium on Adaptive Systems. Bochum

Jongkind, W. (1977). Model Reference Adaptive Satellite Attitude Control using Euler's Theorem on Rotations. Master's Thesis. Laboratory for Control Engineering, Delft University of Technology.

Landau, Y.D. (1979). Adaptive Control. Marcel Dekker Inc. New York.

Mortensen, R.E. (1968). A Globally Stable Linear Attitude Regulator. International Journal of Control, Vol. 8, no2, (297-302).

Ten Hacken, G.V. (1976). Adaptive Control with the Model Reference Principle. Master's Thesis. Laboratory for Control Engineering. Delft University of Technology.

IMPLICIT REFERENCE MODEL

AND OPTIMAL AIM STRATEGY

FOR

ELECTRICAL GENERATOR ADAPTIVE CONTROL

E. Irving
Automatic Control Division
Direction des Etudes et Recherches
Electricité de France
International Symposium on Adaptive Systems
Ruhr-University Bochum 20-21 March 1980

ABSTRACT

The "steady state stability" of the French power transmission system has been
greatly improved by supplying additional signals to the voltage regulator. As usual,
the control system parameters, including additional signals, have been chosen in
order to cover all the possible operating conditions and network configurations.
These additional signals lead to an increase in the "steady state stability" domain
even with a regulator having constant parameters. Better results have been obtained
using several different adaptive control methods. The first method which has been
studied and implemented by analog hardware was based on the hyperstability approach
and introduced new concepts as series and partial reference models.

In order to master completely the time constants of the closed-loop system, a second
adaptive control method using a predictor and an implicit reference model has been
studied and implemented numerically by micro-processors. At last, a simpler and
more robust new method has been introduced : the reference model optimal aim
strategy using a very simple predictor.

I. INTRODUCTION

This paper discusses two methods of adaptive control applied to automatic voltage
control. Among the different control problems in power systems, those related to
the stability of power network seemed of great interest for the following reasons.

Studies made some years ago concerning the "steady state stability" (small variations
stability) of the French power network as well as its actual performance in certain
special conditions revealed a need for an improvement in the power network unit
voltage regulation. This fact led to the use of additional stabilizing signals

obtained, for example, from the variations of the active power output (Monville 1978).

Moreover, studies concerning the dynamic behaviour of the future power systems show new stability problems appearing. Such stability problems are essentially due to the weak "steady state stability" of the future network resulting from unbalanced developments of unit power and network short-circuit power. In fact, on the one hand, the spinning power of units is increasing faster than is the short-circuit power of network, and on the other hand, the nuclear plants can be situated far from the consumption centers. These two facts lead to a weakening of the static stability.

General studies concerning voltage regulation equipped with stabilizing devices show the best values of the parameters to be dependent on the operating conditions of the generator and the network structure. Thus, calculating the optimal values of these parameters has to be a compromise which may not be very efficient in certain restricted situations. So as to maintain efficient operation in all situations, it seems of interest to automatically adjust the values of the parameters of the voltage regulation.

At last, after the French electrical failure of last December, the controllers of the network became aware that stability of the system they are in charge of was not any more a future but a present problem. Following the decision which has been taken to equip all the future 1300 MW units with voltage adaptive control (Irving et al., 1979), the recent events involved another decision : to equip the present most powerful units of South East of France with this new type regulator.

The benefit of the last decision has been estimated to be an additional transmitted power available larger than 3000 MW from the South East to the Parisian area.

II. A DISCRETE TIME MODEL OF THE CONTROLLED SYSTEM

From the basic electrical and mechanical equations of the generator linked to the network and for each working point defined by

$V(t)$: stator voltage

$P_e(t)$: electrical active power

$Q(t)$: electrical reactive power

$X(t)$: linkage reactance of the transmission line

if the following phenomenas are neglected

- amortisseur effects

- solid iron eddy current in both axes

- eddy current losses in windings, case and associated hardware it is possible to deduce the following simplified dynamical linearized equations (see Irving et al., 1979) :

$$dy(t)/dt = Ay(t) + Bu(t) \tag{2.1}$$

$$y^T(t) = [\Delta V(t), \Delta\Omega(t), \Delta P_e(t)] \tag{2.2}$$

$$u^T(t) = [\Delta V_f(t), \Delta P_m(t)] \tag{2.3}$$

$\Omega(t)$ being the rotating speed of the unit, $P_m(t)$ the mechanical power supplied to the generator by the turbine being a disturbance input.

$\Delta V_f(t)$ the field voltage being the control input.

The notation $\Delta x(t)$ means a small variation of the variable $x(t)$ from a permanent value.

From the latter continuous time simplified model, the following discrete-time simplified model can be deduced

$$y(t + \Delta t) = \Phi y(t) + \Gamma u(t) \tag{2.4}$$

$$y^T(t) \quad = [\Delta V(t), \Delta\Omega(t), \Delta P_e(t)] \tag{2.5}$$

$$u^T(t) \quad = [\Delta V_f(t), \Delta P_m(t)] \tag{2.6}$$

$\Delta V(t)$, $\Delta\Omega(t)$, $\Delta P_e(t)$, $\Delta V_f(t)$, $\Delta P_m(t)$ are the discrete-time equivalent of the small variations defined in the continuous time case.

We recall that the purpose of this way of defining variations of the different variables is to make the linear controller work with an adequately defined linearized system. It should be noted that one of the main reasons making the voltage adaptive controller efficient, in addition to the full mastering of the dynamic of the closed-loop, is this way of defining variations.

As in the continuous time case, $\Delta V(t)$ is defined by

$$\Delta V(t) = V(t) - V_a(t) \tag{2.7}$$

$V_a(t)$ is the "moving average" of $V(t)$ defined by

$$V_a(t) = e^{-\Delta t/T} V_a(t-\Delta t) + (1-e^{-\Delta t/T}) V(t-\Delta t)$$

with the numerical choice

$$\Delta t = 0.1 \text{ s} \qquad\qquad T = 10 \text{ s}$$

$\Delta\Omega(t)$, $\Delta P_e(t)$, $\Delta P_m(t)$ are defined in the same way.

It is worth noticing two particular points.

First, $\Delta V_f(t)$ is the output of the adaptive controller, but it is $V_f(t)$ which must be injected in the excitation system of the unit. This point is solved by applying (2.7) in a reverse sense, so $V_f(t)$ can be deduced by

$$V_f(t) = \Delta V_f(t) + V_{fa}(t)$$

$V_{fa}(t)$ being defined as $V_a(t)$.

Second, $r(t)$ the reference input of the linearized system is defined by

$$r(t) = \Delta V_c(t)$$

$V_c(t)$ being the set point input of $V(t)$.

If $\Delta V_c(t)$ is defined by

$$\Delta V_c(t) = V_c(t) - V_{ca}(t)$$

$V(t)$ will present a slow drift.

To avoid this slow drift, $\Delta V_c(t)$ is defined differently by

$$\Delta V_c(t) = V_c(t) - V_a(t)$$

III. THE PRACTICAL CONTROL PROBLEM

When considering the behaviour of the unit, we have to solve several control problems we recall in the following requirements.

First requirement

In spite of the variations of $V(t)$, $P(t)$, $Q(t)$, $X_t(t)$ the behaviour of the state variables of the unit must be sufficiently damped (static stability).

Second requirement

Control the static value of $V(t)$ since its dynamic behaviour is acceptable if the first requirement is fulfilled.

Third requirement

In spite of the sudden variations of the mechanical power $P_m(t)$, such as the closing down of the unit valve after a network fault, the unit speed $\Omega(t)$ must present the least possible variations (transient stability).

Fourth requirement

The control system must work with the same efficiency for different static values of

$V_c(t)$ (set point input of $V(t)$) and $P_m(t)$ which implies both a controller working with variations of the differents signals and rejection of continuous components (linearization requirement).

Fifth requirement

As far as possible, do not use the valve unit to maintain stability because, firstly, it is not an efficient way to tackle the previous problems for thermal time constants reasons, secondly, the reliability of the unit may be endangered (nuclear units).

For these reasons, it is proposed to act uniquely on the field voltage $V_f(t)$ to fulfil the previous requirements.

IV. THE DYNAMIC BEHAVIOUR MONITORING PROPERTY

Now, examine the question of mastering the dynamic of the closed-loop : as in the continuous time case, is there a structure of the discrete time controller which permits to master completely the poles of the discrete time transfer function of the closed-loop system ?

The basic difficulties of the discrete time case are the fact that the controlled system as well as the controller present, both, a sampled time delay.

It will be seen, that in spite of these basic delays, the same arguments as in the continuous time case can be used in the discrete time case (the existence proof of a controller mastering the time constants of the closed-loop system).

Define q the shifting operator by

$$qy(t) = y(t+\Delta t)$$

and q^{-1} by

$$q^{-1}y(t) = y(t-\Delta t)$$

Equation (7.1) can be written in the form

$$y(t) = (I-q^{-1}\Phi)^{-1} q^{-1} \Gamma u(t) \tag{4.3}$$

or equivalently by

$$
\Delta(q^{-1}) \begin{bmatrix} y_1(t) \\ y_2(t) \\ y_3(t) \end{bmatrix} = \begin{bmatrix} B_1(q^{-1}) & D_1(q^{-1}) \\ B_2(q^{-1}) & D_2(q^{-1}) \\ B_3(q^{-1}) & D_3(q^{-1}) \end{bmatrix} \begin{bmatrix} u_1(t) \\ u_2(t) \end{bmatrix}
$$

$\Delta(q^{-1})$, $B_i(q^{-1})$, $D_i(q^{-1})$, $i = 1,2,3$, are polynomials given by

$$
\Delta(q^{-1}) = 1 + \sum_{i=1}^{3} \Delta_i \, q^{-1}
$$

$$
B_i(q^{-1}) = \sum_{i=1}^{3} b_{ij} \, q^{-j} \qquad\qquad i = 1,2,3
$$

$$
D_i(q^{-1}) = \sum_{i=1}^{3} d_{ij} \, q^{-j} \qquad\qquad i = 1,2,3
$$

As in the continuous time case, the most general form of linear controller is given by

$$
S(q^{-1}) \, u_1(t) = - \sum_{i=1}^{3} R_i(q^{-1}) \, y_i(t) + C(q^{-1}) \, r(t) + T(q^{-1}) \, u_2(t) \quad (4.8)
$$

with $C(q^{-1})$, $S(q^{-1})$, $T(q^{-1})$, $R_i(q^{-1})$ polynomials of the following form

$$
S(q^{-1}) = 1 + \sum_{i=1}^{n_s} s_i \, q^{-1}
$$

$$
C(q^{-1}) = \sum_{i=1}^{n_c} c_i \, q^{-1}
$$

$$
T(q^{-1}) = \sum_{i=1}^{n_T} \theta_i \, q^{-1}
$$

$$
R_i(q^{-1}) = \sum_{j=1}^{n_{ri}} r_{ij} \, q^{-1} \qquad\qquad i = 1,2,3
$$

We notice that the polynomials $C(q^{-1})$, $T(q^{-1})$, $R_i(q^{-1})$, $i = 1,2,3$, have all the term q^{-1} in factor which means that we have represented in the equation of the controller the computing time delay of one sampling interval.

As in the continuous time case, the q^{-1} transfer function is given by the following equation

$$
(\Delta(q^{-1}) \, S(q^{-1}) + \sum_{i=1}^{3} B_i(q^{-1}) \, R_i(q^{-1})) \, y_i(t) = B_i(q^{-1}) \, C(q^{-1}) \, r(t) +
$$

$$
(D_i(q^{-1}) + B_i(q^{-1}) \, T(q^{-1})) \, u_2(t) \qquad\qquad (4.13)
$$

The polynomial $\tilde{A}(q^{-1})$ defined by

$$\tilde{A}(q^{-1}) = A(q^{-1}) \, S(q^{-1}) + \sum_{i=1}^{3} B_i(q^{-1}) \, R_i(q^{-1}) \qquad (4.14)$$

representing the dynamical properties of the closed-loop, we have to find the different minimal degrees n_s, n_{ri}, $i = 1,2,3$ and the values of the terms of the polynomials $S(q^{-1})$ and $R_i(q^{-1})$, $i = 1,2,3$ such that the terms of the polynomial $A(q^{-1})$ be given, $A(q^{-1})$ being chosen a "stable polynomial".

To clarify this choice, it has been found very convenient to write the equation (4.14) in the form of a table (table 1), with arranging the different products of polynomials by monomials on separate lines, the terms of the same degrees being in the same column.

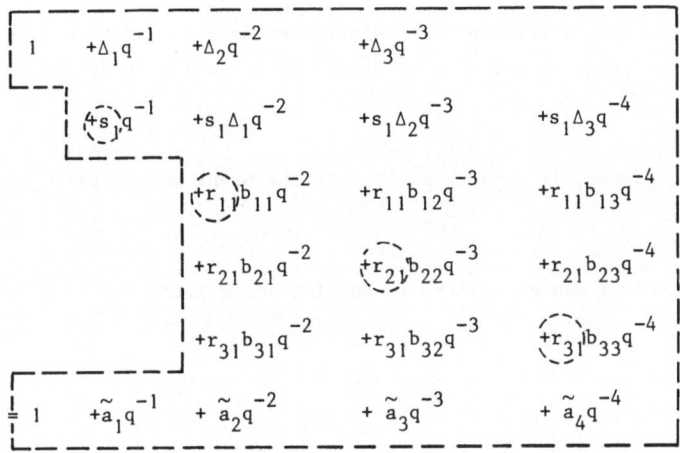

Table 1 : Choice of the controller structure

From Table 1, it appears that the following choice of the controller polynomials

$$S(q^{-1}) = 1 + s_1 \, q^{-1}$$
$$R_i(q^{-1}) = r_{i1} \, q^{-1}$$

gives enough degrees of freedom to master the dynamic properties (as far as concerning poles) of the closed-loop. Moreover, the numerical values of s_1, r_{i1}, $i = 1,2,3$ are given by the following linear system of equations

$$s_1 = a_1 - \Delta_1 \qquad (4.17)$$

$$
\begin{bmatrix} b_{11} & b_{21} & b_{31} \\ b_{12} & b_{22} & b_{32} \\ b_{13} & b_{23} & b_{33} \end{bmatrix} \begin{bmatrix} r_{11} \\ r_{21} \\ r_{31} \end{bmatrix} = \begin{bmatrix} \tilde{a}_2 - s_1 & \Delta_1 - \Delta_2 \\ \tilde{a}_3 - s_1 & \Delta_2 - \Delta_3 \\ \tilde{a}_4 - s_1 & \Delta_3 \end{bmatrix} \tag{4.18}
$$

In conclusion, knowing estimates $\hat{\Phi}$, $\hat{\Gamma}$ of Φ, Γ given by an adjustable predictor, it is possible to obtain $A(q^{-1})$ and $B_i(q^{-1})$ $i = 1,2,3$ by using (4.3) and finally, solve (4.17) and (4.18).

V. AN ADAPTIVE PREDICTOR ALGORITHM

A predictor algorithm using recursive least squares needs some additional features to avoid the different dead-locks it may encounter if it is supposed to work in an adaptive system.

Let us examine these particular aspects applied to the predictor of the simplified model of the unit.

The simplified model (7.1) can be written in the following form

$$
\begin{bmatrix} y_1(t) \\ y_2(t) \\ y_3(t) \end{bmatrix} = \begin{bmatrix} \Phi_1 \\ \Phi_2 \\ \Phi_3 \end{bmatrix} \begin{bmatrix} y_1(t-\Delta t) \\ y_2(t-\Delta t) \\ y_3(t-\Delta t) \end{bmatrix} + \begin{bmatrix} \Gamma_1 \\ \Gamma_2 \\ \Gamma_3 \end{bmatrix} \begin{bmatrix} u_1(t-\Delta t) \\ u_2(t-\Delta t) \end{bmatrix} \tag{5.1}
$$

The previous system of equations shows that the predictor problem can be split into three decoupled predictor problems which may be solved using parallel computation. So each predictor must be designed such that its output converges in the quickest way possible to the output of the following model

$$
y_i(t) = \Phi_i \, y(t-t) + \Gamma_i \, u(t-t) \tag{5.2}
$$

$$
1 \leq i \leq 3
$$

As the whole simplified model (5.1) may be unstable, it is necessary to utilize a series-parallel predictor of the following form

$$
\hat{y}_i(t) = \hat{\Phi}_i(t) \, y(t-\Delta t) + \hat{\Gamma}_i(t) \, u(t-\Delta t)
$$

$$
= x^T(t) \, \hat{p}(t)
$$

This choice of a series-parallel predictor avoids the stability problem because the equation of the output behavior error of this predictor has no dynamics dependent on the dynamics of the identified system as in the purely parallel case, see Landau (1979). with

$$x^T(t) = (y^T(t-\Delta t), \ u^T(t-\Delta t))$$

$$\hat{p}^T(t) = (\hat{\Phi}_i^T(t), \ \hat{\Gamma}_i^T(t))$$

as $p(t)$ defined by

$$p^T(t) = (\Phi_i^T, \ \Gamma_i^T)$$

can be supposed to satisfy the following state space equation

$$p(t+\Delta t) = p(t) + v(t)$$

$$y_i(t) = x^T(t) \ p(t) + e(t)$$

with $v(t)$ the parameters noise and $e(t)$ the measurement noise with the following statistics

$$E(v(t) \ v^T(t)) = \Xi(t)$$

$$E(e^2(t)) = \sigma^2$$

the predictor algorithm can simply be a Kalman predictor (Jazwinski (1970)) satisfying the following equations

$$\hat{p}(t) = \hat{p}(t-\Delta t) + (\sigma^2 + x^T(t) \ \Pi(t) \ x(t))^{-1} \ . \ \Pi(t) \ x(t)$$

$$(y_i(t) - x^T(t) \ \hat{p}(t-\Delta t))$$

$$\Pi(t+\Delta t) = \Pi(t) + \Xi(t)$$

$$- \Pi(t) \ x(t) \ (\sigma^2 + x^T(t) \ \Pi(t) \ x(t))^{-1} \ x^T(t) \ \Pi(t) \tag{5.13}$$

The particular aspect of adaptive control is the fact that the statistics $\Xi(t)$ may vary from time to time at unknown instants.

$\Xi(t)$ being unknown is generally set to zero and the (5.13) equation shows that $\Pi(t)$ will decrease continuously in this case and attain very small values.

The small values of $\Pi(t)$ involve too slow an adaptation which is a severe drawback, especially after a sudden variation of the parameters of the controlled system.

The well-known remedy which is generally applied in such a situation is the multiplication of $\Pi(t)$ by a factor $\lambda(t)$ called a "forgetting factor" satisfying the inequality

$$1 \le \lambda(t) \le \lambda_{max}$$

It has been found simple to choose $\lambda(t)$ in such a way that the trace of $\Pi(t)$, related to the speed of adaptation, remains constant.

Unfortunalety, the use of a forgetting factor may involve quasi-singularity of $\Pi(t)$ in case the data of the predictor become singular which means that the matrix $W(t)$ defined by

$$W(t) = \sum_{i=0}^{t} x(t-i) \; x^T(t-i)$$

becomes singular.

The dead-lock aspect of the last situation is the fact which can be proved that, even if the data are going back to regularity, the $\Pi(t)$ matrix remains singular. Clearly, it is not the multiplication of all terms of $\Pi(t)$ by some factor $\lambda(t)$ which can transform a singular matrix into a regular matrix.

To regularize the $\Pi(t)$ matrix, it has been found convenient to multiply its diagonal elements by a factor such as $\lambda(t)$ chosen so that the trace of $\Pi(t)$ remains constant and to multiply the other terms by a less than $\lambda(t)$, for example $0.95\,\lambda(t)$.

This particular algorithm with constant trace forgetting factor and regularization operation has been verified to be very efficient in different adaptive control problems.

VI. IMPLICIT REFERENCE MODEL ADAPTIVE CONTROL OF THE GENERATOR

Having defined ways of estimating the controlled system and computing a controller monitoring the dynamic properties (as far as poles and static gain are concerned) of the closed loop system, it is possible to use an implicit reference model adaptive control method.

Utilization of an explicit reference model adaptive control method is impossible because the zeroes of the stator voltage on field voltage transfer function are complex and poorly damped.

Before introducing the implicit reference model adaptive control methods, it is worth recalling that explicit reference model adaptive control methods avoid explicit identification of the controlled system in order to obtain convergence of the behaviour error without assuming the convergence of the internal parameters of the identifier.

The hypothesis of convergence of internal parameters has been eliminated in order to avoid the necessity, to satisfy it, of injecting an extra excitation in the closed-loop system. This excitation being harmful to unit reliability, would rarely be accepted by people in charge of the plants (especially in nuclear plants).

The implicit model reference adaptive control methods reintroduce identification as a basic tool for designing adaptive systems, but without assuming convergence of the internal adjustable parameters of the identifier. It is only assumed that the output of the identifier converges to the output of the identified system. This last restricted hypothesis does not imply the necessity of an extra excitation. In short, what we assume now, is convergence of what we may call the "external behaviour error" of the identifier. For this last reason, we call "predictor" this special kind of identifier.

The interest of this new scheme is due to the fact that we can prove convergence of the output of the controlled system to the output of an implicit reference model with only the previous restricted hypothesis, so without assuming extra excitation.

Before carrying out this proof of convergence, let us examine on Fig. 1. the scheme of the implicit model reference adaptive control methods.

From Fig. 1, it is worth noticing that the loop composed of the adjustable controller is not closed on the controlled system as it is usually done, but on the adjustable predictor. This last feature will allow to obtain a stability proof of the output of the controlled system.

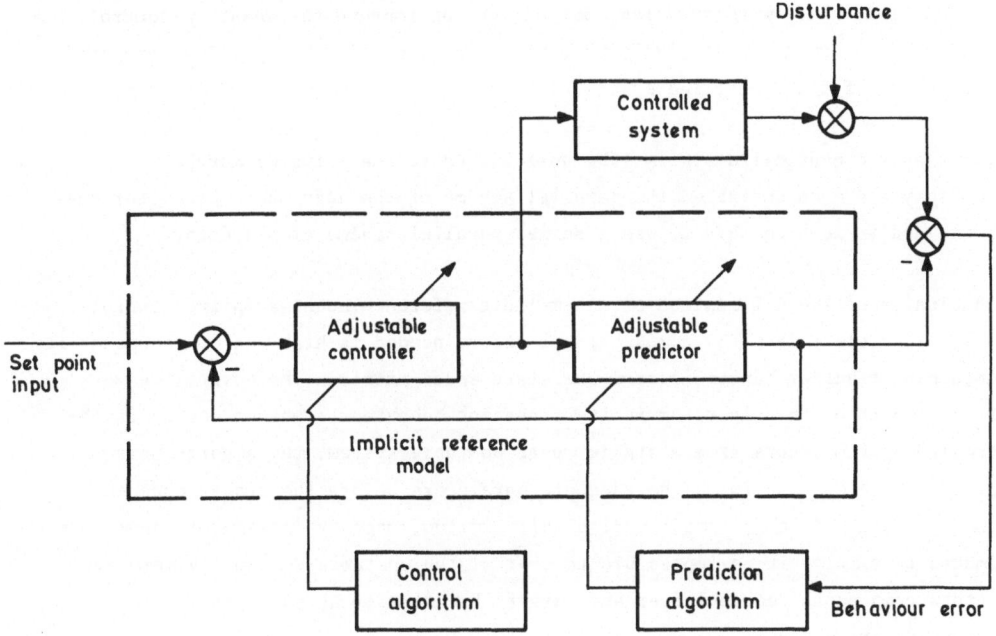

Fig. 1 Implicit reference model adaptive control system

This proof of stability of the implicit reference model methods is sketched next in the two following parts.

First, we assume that the predictor algorithm is such that the output of this predictor converges to the output of the controlled system.

Second, we assume that the controller adjustment algorithm, taking the internal parameters of the predictor in consideration, will give to the adjustable controller such parameters that the closed-loop composed of the adjustable predictor and the adjustable controller be a stable system with given time constants, which is always possible as been proved previously. Then, this closed-loop is equivalent to a reference model, which being not effectively realized is called "implicit". From this second hypothesis, it can be deduced that the output of this implicit reference model is stable.

The conclusion is straightforward : as the output of the controlled system is equal to the sum of the output of a stable system (output of the implicit reference model) and of an another variable converging to zero (behaviour error of the predictor), so this output is stable with given time constants.

It is now clear that, with this last method, we can improve the adaptive control methods used in Irving et al (1979) by maintaining constant the dynamical properties of the output of the controlled system.

In fact, as the controlled system (the unit linked to the network) may be oscillatory and even unstable, the parallel scheme of the adjustable predictor does not work and it is necessary to use a series-parallel scheme of predictor.

As conclusion of the utilization of an implicit reference model adaptive control method, it is clear that the computational effort needed is high due to the necessity of obtaining transfer functions from the state space matrices. Moreover, the most worrying aspect of the algorithm is its transient behaviour : when a change in the controlled system occurs from a stable to an unstable system, the algorithm is completely unable to estimate the adequate parameters before the state components become too large. As the control variable is limited, when the parameters have been estimated to slowly, it is impossible to control the unstable system any more and the state components become larger and larger. Practically at this instant, the generator is tripped off.

The remedy to such situation has been found to use what is defined as a "conservative politics" : using the fact that the impulse response at the origin of time is always positive as indicated in Fig. 2, if a derivative of the controlled variable is

231

measured bigger than the maximum derivative in the case the closed-loop system is
perfectly adapted, it is decided to counteract this dangerous situation in the
quickest way possible. Then, the adequate sign bang-bang control is produced. With

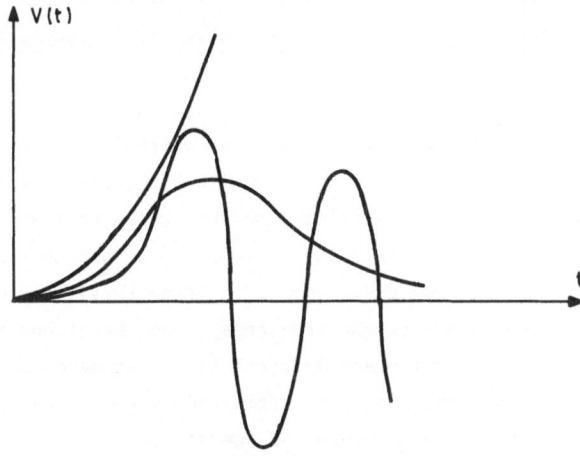

Fig. 2 Impulse responses of the stator voltage

this bang-bang control two interesting effects are obtained :

- first, the state-space components are maintained in a controllable domain,

- second, as the input of the estimated system becomes "richer", the system is
quickly estimated.

As is shown in the numerical results, the implicit reference model adaptive control
method with the latter "conservative politics" works pretty well. Nevertheless,
it happens that in some restricted situations even this "ad hoc" combination fails.
So the need for a simpler and more robust control algorithm appears.

VII. AN ADAPTIVE REFERENCE MODEL OPTIMAL AIM STRATEGY

The optimal aim strategy, fully described in Barnard (1976), seems very attractive as
an alternative control algorithm for the unit voltage adaptive control due to the
following reasons.

Why the implicit reference model control method has an unsound behaviour in the
transient phases is due to the following aspect of the method: as there are no
external excitations on the controlled system, only the convergence to zero of the

prediction error can be proved and not the convergence of the estimated parameters
to the true ones. In the transient phase, it may happen that the estimated parameters,
being completely false, give a very quickly increasing control variable and,
moreover, with the wrong sign. For the reasons explained previously, this latter
situation is not acceptable in case of an unstable controlled system.

In short, the implicit reference model adaptive control algorithm is unsound because
the quality criterium of the estimator algorithm which is the convergence of the
external prediction error, is not compatible with the quality criterium of the
controller : determination of internal parameters chosen so that the parameters of
the closed-loop system composed with the estimator and this controller has the
parameters of a stable and damped system. As we have never assumed that internal
estimated parameters converges to the true ones, good operations of the implicit
scheme happen after a transient phase when the false estimated parameters induce
rich enough excitation on the controlled estimated system so that these estimated
parameters arrive in the adequate region of operation.

Contrary to the implicit scheme, the quality criterium of optimal aim strategies is
directly compatible with the quality criterium of the predictor : a control is chosen
directly so that the future state space vector $y(t+\Delta t)$ be at a minimum distance $\mathcal{D}$
from a given target $y_c(t+\Delta t)$.

In the case of the discrete time model of the generator defined by (2.4), the
distance $\mathcal{D}$ can be chosen as the euclidian norm of $y_c(t+\Delta t) - y(t+\Delta t)$, so $\mathcal{D}$ is given
by

$$\mathcal{D} = [y_c(t+\Delta t) - y(t+\Delta t)]^T [y_c(t+\Delta t) - y(t+\Delta t)] \tag{7.1}$$

$\mathcal{D}$ is minimized simply by the following choice $\hat{u}(t)$ of $u(t)$

$$\hat{u}(t) = (\Gamma_1^T \Gamma_1)^{-1} \Gamma_1^T [y_c(t+\Delta t) - \phi y(t)] \tag{7.2}$$

Indeed, the choice of the control variable is slightly more complicated.

First, as $u_1(t)$ is the control input (field voltage) and $u_2(t)$ is the disturbance
input (mechanical power) the equation (2.4) can be written in the following form

$$y(t+\Delta t) = \phi y(t) + \Gamma_1 u_1(t) + \Gamma_2 u_2(t) \tag{7.3}$$

Second, as in the 4 §, it is necessary to consider the computation delay of one
sampling time interval Δt. The consequence of the computational delay being the
necessity of aiming at the target $y_c(t+2 \Delta t)$ with the control variable $u_1(t+\Delta t)$.

Then, $y(t + 2 \Delta t)$ is given by

$$y(t+2 \ \Delta t) = \phi y(t+\Delta t) + \Gamma_1 u_1 (t+\Delta t) + \Gamma_2 u_2 (t+\Delta t) \tag{7.4}$$

Suppose

$$u_2 (t+\Delta t) = u_2 (t) \tag{7.5}$$

$\mathcal{D}$ being defined by

$$\mathcal{D} = [y_c (t+2 \ \Delta t) - y(t+2 \ \Delta t)]^T \ [y_c (t+2 \ \Delta t) - y(t+2 \ \Delta t)] \tag{7.6}$$

$\hat{u}_1 (t+\Delta t)$ is given by

$$\hat{u}_1 (t+\Delta t) = (\Gamma_1^T \ \Gamma_1)^{-1} \ \Gamma_1^T [y_c (t+2 \ \Delta t) - \phi y(t+\Delta t) - \Gamma_2 u_2 (t)] \tag{7.7}$$

with $y(t+\Delta t)$ given by (7.3).

To define completely the algorithm, it remains to define the "moving target" $y_c (t+2 \ \Delta t)$. The idea of choosing a moving target is due to the fact that a fixed target may give some dead-lock situations as will be illustrated by the following second order state space discrete-time system

$$\begin{cases} x_1 (t+\Delta t) = x_1 (t) + x_2 (t) & (7.8) \\ \\ x_2 (t+\Delta t) = x_2 (t) + u(t) & (7.9) \end{cases}$$

$x_1 (t)$ may be considered as the position of a vehicle, $x_2 (t)$ the speed and $u(t)$ the moving force as a control input.

Suppose

$$t = 0 \qquad x_1 (t) = 0 \qquad x_2 (t) = 0$$

the control variable $u(t)$ is chosen so that the state at $t+\Delta t$ be at the minimum distance of the target x_c defined by

$$x_c^T = (L \ , \ 0) \tag{7.11}$$

from (7.2)

$$\hat{u}(t) = [0 \quad 1] \begin{bmatrix} L \\ 0 \end{bmatrix} = 0 \tag{7.12}$$

which shows that the previously defined target will never be attained.

This dead-lock situation is explained geometrically on Fig. 3 where the variation vector $\Delta x(t)$ of the state vector $x(t)$ defined by

$$\Delta x^T (t) = [x_1 (t+\Delta t) - x_1 (t), \ x_2 (t+\Delta t) - x_2 (t)] \tag{7.13}$$

is represented.

Evidently, at the time origin, the variation vector is perpendicular to the aiming vector $\Delta x_c(t)$ defined by

$$\Delta x_c^T(t) = [x_{1c} - x_1(t), \ x_{2c} - x_2(t)] \tag{7.14}$$

As the control amplitude is the modulus of the variation vector, the fact that the vehicle cannot start is explained.

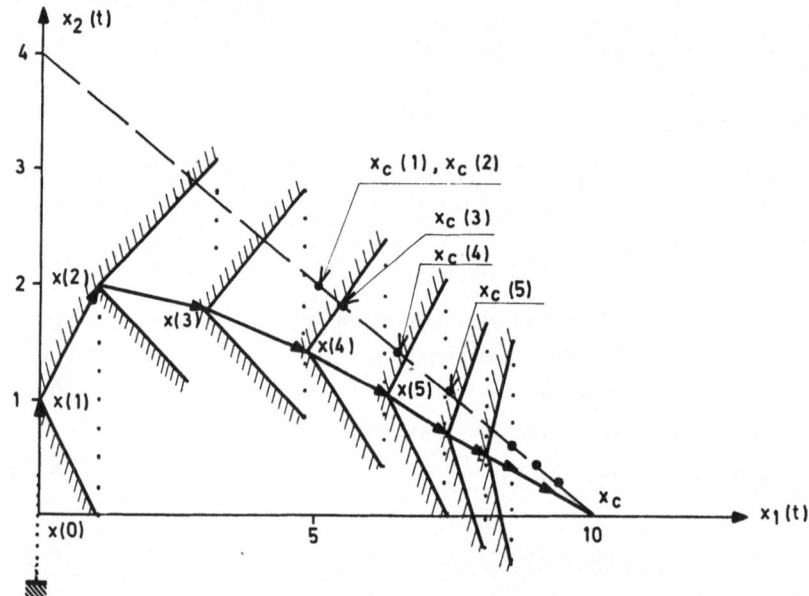

Fig. 3 A reference model moving target optimal aim strategy

A way to get out from this latter dead-lock situation is to define x_c as a moving target $x_c(t)$ going towards the final target x_c defined by (7.11). In the case of Fig. 3, the moving target is located on the dashed straight line containing the final x_c, its abscissa being given by

$$x_{c1}(t+\Delta t) = (L + x_1(t))/2 \tag{7.15}$$

As the moving target is on the dashed line, $x_{c2}(t)$ is given by

$$x_{c2}(t) = - \alpha(x_{c1}(t) - L) \qquad (\alpha = 0.4) \tag{7.16}$$

Then $\hat{u}(t)$ is given by

$$\hat{u}(t) = x_{c2}(t) - x_2(t)$$

$$= - \alpha[(L + x_1(t))/2. - L] - x_2(t) \tag{7.18}$$

Replacing $u(t)$ by $\hat{u}(t)$ in (7.9) gives the following matrix equation

$$
\begin{bmatrix} x_1(t+\Delta t) \\ x_2(t+\Delta t) \end{bmatrix} = \begin{bmatrix} 1 & 1 \\ \alpha/2 & 0 \end{bmatrix} \begin{bmatrix} x_1(t) \\ x_2(t) \end{bmatrix} + \begin{bmatrix} 0 \\ \alpha L/2 \end{bmatrix} \tag{7.19}
$$

This latter system is stable if

$$
0 < \alpha < 2 \tag{7.20}
$$

At the equilibrium point

$$
x_1(t) = L \tag{7.21}
$$

$$
x_2(t) = 0 \tag{7.22}
$$

As a conclusion of the results obtained in the latter example, it is proved that a moving target optimal aim strategy is equivalent to a constant state-space feedback (equation 7.18), when the limitations on the control variable are not attained, so can be made stable by an adequate choice of the characteristics of the moving target. Moreover, the choice of the equation (7.15) for the abscissa of the moving target is equivalent to the technique of the series reference model with time constant 0.7 Δt and of a parallel reference model with a unity gain and an input L.

The difference with the reference model technique is the fact, which is the key to the efficiency of the method, that an acceptable behaviour is obtained simultaneously for both components of the state-space vector by the least squares minimization of the "aiming error".

As a reference model moving target optimal aim algorithm can be made equivalent to a constant state-space feedback, it is understandable that this control method works also in the case of non-minimum phase system or system with oscillatory zeroes.

As a consequence of the last remark, the idea of a reference model moving target optimal aim strategy can be applied for the generator voltage control.

The following choice of the components of the moving target has been shown to be particularly adequate in the case of this latter problem.

This choice is defined by

$$
y_{c1}(t+2\,\Delta t) = \alpha y_1(t) + (1-\alpha)\, r(t) \tag{7.23}
$$

$$
y_{c2}(t+2\,\Delta t) = -\,(r(t) - \alpha y_1(t)) \tag{7.24}
$$

$$
y_{c2}(t+2\,\Delta t) = u_2(t) \tag{7.25}
$$

$$\alpha = e^{-2 \Delta t/\tau_R} \tag{7.26}$$

Equation (7.23) is simply a series-parallel reference model on the first component, τ_R being the desired time constant of the reference model.

Equation (7.24) is the idea of forcing the moving target to be on a plan joining the final target y_c defined by

$$y_c^T = [r(t), \quad 0, \quad u_2(t)]$$

Equation (7.25) is the realisation of the equilibrium in permanent operations where the electrical active power $(y_3(t))$ is equal to the mechanical power $(u_2(t))$.

It is interesting to note that the unity factor of the term $r(t) - \alpha y_1(t)$ of equation (7.24) has been found by comparing the respective evolutions of $y_1(t)$ towards its set-point value and of $y_2(t)$ towards its equilibrium value in the case of the implicit reference model perfectly adjusted, see Fig. 4 and 5.

Indeed, the equations of the moving target which has been implemented are slightly different from the (7.23), (7.24) and (7.25) equations due to the linearization procedure described in the second chapter.

In fact, the variables $y_1(t)$, $y_2(t)$, $y_3(t)$, $u_1(t)$, $u_2(t)$ represent small variations of the variables $V(t)$, $\Omega(t)$, $P_e(t)$, $V_f(t)$, $P_m(t)$ relatively to their "moving average" values, but it is desirable that all the optimal aim strategy be applied on the actual values of the different variables instead of their variations. Then, the (7.23), (7.24) and (7.25) equations must be replaced by

$$V_c(t+2 \Delta t) = \alpha V(t) + (1-\alpha) V_c(t) \tag{7.28}$$

$$\Omega_c(t+2 \Delta t) = - (V_c(t) - \alpha V(t)) \tag{7.29}$$

$$P_{e_c}(t+2 \Delta t) = P_m(t) \tag{7.30}$$

Define by $V_{ca}(t)$, $\Omega_{ca}(t)$, $P_{ea}(t)$, $V_{fa}(t)$, $P_{ma}(t)$ the moving average values of the corresponding variables, using the same idea as previously, it is necessary to aim at the target defined by (7.28), (7.29) and (7.30) in the least squares sense by the predicted values of the actual variables so by $y(t+2 \Delta t) + y_a(t+2 \Delta t)$. As the time constant T of the moving average filter is large

$$y_a(t+2 \Delta t) \sim y_a(t)$$

and $\hat{u}_1(t+\Delta t)$ is defined by

$$\hat{u}_1(t+\Delta t) = (\Gamma_1^T \Gamma_1)^{-1} \Gamma_1^T [x_c(t+2\ \Delta t) - \phi y(t+\Delta t) - \Gamma_2 u_2(t) - y_a(t)] \quad (7.32)$$

with

$$x_c^T(t+2\ \Delta t) = [V_c(t+2\ \Delta t), \Omega_c(t+2\ \Delta t), P_{e_c}(t+2\ \Delta t)] \quad (7.33)$$

instead of (7.7)

VIII. NUMERICAL EXPERIMENTS

To appreciate the improvements brought by the reference model adaptive optimal aim strategy on the implicit reference model adaptive control method, numerical simulations have been performed with both methods in the same conditions for variations of the parameters of the controlled system. These environmental conditions can be chosen as the hardest to be fulfilled by an adaptive system : the controlled system becomes suddenly unstable with a very quick increase of the state components, the actual technological limitations on the control variable being effectively represented. Fig. 4 shows how an implicit reference model adaptive control using conservative politics behaves. The first curve is the control variable $V_f(t)$ and the other curves, the variables $V(t)$, $\Omega(t)$, $P_e(t)$. $P_m(t)$ being kept constant. This conservative politics is slightly more complicated than a bang-bang control and operates for 2 seconds before adjustment by the implicit reference model method gives acceptable derivatives.

Fig. 5 shows that the reference model adaptive optimal aim strategy behaves naturally more efficiently : a good control variable is obtained in 0.5 second without the latter conservative politics.

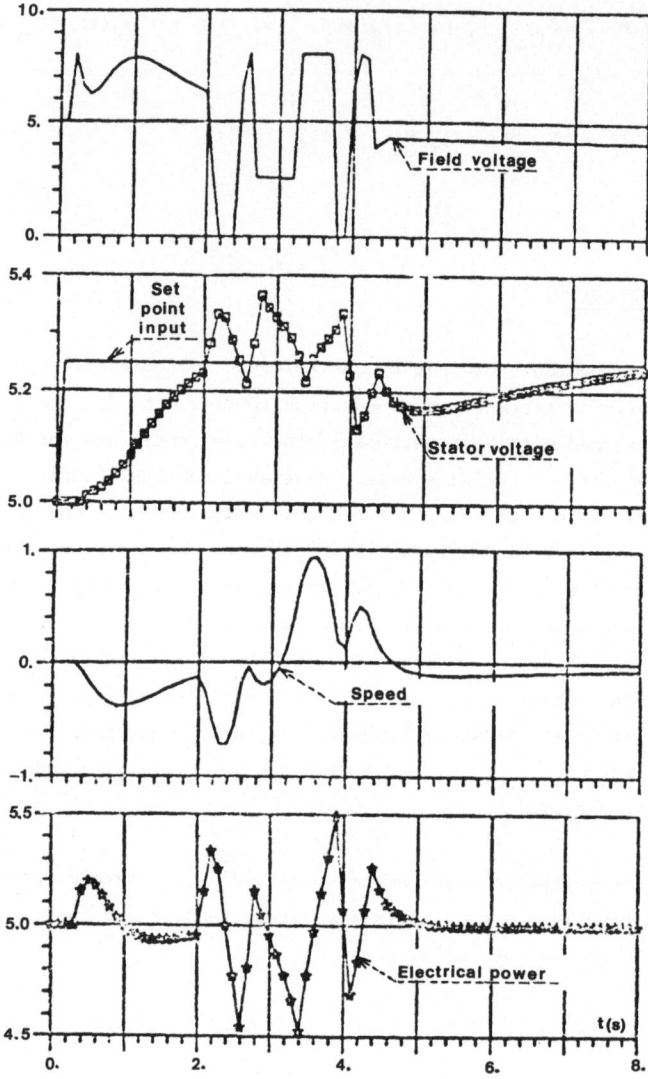

Fig. 4 Performances of the implicit reference
model adaptive control method with conservative politics
(Sudden change from stable to unstable controlled system
occurs at time 2 seconds)

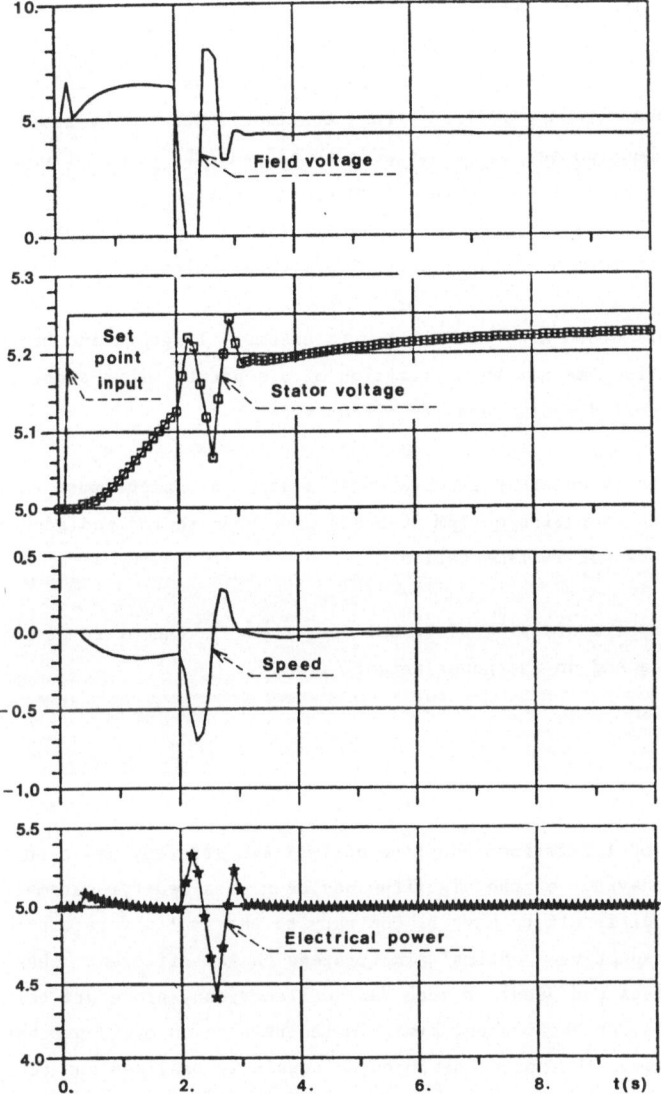

Fig. 5 Performances of the model reference
adaptive optimal aim strategy without conservative politics
(Sudden change from stable to unstable controlled system
occurs at time 2 seconds)

IX. MICRO-PROCESSOR IMPLEMENTATION

The model reference adaptive optimal aim strategy method has been implemented on a 16 bit micro-processor (Texas 9900),see Benejean (1980). Due to the sampling time interval which has been chosen to be 0.1 of a second and the computational time necessary to execute programmed 32 bit arithmetical floating-point operations which varies from 200 to 400 microseconds, parallelization of the estimation algorithm has been performed in three central processor units.

The adaptive control method has been written in assembly language and has been checked in a closed-loop manner by utilization of a separate micro-processor simulating the controlled system and its disturbances.

The numerical results which are presented in this paper are programmed in FORTRAN 32 bit floating-point operation on IBM 3033 and have been reproduced accurately by the latter micro-processor implementation.

The next steps in experiments will be actual operations on the micro-network of Electricity of France and on the power network itself.

X. CONCLUSIONS

Utilization of the model reference adaptive optimal aim strategy has been very successfull. Safe behaviour of the algorithm has been obtained for parameters and output disturbances difficult to master. Contrary to the implicit reference model control method, the adaptive optimal aim strategy has an efficient behaviour during transient phases and does not need "ad hoc" emergency procedure as conservative politics. At the present time, implementation on micro-processor of the methods described in this paper has been successfully realized and are to be experimented on the real French power network giving the hope to another improvement in utilization of electrical power.

XI. REFERENCES

Barnard, R.D. (1976). Optimal Aim Control Strategies Applied to Large-Scale Non-linear Regulation and Tracking Systems. IEEE Trans. on circuits and systems, vol. CAS. 23 n° 12 December 1976.

Benejean, R. (1980). Realisation sur micro-processor de méthodes de commande adaptative pour les groupes turbo-alternateur. Note E.D.F. (to be published).

Bonnard, G. (1973). Micro-réseau. Techniques de l'Ingénieur art. D 633, January 1973.

Irving, E. et al. (1979). Improving Power Network Stability and Unit Stress with Adaptive Generator Control Automatica Vol. 15, pp. 31-46.

Jazwinski, A.H. (1970). Stochastic Processes and Filtering Theory New York Academic Press 1970.

Landau, I.D. (1979). Adaptive Control : the Model Reference Approach M.D.I. New York.

Monville, J.P. (1978). Dispositif permettant d'améliorer la stabilité statique des turbo-alternateurs. Bull. D.E.R. Electricité de France, Série B - réseaux électriques.

ADAPTED REGULATOR FOR THE EXCITATION OF LARGE TURBOGENERATORS

P. Bonanomi, G. Güth
Brown Boveri, Baden/Dättwil, Switzerland

ABSTRACT

An adapted regulator for the control of large turbogenerators is presented. The compensator consists of three feedback channels, the resulting signal being applied to the rectifier bridge of the excitation system. Because of the varying operating conditions and the possible disturbances in the power system, improved performance is obtained by adapting the gains of the feedback loops. The proper adjustments are selected using a look-up table of precomputed values. This adaptation scheme is implemented on a microprocessor. Experiments on a model machine show the performance of the regulator. Owing to the robust design of the control scheme, the regulator may readily be applied in practice.

INTRODUCTION

The excitation of a generator may in principle be controlled manually by the plant operator. Automatic regulators have been introduced a long time ago, however, and improvements in several respects have been achieved. The original purpose of the feedback arrangement was voltage regulation. This basic requirement is important for maintaining stable synchronous operation of the generators in the power system, and for controlling the voltage supplied to customers. Further advantages have been obtained since static (power semiconductor) excitation systems have been used. Through these fast acting devices, the regulator may contribute to the transient stability after faults (keeping the generator from falling out of step), and to the damping of electromechanical rotor oscillations. These features are provided by many commercial regulators nowadays [1].

The adapted regulator presented in this paper improves the performance one step further. The main result is the good damping provided for a wide region of operating points, and the smooth voltage regulation. The improvement is particularly meaningful for turbogenerators, as they are more difficult to control than hydraulic units.

GENERATOR AND POWER SYSTEM MODELS

A variety of mathematical representations for the generator may be found in the literature. For regulator design, a low-order model is adequate. The 7th order (non-linear) representation based on Park's equations is adopted here [2]. Because of the high order and excessive complexity of a complete power system, a model with one machine is considered at the design stage. Multi-machine models are used later for validating the regulator design with time simulations or eigenvalue analysis. The one-machine system is shown in Fig. 1. The generator is connected to the "infinite bus" V_∞ over two transmission lines with circuit breakers. V_∞ is an ideal 3-phase voltage source, and the lines are represented by pure reactances. X_e denotes the equivalent reactance between the generator and V_∞. This configuration corresponds to the realistic case of a remote power station supplying a distant area like a city.

Combining the 7th order representation of the generator with the network equations (Fig. 1), the order of the model can be reduced to 5 by a realistic simplification. During short circuits in the network, the appropriate changes must be made on the reduced model.

The equations of the power system models may be linearized, which is useful for analysing steady state operating situations.

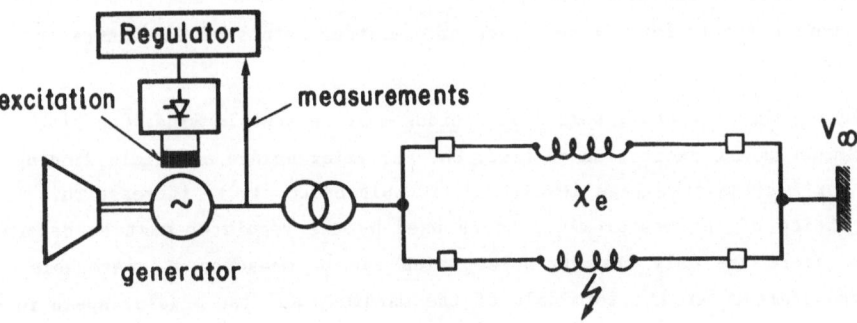

Fig. 1: Power system model

REGULATOR REQUIREMENTS [3]

The control quality is assessed in terms of the steady state behaviour and the dynamic performance in the presence of disturbances. Two kinds of disturbances must be consid-ered:

a) short circuits in the network, line switchings.

b) changes of operating point (Power, voltage, network parameters).

A typical short circuit disturbance consists of a sequence of events: short circuit on the line - circuit breakers open to isolate the faulty line - circuit breakers reclose after about 0.4 second and remain closed if the fault no longer exists. Line switchings may occur independently from short circuits, to control voltage and power flow in the network. Changes in operating point are due to either operator action or events in the power system. The operator may vary the input of mechanical power or change the voltage setpoint. Voltage and power are influenced by external factors as well (change of network topology, load variations).

Having outlined the disturbances to be considered in the design of the regulator, the <u>performance criteria</u> are now given:

1) Regulation of the generator terminal voltage:
   - smoothness (no ripple during steady state operation)
   - speed of response (in order to avoid overvoltages after load or topology changes)
   - accuracy
2) Ability to keep synchronism after a fault. This requirement means simply that the excitation voltage should be at the maximum during faults and during dangerous rotor accelerations. The excitation may return to normal after the first peak of the rotor swing.
3) Damping of rotor oscillations
4) All above criteria for a wide region of operating points of the generator (power loading, voltage)

When designing the regulator, some <u>restrictions</u> must be kept in mind. One of these is the limitation of the excitation voltage. The rectifier bridge and field windings are designed for a certain voltage level, and it would be costly to increase this limit.

The choice of the measurements to be used by the regulator must be determined carefully. There are only few variables which can be measured at reasonable cost, such as measurements at the terminals of the machine, and the angular speed in some cases.

It should be noted also that the generator-turbine shaft is an elastic body with weak damping. Any torsional oscillations in this shaft are dangerous, and can lead to a costly shaft failure (one such incident happened recently). It has been demonstrated in practice that the voltage regulator has the potential of exciting these torsional modes, in particular when measurements of angular speed are used.

Further aspects to be considered are reliability and cost of the regulator. Reliability does not only apply to the hardware but also to the actual control strategy. The cost is made of three components: the initial design, the hardware modules and the final design for each specific application.

REGULATOR DESIGN

The manifold requirements and conditions imposed on the regulator put severe restrictions on the design of the control system. Careful investigations are required when using modern control schemes. Many attempts to apply such new techniques have resulted in just another impractical regulator. Extravagant claims on the performance of a regulator should be considered with care, often being obtained from false assumptions about generator model or regulator requirements.

The adapted regulator presented here is based on the presentday conventional regulator manufactured by BBC (Fig. 2). By a judicious design of the parameters |4,6| very effective performance can be obtained with this analog regulator. For reliability reasons, the voltage feedback is designed to produce adequate control even in the absence of the power and frequency feedback. P- and ω-feedback are regarded as additional signals, the purpose of which is to produce damping (stabilization) of rotor oscillations.

Since the dynamic interaction between the voltage feedback and the stabilizing feedback loops is low, the PID compensator may be designed separately. The PID parameters are determined by making time simulations (step responses) and using Bode Plots. The gains for the power and frequency feedback can be designed by optimizing the time response after a selected disturbance [6] or using the D-decomposition technique [4,5]. The control performance obtained by the first method is nearly equal to an optimal bang bang strategy [6]. D-decomposition (or domain separation) is a method based on the characteristic equation of the linearized model. It produces curves of constant damping in the plane of the compensator gains G_1 and G_2. The method can be combined with an optimization procedure to generate optimal gains (optimal damping of the dominant poles) |1|. The gains designed with the linearized model are also adequate for controlling large disturbances, as demonstrated in [6].

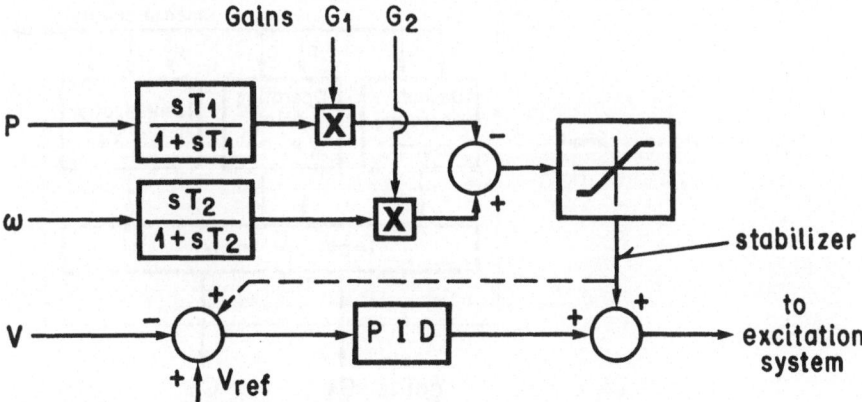

Fig. 2: The analog regulator (presentday BBC regulator).
          --- : alternative connection required by some customers

ADAPTATION

Adaptation of the gains G_1 and G_2 (Fig. 2) is motivated by the following observations:

1) The adequate gains for the damping of rotor oscillations depend on the <u>operating point</u> of the generator, due to the nonlinearities in the system. The operating point itself may vary in time, depending on the loading, voltage and network topology.

2) The power and frequency feedback disturb the control of the terminal voltage in the steady state. The gains G_1 and G_2 should be kept as low as possible at any time.

It appears that the gains should be selected depending on the following two factors:

1) <u>Steady state operating point</u>

Because of the nonlinear characteristic of the generators, the dynamic behaviour around the equilibrium point depends on the operating point, as mentioned above. Fortunately, only three quantities suffice to define the operating point: active and reactive power <u>P and Q</u> respectively, and the <u>reactance X_e</u> of the network. P and Q are measurable directly, while X_e can be identified from local measurements. When the three quantities are known, the linearized model of the generator is fully defined (assuming $V_\infty = 1$ p.u.).

2) <u>Operating regime</u>

Two situations are considered: <u>steady state operation</u> and <u>transient operation</u> (short circuit, rotor oscillations). In steady state, the gains G_1 and G_2 should be <u>reduced</u> for better voltage control, while they may be larger during transient operation.

Fig. 3:   The adaptation scheme

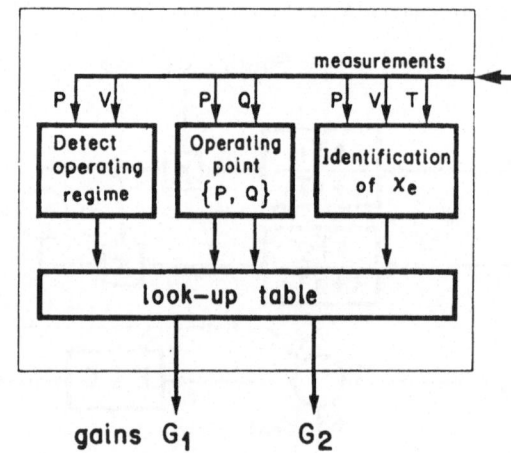

The resulting adaptation scheme is shown in Fig. 3. The entries of the look-up table are computed off-line. The design techniques described in the previous section are used. A set of about 5 gain pairs appears to be adequate in practice. G_1 and G_2 are adjusted whenever major events occur or the parameters have drifted significantly. Updating of the gains is only permitted once every few seconds in order to secure the stability of the adaptation. The identification of X_e is based on a simple curve fitting procedure. The algorithm, which has been tested in the field, is described in [7].

The adaptation scheme (Fig. 3) is implemented on a digital microprocessor. The gains G_1 and G_2 are applied to the analog regulator of Fig. 2.

RESULTS

A small scale power system (configuration of Fig. 1) was built at Brown Boveri, Baden, for designing and testing the regulator. The model generator is a 7.5 KVA replica of a 600 MVA turbogenerator. The following results were obtained.

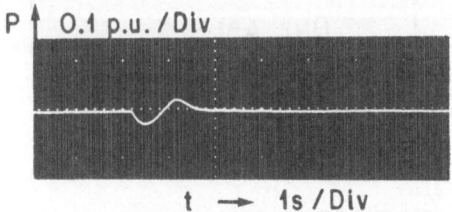

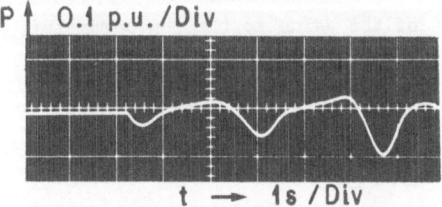

Fig. 4: stable case   Fig. 5: unstable case
$G_1 = 20$ $G_2 = 2.5$ $G_1 = 20$ $G_2 = 0$

Disturbance: Step disturbance on the voltage setpoint.
Operating point: $P = 0.7$ p.u., $Q = 0.3$ p.u., $X_e = 1.2$ p.u.

Fig. 6:
Self-induced oscillations.
Event: gains switched from $G_1 = 20$,
$G_2 = 0$ to $G_1 = 20$, $G_2 = 2.5$.
Operating point: $P = 0.7$ p.u.,
$Q = 0$ p.u., $X_e = 0.5$ p.u.

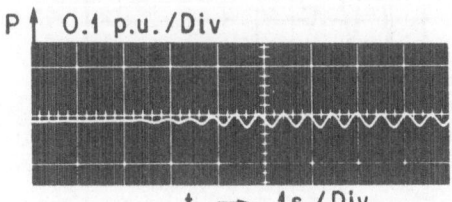

The need for adaptation at <u>critical operating points</u> is demonstrated in Fig. 4,5 and 6. Two sets of gains are used, each designed for a specific operating point. Fig. 5 and 6 show that the gains designed for one case may not be used in the other. When the wrong gains are used, instability occurs after a disturbance (Fig. 5) or in the form of self-induced oscillations (Fig. 6).

Fig. 7 shows the region of <u>stable operating points</u> for two typical values of X_e. The stability boundary is evaluated using the linearized model. The gains of the unadapted regulator are the result of a <u>practical compromise between the various performance criteria</u>.

Fig. 8 shows a boundary situation of Fig. 7. The opening of a transmission line (Fig. 1) causes the reactance X_e to double from 0.3 p.u. to 0.6 p.u. The steady state operating point after switching is P = 0.8 p.u., Q = 0.2 p.u. The response is badly oscillatory. Adaptation of the gains to the new parameter X_e produces the well-damped response in Fig. 9.

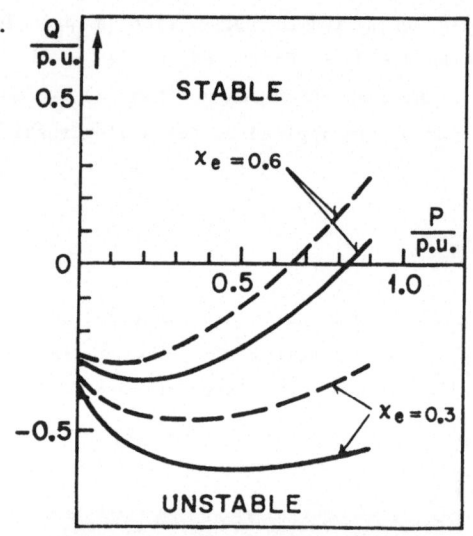

Fig. 7:
Steady state stability limits of the generator
---- unadapted regulator
—— adapted regulator

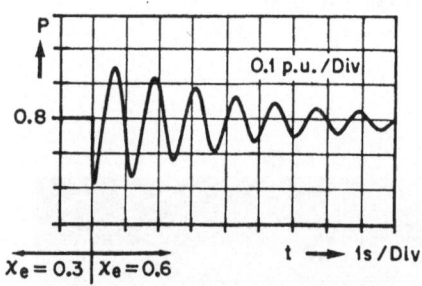

Fig. 8: Response to the opening of a transmission line, unadapted regulator.

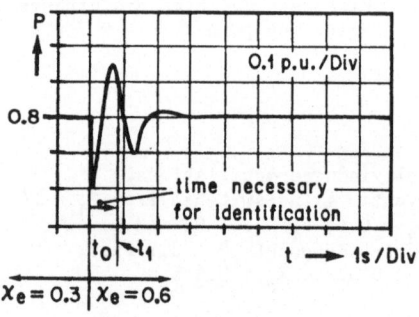

Fig. 9: Same as Fig. 8, but with adapted regulator.

The response after a 3-phase fault is shown
in Fig. 10 for the adapted regulator. Using
the regulator with fixed compromise gains,
only the damping would be reduced. Stabilization of the first swing would be the same.
In fact, for any fault where the generator
does not fall out of step immediately ("first
swing instability") the subsequent oscillations
are stabilized properly with this regulator.
Thus, the conventional regulator of Fig. 2
appears to be quite optimal in this case.

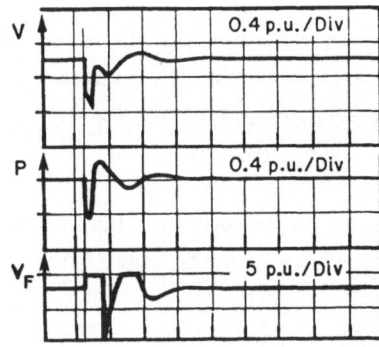

Fig. 10: Short circuit of 150 ms.
on one line, without reclosure of the line.
X_e = 0.3 p.u. before, and 0.6 p.u. after
the fault. V_F = Field voltage.

IMPLEMENTATION IN PRACTICE

An experimental version of the adapted regulator was tested in a hydraulic power
station in Spain. The main results are:
1) The identification procedure for the reactance X_e functions as desired.
2) Adjustment of the gains from the discrete look-up table is quite adequate.
3) Operation of the computer in the hostile environment posed no problem
    (despite the presence of the author!).

The practical circumstances did not permit a demonstration of the improved stability
region provided by the adapted regulator. The machines had to be operated at fairly
secure operating points, and the hydrogenerators were easy to control anyway. It is
true in general that the significance of the adapted regulator with respect to steady
state stability is moderate. Utilities tend to operate their units at conservative
operating points, such that the contingency of a large disturbance may not endanger
synchronous operation.

The main advantage of the adapted regulator lies in the smooth voltage regulation. The implementation of additional features (adjustment of voltage feedback gain
and stabilizer limits) would also improve the damping after large disturbances. Such
extensions may be considered in the future.

Computer reliability is not a stringent problem in the present application. The
stabilizing signal (Fig. 2) has a limited influence and is not always vital to the
operation of the generator. The signal is monitored by a protective device and disconnected automatically in case of abnormal behaviour. If such a failure happens and the
generator becomes potentially unstable, the station operator must move the machine to
a safe operating point (less power, higher voltage, according to Fig. 7).

CONCLUSION

An improved regulator for the excitation of large turbogenerators has been presented. The design is based on a conventional analog regulator, which satisfies the main requirements and provides reliable control. An adapting device is added, which adjusts the feedback gains according to the operating point and operating regime of the generator. The procedure uses a look-up table with predetermined gains and is implemented on a microprocessor. The adapted regulator provides smooth voltage control and an improved steady state stability region for the generator.

REFERENCES

[1] G. D'Ans, H. Glavitsch, L.M. Panis, "An Evaluation of Present-Day Excitation System Control of Turbogenerators". IFAC Symposium on automatic control and protection of electric power systems, Melbourne, Australia. Preprints of papers. pp. 199-203, February 1977.

[2] T. Laible, Die Theorie der Synchronmaschine im nichtstationären Betrieb. Springer Verlag, Berlin/Göttingen/Heidelberg, 1952

[3] "IEEE Guide for Identification, Testing and Evaluation of the Dynamic Performance of Excitation Control Systems", IEEE Std. 421A-1978.

[4] F. Peneder, R. Bertschi, "Optimizing the Setting of Slip Stabilization Equipment by Digital Computers and Special Computing Procedures", Brown Boveri Review, Vol. 65, No. 9, pp. 590-597, 1978.

[5] J. Nanda, "Optimization of Voltage Regulator Gains by the D-Decomposition Technique for best Steady-State Stability", IEEE Transactions on Power Apparatus and Systems, Vol. PAS-90, pp. 2488-2494, Nov./Dec. 1971.

[6] R. Craven, H. Glavitsch, "Large Disturbance Performance of an Adapted Linear Excitation Control System", IEEE paper no. A 79 452-4, presented at the IEEE PES Summer Meeting, Vancouver, BC, Canada, July 15-20, 1979

[7] P. Bonanomi, R. Bertschi, "On-Line Identification of an Equivalent Reactance for Stability Applications", paper submitted for presentation at the IEEE PES Summer Meeting, Minneapolis, Minnesota, USA, July 13-18, 1980.

[8] P. Bonanomi, G. Güth, F. Blaser, H. Glavitsch, "Concept of a Practical Adaptive Regulator for Excitation Control", IEEE paper no. A 79 453-2, presented at the IEEE PES Summer Meeting, Vancouver, BC, Canada, July 15-20, 1979.

ADAPTIVE CONTROL BY SELF-SELECTION
- AN APPLICATION TO HYDROPOWER CONTROL

A.H.Glattfelder
Research Department
ESCHER WYSS Limited

CH-8023 Zürich/Switzerland

W.Schaufelberger
Fachgruppe für Automatik der ETH
Swiss Federal Institute of Technology

CH-8092 Zürich/Switzerland

ABSTRACT

Adaptive regulator structures such as self-selecting controllers provide a practical
solution to many control problems with strong constraints on control and state variables. First results on stability investigations are presented. The design elements are
applied to the aperture control of a hydraulic turbine. The regulator is implemented
on a microcomputersystem and test results on an analog model are given.

1. INTRODUCTION

Saturations on control and state variables appear in most practical control problems during start-up or large load-shifts. A convenient solution for such problems is
to adjust the weight factors in the cost functional in order to avoid the constraints
and thus linearize the control problem /1/. In many practical cases however, the constraints are such that this procedure will not yield an acceptable control performance. Then the well known optimization problem with constraints /1/ must be solved.
It requires a relatively large additional design effort.

On the other hand, industrial control system design especially for thermal power
plants has led to a modular design technique /2/ for control problems with multiple
constraints, which has been successfully applied to practical cases in different
fields /3/, /4/, /5/.

The design technique leads to an adaptive controller structure as shown in fig. 1
for the basic case of one pair of constraints y↑ and y↓ on y. Its aim is to yield
trajectory control along the constraints y↑ or y↓, whereas the basic purpose of the
"variable structure" systems (used in /6/ on the same process) is insensitivity to
plant parameter variations. A self-selecting MIN-MAX-module automatically and continuously selects from several control algorithms the most constraining one for the
control variable. The control algorithms may be e.g. linear with constant coefficients or of the parameter self-tuning type.

From experience this design procedure is easy to apply to a given problem. It produces a transparent solution, which is convenient to implement and control performance is acceptable by practical standards in many cases.

However the design procedure is until now a heuristic one, which raises many interesting theoretical problems. First results in the area of stability analysis are given in the next section, but much remains to be done and further investigations in this field are encouraged.

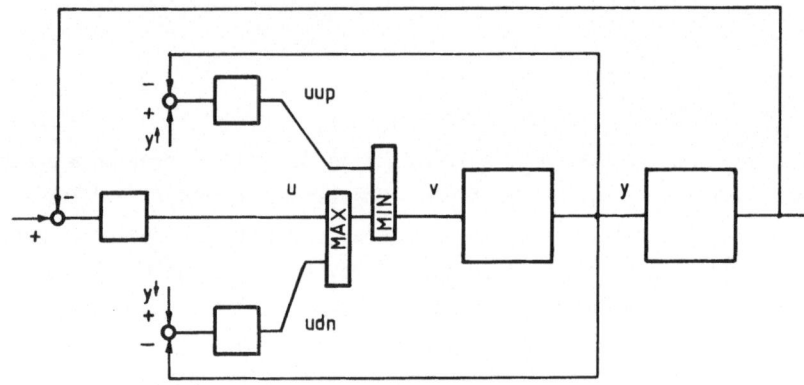

<u>fig. 1</u>
The basic structure-adaptive system.

2. STABILITY ANALYSIS OF THE SELF-SELECTING CONTROLLERS

2.1. Introduction

The stability analysis is not trivial. This results from the fact, that the proposed control systems are nonlinear and that they cannot be broken down in the usual way into a linear dynamic part and a nonlinear static part. Only one variable will be constrained in the following as shown in fig. 1.

A first fact to be noticed is the following: between the switching events the system is linear and time invariant and has all eigenvalues of the controlled part in the left half plane. This does not guarantee stability, but it is nonetheless an indication that a certain kind of stability can be expected. For stability investigations, it may be useful to replace the limiting circuit in fig. 1 by the following slightly modified algorithm:

(* us and ys are the stationary values of u and y,
   uup   and udn
   are precomputed limiting values *)

```
v: = u ;
if (y > ys) ∧ (u > uup) then v: = uup ;
if (y < ys) ∧ (u < udn) then v: = udn ;
```

No general theory for stability analysis is available at this time, but the following ideas and facts may help in many situations:

2.2. Second order systems

Second order systems can be investigated in the phase plane. Stability investigations are usually simple in this case. This yields also insights which are useful in the higher order case.

2.3. Limiting input or output only

These two special cases lead to nonlinear control systems which can be separated into a nonlinear static part and a linear dynamic part. The analysis methods that have been developed for this case can be applied.

2.4. Hyperstable blocks

It can easily be shown that a hyperstable block remains hyperstable when the above stated algorithm is used to design a limiting circuit around it. Designs based on hyperstability theory are thus possible.

2.5. General case

The stability problem is difficult in this case. In simple cases, the solution of the system may actually be computed since the system is linear and time-invariant between the switching events. This fact may also help to develop fast simulation methods. Computations can be considerably simplified if it is assumed that the limiting control systems reach their steady state fast. The total system may also be represented by a system with a timevarying gain. The second method of Ljapunov used in /7/ for this case may be used to find a range of the gain for which the system is asymptotically stable. The stability problem is then solved for transients for which the gain remains in this range. Ideal Ljapunov functions /8/, /9/ may also be helpful.

It is hoped that further work in this area will lead to more conclusive results.

3. DESIGN

The design elements of fig. 1 shall now be incorporated into a more complex structure with five pairs of constraints.

The main controlled variable in the hydropower plant fig. 2 (if the generator is connected to the grid) is the turbine guide vane aperture, /6/. Its setpoint is manipulated either manually or by the load-frequency-controller. A standard linear regulator is used. It performs acceptably for small setpoint variations, but poorly for large ones because restrictions appear. For instance the actuator subsystem has saturations on the servomotor stroke and the pilot valve stroke which appear only for large or fast setpoint variations. Also the hydraulic subsystem produces a transient pressure rise due to the inertia of the water column when the aperture is sharply reduced /10/, but the pressure p_T at the turbine must not exceed certain safety limits. Finally the storage capacities of the upstream and downstream reservoirs will allow transient load following only within their level bounds.

The process model in fig. 2 visualizes how these variables are interrelated for small deviations from nominal operating conditions /11/, /12/. Note the large spread of eigenvalues by a factor of $2 \cdot 10^3$ in the model. On a real plant a factor $> 10^4$ should be expected. The gains and time constants are functions of the operating conditions /6/, /12/. In a first design step however they are considered to be constant. The simplified model can be used to check the controller function and yield first approximates for the controller parameters prior to testing it on-line on the more complex real process.

On this basis the adaptive regulator structure fig. 3 is designed. It consists of the main PD-controller for the servomotor stroke h_{SM} (that is guide vane aperture) connected to an integrator ($u \rightarrow v$) to eliminate steady state errors, and of one self-selecting module for each constrained variable. The individual setpoints are given in fig. 3 as well as the gain and rate factors holding for the plant model in fig. 2. They can be determined by standard design methods. Note that both the upstream level limit controller and the pressure limit controller have negative gains and are connected crosswise to the MIN-MAX-selector modules in order to achieve overall negative feedback and the correct limiting actions on u.

Because both the upstream and downstream level limiter loops contain three integrators in series, additional feedback signals are necessary to stabilize them. Time derivatives of the process variables must be utilized, as adding the signals directly would introduce steady state offsets.

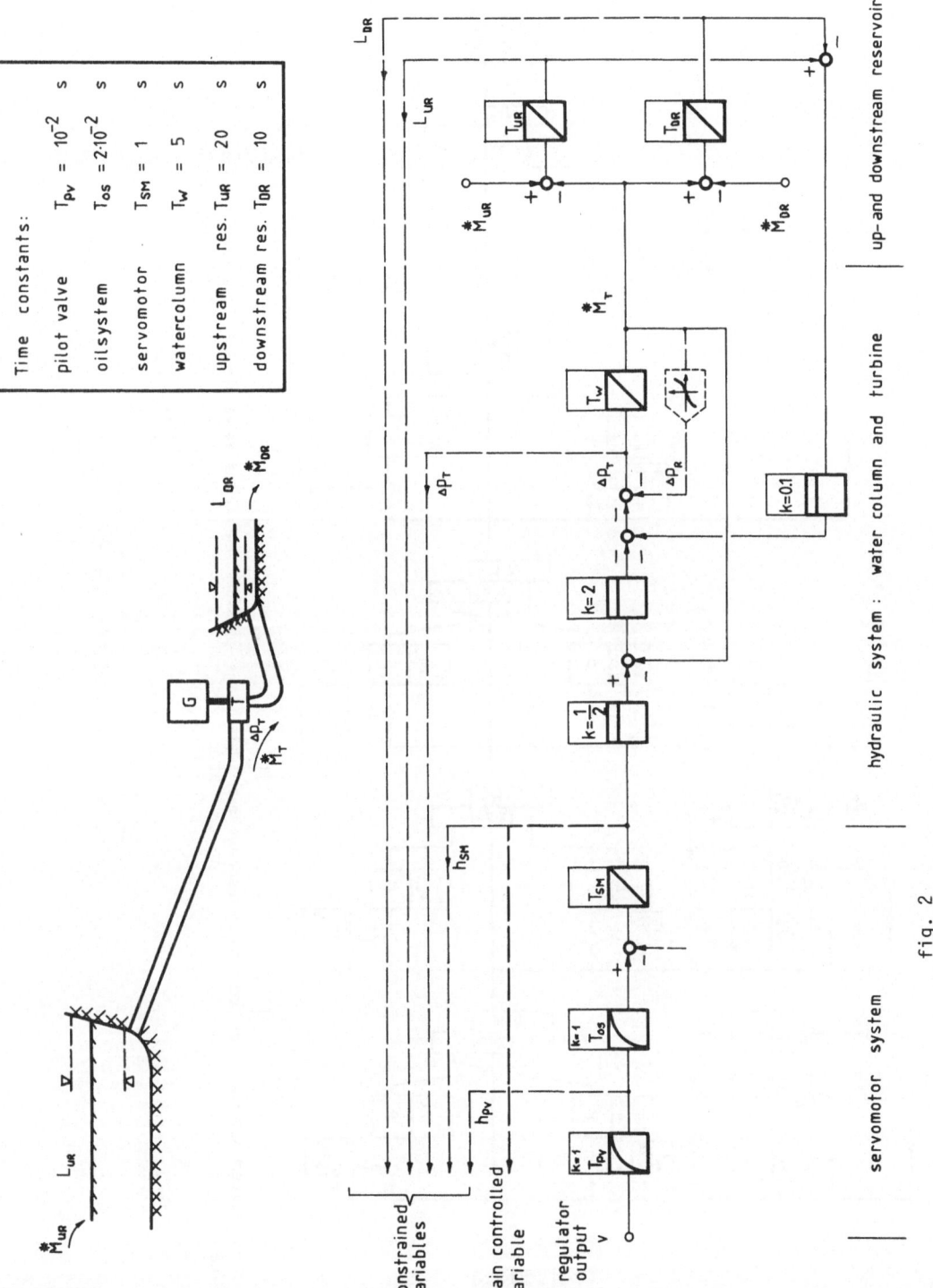

fig. 2

Sketch of the hydropowerplant and block diagram of the model system.

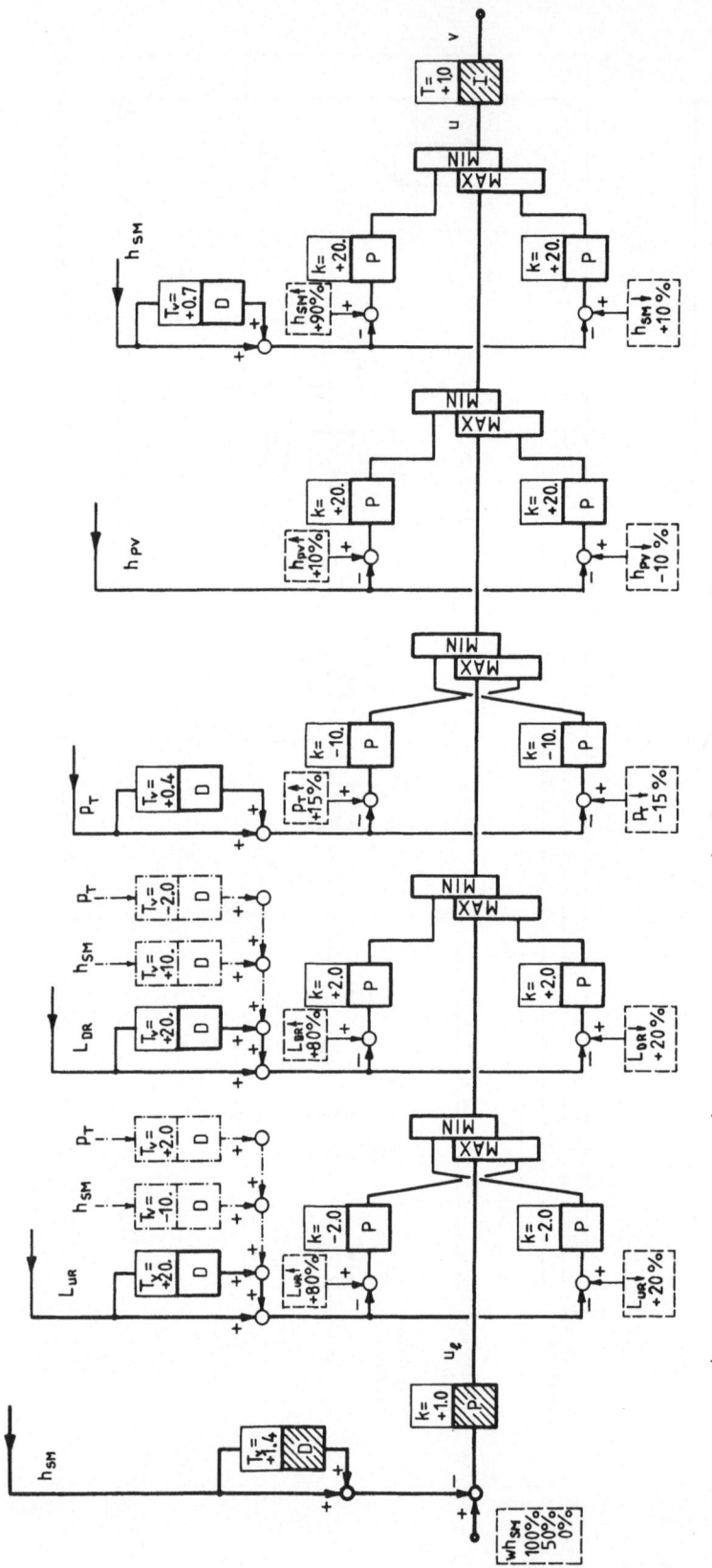

fig. 3
The self-selecting regulator.

4. IMPLEMENTATION

From previous experience with this type of problem, the choice of a high level language and of floating-point representation of all internal variables to avoid scaling problems was considered crucial.

Therefore the program for the regulator shown in fig. 3 is written in PASCAL. It contains only four statements at assembly level to synchronize the computer operation to the external clock.

The program consists of two main parts. The first part contains the parameter input via TTY and the initialization of the interface system. The regulator is executed in the second part which is periodically triggered by the external clock. The limiting controller pairs with their respective self-selecting modules in fig. 3 are implemented by calling a separate procedure for each of them consecutively. Each procedure can be disabled by an external switch for test purposes. A real function MIN-MAX (uin, uup, udn) is used for the self-selection module

$$u = Min (uup, Max(uin, udn))$$

Otherwise the program is straight-forward.

The program was developped and executed on an LSI-11 microcomputer system with a standard process control interface at the Control Laboratory of the Swiss Federal Institute of Technology. Program development took one day of work. The program has roughly 180 lines of code. The regulator part contains about 60 floating point operations and executes in approx. 17 ms on the system used. Therefore the experiments were performed with a sampling time of 20 ms, that is twice the smallest time constant T_{pv} of the plant model fig. 2. The sampling time should not be increased much further or the pilot valve limiting response will deteriorate. This is an important point when selecting a suitable hardware configuration.

5. RESULTS

The regulator implemented on the microcomputer system was tested on an analog model of the plant given in fig. 2. An EAI-Mini AC of the Control Laboratory was used.

Typical system transients are shown in fig. 4 and 5. By inspection they are suboptimal with respect to minimum-time transients. However from a practical point of view they may be considered as acceptable.

If the regulator parameters are precalculated (e.g. by standard linear methods) and the constraint control loops are incorporated one after the other (switch enabling each one proved to be very helpful), then the start-up procedure is completed in a few hours of real time tests. This confirms prior experience with analogue regulator implementations /4/, /5/.

The large spread of process eigenvalues caused a minor difficulty. - In a first phase, only the relatively fast loops for the pilot valve and servomotor strokes and the transient pressure rise were considered. Their response to a large main setpoint step is shown in fig. 4. On the h_{pv}-trace it is indicated (by their respective setpoints) which of the controller outputs is selected as control variable at each moment. - In the second phase, the level constraints were included. Their limiting loops require the same sampling rate for the stabilizing feedback signals. This sampling time applied to the very slowly changing level signals produces very small differences and thus a poor rate signal. To avoid this, the rate signals were evaluated only after a 50-samples-period, that is after approx. 5% of the large time constant. This should be taken into account when considering even larger time constants on real plants. Fig. 5 shows a transient from the upward to the downward limit on the upstream reservoir level, when the aperture setpoint is switched from fully closed to fully open. The Δp_T-trace shows which output is selected. The integral $\int h_{SM} \cdot dt$ is a measure for the electric power drawn from the upstream storage capacity. Note that the right most flank of $h_{SM}(t)$ due to the $L_{UR\downarrow}$-controller action will turn out comparatively steeper for a larger time constant T_{UR} and thus yield a transient closer to minimum-time behaviour.

On this basis the application to a real plant can be considered. If additional restrictions should appear or others turn out to be unnecessary then the corresponding modifications to the regulator structure can be done by the same design procedure. Moreover it may be useful to apply self-tuning methods to the regulator parameters. Both measures can be implemented more easily on a microcomputer system than on an analogue system.

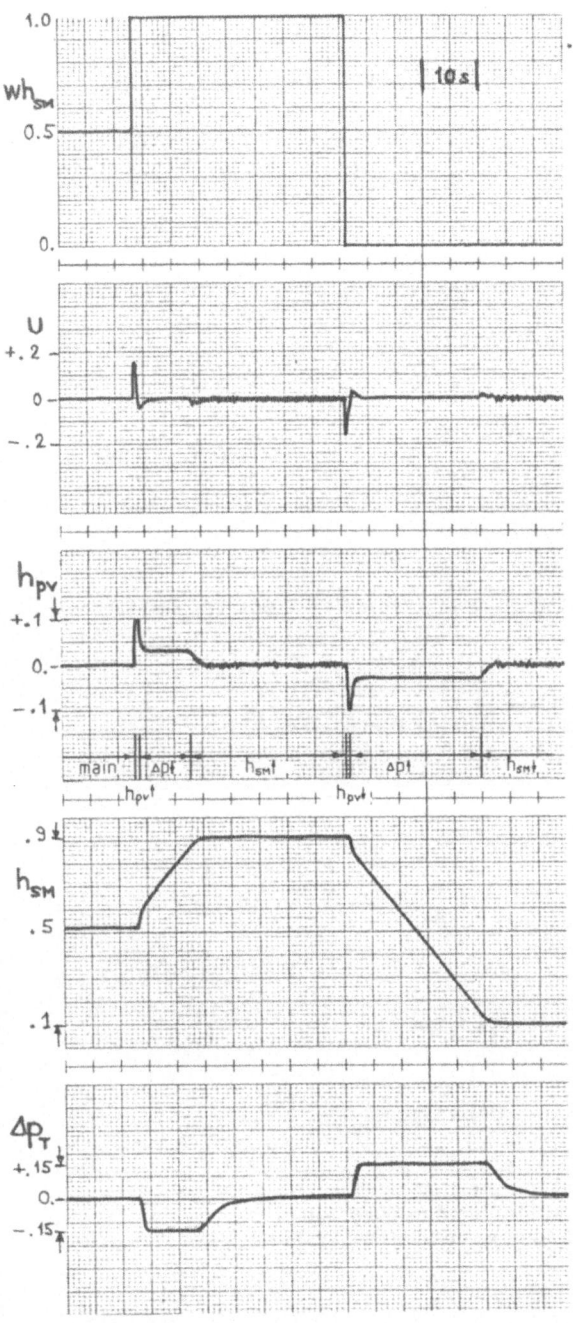

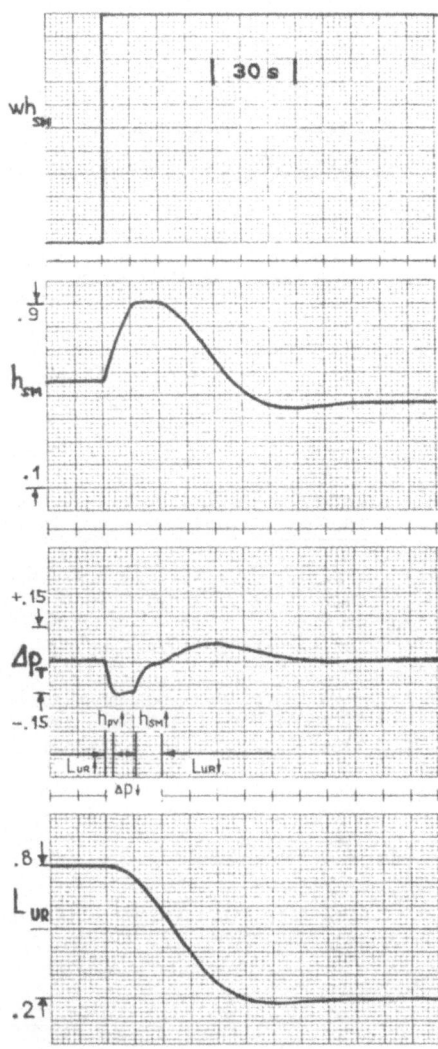

fig. 5

System response to upper and lower bounds on the upstream reservoir level (L_{UR}) in addition to the system of fig. 4.

fig. 4

System response to large main setpoint (wh_{SM}) steps. Note the self-selection sequence indicated on the pilot valve stroke (h_{pv}) trace.

REFERENCES

/1/ Sage A.P.
     Optimum Systems Control, Prentice Hall, 1968

/2/ Glattfelder A.H.
     Regelungssysteme mit Begrenzungen
     R.Oldenbourg Verlag, München, Wien, 1974

/3/ Glattfelder A.H.
     Optimal coupling of an incinerator plant to a district central heating system
     Proc. 4th IFAC/IFIP on Digital Computer applications to process control/Vol II,
     Springer Verlag, Berlin-Heidelberg-New York, 1974

/4/ Glattfelder A.H./Gross L.
     Weitbereich-Leistungsregelung eines Trommelkessels, Entwurf, Realisierung
     und Betriebserfahrungen
     "Brennstoff-Wärme-Kraft", 27 (1975), pp. 379-382

/5/ Glattfelder A.H./Gross L.
     Weitbereichregelung eines Wärmeverbundnetzes
     "Brennstoff-Wärme-Kraft", 29 (1977), pp. 27-34

/6/ Erschler J./Roubellat F./Vernhes J.P.
     Automation of a hydroelectric power station using variable-structure control
     systems
     Automatica 10 (1974), pp. 31-36

/7/ Glattfelder A.H./Huguenin F./Schaufelberger W.
     Microcomputer based Self-Tuning and Self-Selecting Controllers
     Automatica 16 (1980), in print

/8/ Itschner B.
     Einführung idealer Ljapunovfunktionen
     Regelungstechnik 25 (1977), pp. 216-221

/9/ Itschner B.
     Stabilitätsanalyse mit Hilfe idealer Ljapunovfunktionen
     Regelungstechnik 25 (1977), pp. 251-257

/10/ Jaeger Ch.
     Fluid transients in hydroelectric engineering practice
     Blackie, Glasgow, London, 1977

/11/ Borel L.
     Stabilité de réglage des installations hydroélectriques
     Payot, Lausanne; Dunod, Paris; 1960

/12/ Glattfelder A.H./Jeanneret F./Glauser A./Huser L.
     Computer-Aided Design of Adaptive Hydraulic Turbine Governors
     Proc. IFAC Symposium on Computer Aided Design of Control System
     Zürich, 1979

ADAPTIVE TIME-OPTIMAL POSITION CONTROL
WITH MICROPROCESSOR

U. Claussen, Gartenstrasse 21a, D-8551 Roettenbach, Germany

1. INTRODUCTION

Often the aim of control is to transform any initial state of a dynamic
system into a desired stationary state in minimal time. If the controller
output is limited, as is the case in most real processes, time optimal
control can be usefully applied to the system. Control theory states that
an aperiodic system of n-th order demands exactly one switch-on, n-1
switch-overs and one switch-off of the maximum controller output signal,
in order to lead the system into the stationary state as fast as possible.
Each switching must occur at a definite predetermined instant (fig.1).

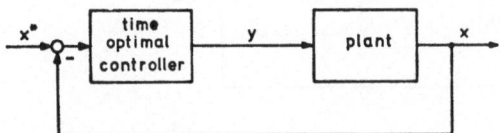

Fig.1

In many real applications the system parameters are not well-known. They
also drift slowly away from their initial values or they vary abruptly
due to external influences. This is critical in time optimal loops as a
slightly misaligned controller can cause considerable overshoot and oscillation, i.e. by no means time optimal system adjustment. As such a system
response is not tolerable in most cases, the time optimal controller has
to be set to the worst-case combination of system parameters, that can
be expected. The real transient response consequently takes longer time
and requires a greater number of switchings than theoretically necessary.

In these cases the use of an adaptive time optimal control scheme is advantageous. Such a controller is able to adapt its parameters automatically to changes in the system, so that time optimal transients will always be accomplished. It is thereby assumed, that the system's order and
structure are well-known and the controller only has to be adapted to
variable gains, time constants and disturbances.

Monitoring these influences directly, which is the easiest way, is often
not practicable in real applications. Sensors are expensive, susceptible

to faults and easily damaged, and the desired values may not be measurable at all. However it is possible to identify these missing quantities from the system's performance. For this only information is used,
which is in any case necessary for time optimal control (fig.2). The
considerable amount of calculation required for identification and adaption demands the use of a digital computer. The resulting hardware costs
however are moderate, if a microprocessor is used. This is especially the
case, if the time optimal control itself is executed by the same processor, like in the example presented below.

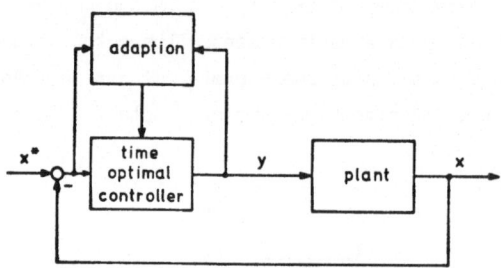

Fig. 2

2. TIME OPTIMAL POSITION CONTROL

In the numerous industrial processes, which involve transport and motion,
position control is of special practical importance. Time optimal control
was introduced early in this field /2/, while attempting to increase the
exploitation of the sophisticated machinery. The method is well-known and
will only be described so far as it is important for the following.

Fig.3 shows a 2-nd order system with a time optimal controller. The quantities referring to linear motion (F force, x position) can also be substituted by their rotary counterparts (T torque, ε angle of revolution).
Instead of the usual two-step controller for the velocity loop /1/, in
this case a proportional plus integral controller with limited output
was used. Its performance is the same for large control errors, but after
approaching the desired position there will be no infinite oscillations,
which cause unnecessary wear. There is no special type of drive required
by this method. For example a d.c. or a synchronous motor with permanent
excitation is suitable. In this case the latter type was used for the experiments.

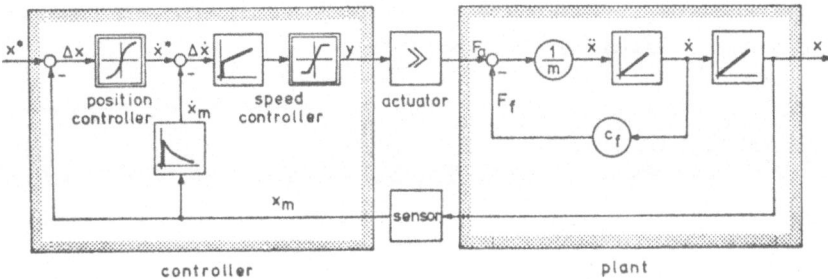

Fig. 3

The equation of motion for the system (fig.3) is

$$m\ddot{x} + c_f\dot{x} = F_a \tag{1}$$

Using the abbreviations

$$\frac{m}{c_f} = T \quad , \qquad \frac{\hat{F}_a}{c_f} = \hat{\ddot{x}} \tag{2}$$

the following time functions can be derived
($F_a(t<0) = 0$, $F_a(t\geqslant 0) = F_a$, $\ddot{x}(t=0) = \dot{x}(t=0) = 0$)

$$\frac{x(t)}{\hat{\ddot{x}}\,T} = \frac{t}{T} - (1 - e^{-t/T}) \tag{3}$$

$$\frac{\dot{x}(t)}{\hat{\ddot{x}}} = 1 - e^{-t/T} \tag{4}$$

The equation of state, $x = f(\dot{x})$, is yielded by elimination of time

$$\frac{x(\dot{x})}{\hat{\ddot{x}}\,T} = -(\frac{\dot{x}}{\hat{\ddot{x}}} + \ln \left| 1 - \frac{\dot{x}}{\hat{\ddot{x}}} \right|) \tag{5}$$

To get the desired characteristic of the time optimal controller
$\dot{x}^* = f(\Delta x)$, the variable $\dot{x}$ has to be separated. This is not possible analytically. But an iterative solution of (5) supplies a corresponding
table.

This set-up implies, that the iterative calculation has to be carried out
for every new set of parameters. This is disadvantageous, particularly if

one considers extending the system to an adaptive controller for permanently
changing load conditions as described below.

A simplified equation of motion is valid for the special case of a system
without friction ($c_f \to 0$).

$$m \ddot{x} = F_a \tag{6}$$

(3),(4) and (5) merge into

$$\frac{x(t)}{\hat{\ddot{x}} T} = \frac{1}{2} \left(\frac{t}{T} \right)^2 \tag{7}$$

$$\frac{\dot{x}(t)}{\hat{\dot{x}}} = \frac{t}{T} \tag{8}$$

$$\frac{x(\dot{x})}{\hat{\ddot{x}} T} = \frac{1}{2} \left(\frac{\dot{x}}{\hat{\dot{x}}} \right)^2 \tag{9}$$

Separation of $\dot{x}$ and use of (2) produce the characteristic of the time
optimal controller for a system without friction.

$$\dot{x}^*(\Delta x) = \sqrt{2 \left| \frac{\hat{F}_a}{m} \Delta x \right|} \ \text{sign} (\Delta x) \tag{10}$$

Applying such a simplified controller to the real process with friction
will result in not entirely time optimal transients. The increase in time
depends on the values of mass, coefficient of friction and position displacement required. However calculations have proved, that the additional
amount of time will not exceed 3.5% of the total time for positioning /3/.
If this deviation is tolerable, as will be the case in almost all real
applications, a system like the one shown in fig.3 can be controlled quasi-
time optimally with a simplified controller of the type given in equ.(10).

3. ADAPTIVE TIME-OPTIMAL POSITION CONTROL

Dimensioning a time optimal controller as well as a linear controller
requires knowledge of system parameters. But there are differences in the
sensitivity to mistuning. While a linear system will respond with a slightly changed damping characteristic, a time optimal system can produce considerable overshoot and an oscillating approach to the position set point.
In general such performance is not admissable.

If variable parameters or disturbances are expected, the controller has to be aligned to worst-case conditions. This is the only way to avoid over-shoot in any case. So a normal transient will not be performed time optimally, and positioning will take a longer time than necessary. Let us for example imagine a device for transportation of goods, capable of carrying a mass ten times as large as the no load mass of the drive. If a full-load positioning takes 1 s , the return trip with no load will take 1 s as well, assuming a controller with fixed parameters. Even though the no load trip could be done in only 0.3 s , if the drive would apply full no load acceleration. This makes clear, that an adaptive control scheme can increase the efficiency of a machine.

Additional sensors are undesirable, because of the reasons mentioned at the beginning. So the unknown system parameters have to be calculated on the basis of the time functions of those quantities, that are needed anyhow for control. In this case (fig.3) the controller output y(t) and the measured variable $x_m(t)$ are available. A variation of the mass to be moved and an unknown disturbing force F_d were assumed to be the essential influences affecting the system (fig.4). These quantities are considered to be constant for at least one positioning displacement.

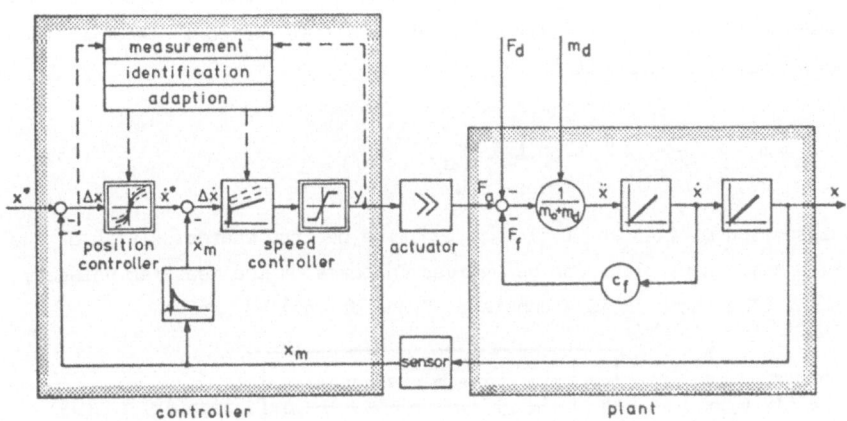

Fig. 4

The characteristic of the time optimal controller is now (cf.(10))

$$\dot{x}^*(\Delta x) \;=\; \sqrt{2 \left| \frac{\hat{F}_a \,\text{sign}(\Delta x) + F_d}{m_o + m_d} \,\Delta x \right|} \;\; \text{sign}(\Delta x) \tag{11}$$

Initially, the disturbing force F_d and the additional mass m_d are the unknown variables.

From the equation of motion

$$(m_o + m_d)\ddot{x} + c_f \dot{x} = F_a - F_d \tag{13}$$

one can derive that the controller output for the stationary state $(\ddot{x} = \dot{x} = 0)$, y_o, is proportional to the disturbing force

$$F_d = K y_o \qquad \text{for } \ddot{x} = \dot{x} = 0 \tag{13}$$

So as long as the drive is at rest, the actual disturbing force can be computed continuously. When a new position set point is commanded, only one further measurement is consequently necessary until the parameters are identified completely.

This measurement is executed immediately after acceleration towards the new position has started. Equ.(7) can be expressed as

$$x(t) = \frac{1}{2} t^2 \frac{\hat{F}_a \text{sign}(\Delta x) - F_d}{m_o + m_d} \tag{14}$$

Measuring the position difference x_t after a test period t_t leads to the unknown value of the mass

$$m_o + m_d = \frac{t_t^2}{2x_t} (\hat{F}_a \text{sign}(\Delta x) - F_d) \tag{15}$$

By insertion of (14) and (15) into (11) the desired characteristic of the time optimal controller can be derived in terms of the measured values y_o and x_t. (The factor K is a constant of the drive.)

$$\dot{x}^*(\Delta x, y_o, x_t) = \sqrt{4 \left| \frac{x_t}{t_t^2} \frac{\hat{F}_a + K y_o \text{sign}(\Delta x)}{\hat{F}_a - K y_o \text{sign}(\Delta x)} \Delta x \right|} \ \text{sign}(\Delta x) \tag{16}$$

A minimum position displacement must occur, after which reliable adaption can be guaranteed. This position displacement depends on the sum of the test period t_t and the computing time t_c. This is because a valid switch-over of the controller output y from accelerating to decelerating cannot be executed earlier than after these two time intervals. This is the first moment at which the actual parameters of the time optimal controller are available. Consequently, immediately after the start the parameters have to be adjusted to the worst-case conditions. This is why very small position shifts are not performed in an optimal way. But as only small changes in setting are concerned by this, the over-all increase in time is negligible.

If the inner loop contains a continuous type of controller, this controller has to be adapted to the variable system parameters as well. In this case the gain of the proportional plus integral element has to be changed, in order to keep the gain of the closed loop constant.

4. RESULTS

An adaptive time optimal controller, as described in the preceeding section, was designed for a 1.2 m long positioning drive with a synchronous linear motor /4/. This was simulated on a digital computer and realized using a microprocessor type Intel 8080 with fast external multiplying unit. The drive can be adequately described as a 2-nd order system. The essential variable quantities of the system are the angle of declination between the track and the horizontal line ($\pm$ 90 deg.) and the additional mass to be transported. Similar operating conditions for example are found in industrial robot systems, which are used for handling of goods. The motor has a maximum acceleration of 48 m/s^2 and takes a time of 0.5 s for a position displacement of 1 m.

The control function including adaption are entirely executed by the microprocessor. Because of the highly dynamic requirements of the drive, the cycle time for control must not exceed 2 ms. Considering that the adaption itself takes about 10...20 ms time of computation it is clear, that the adaption can only be executed step-by-step in a background program. This time shared operation is efficiently supported by the microprocessor's interrupt logic. Due to the continuous interrupts a complete adaption process takes 50...100 ms.

Figs. 5 a,b,c show oscillograms of the time optimally controlled drive, executing position shifts from 0.2 to 0.8 m under different load conditions. The first figure presents a no load trip on horizontal track. Positioning is finished after 0.4 s. The second photograph was taken with 6 kg additional mass (m_o = 1.54 kg). According to the reduced maximum acceleration the gradient of speed is less and the desired position is now reached in 0.8 s. The influence of adaption can be detected by the speed set-point $\dot{x}^*(t)$. It must be noticed, that these oscillograms were taken using a program different from that described above. In this case immediately after starting, the parameters were adjusted to the no load and not the worst-case conditions. Consequently fig. 5b shows, that in the first time interval the position controller demands the same setting for speed as in the figure before (no load), until after about 100 ms the gain of the controller is reduced corresponding to the actual load conditions.

Finally fig. 5c shows an upward motion in a vertical direction with $m_d = 1\,kg$ additional mass. In this case the identification program detects that gravity will support the decelerating force while braking. So compared to a no load horizontal shift the gain of the time-optimal position controller can be increased, although an additional mass has to be transported. Accordingly the time for acceleration is much longer than for braking. Owing to adaptive control under any of the three load conditions a time optimal transient without overshoot is achieved.

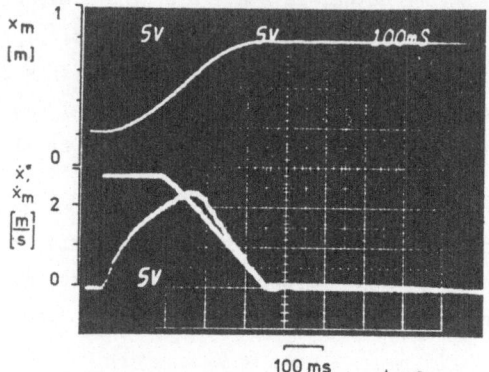

Fig. 5a

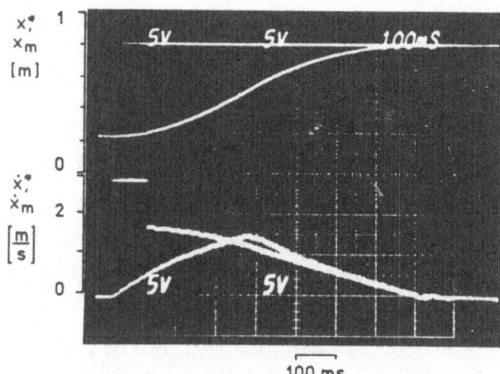

Fig. 5b

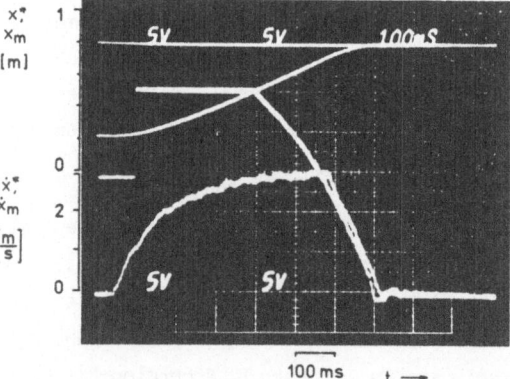

Fig. 5c

5. SUMMARY

Conventional time optimal control schemes can only be applied to systems with well-known, time-invariable parameters. For the example of a positioning drive it is demonstrated, that also systems with variable parameters can be controlled time optimally using an adaptive method. The required on-line computations can advantageously be executed by a microprocessor.

REFERENCES

/1/ Lerner, J.   Schnelligkeitsoptimale Regelungen
                       Oldenbourg, München 1963

/2/ Anke, K.   Digitale Wegregelung
     Ertel, K.   Siemens-Z. 1960, S. 664 ... 671
     Sinn, G.

/3/ Claussen, U.   Adaptive zeitoptimale Lageregelung eines linearen
                       Stellantriebs mit synchronem Langstatormotor
                       Dissertation Techn. Universitaet Braunschweig, 1980

/4/ Claussen, U.   Microprocessor – Controlled linear synchronous Motor
     Leonhard, W.   as Positioning Drive
                       Proc. Int. Conf. on Electrical Machines, Brussels 1978

A SIMPLE ADAPTIVE CONCEPT FOR THE CONTROL OF

AN INDUSTRIAL ROBOT

C.T. Cao

Department of Control Engineering, FB7,

University of Duisburg

Bismarckstr. 81, D - 4100 Duisburg 1

Federal Republic of Germany

1. INTRODUCTION

In recent years manipulators and industrial robots have been used increasingly in industrial automation. With the development of low cost microprocessors the need for improved control techniques will arise (Freund and Syrbe 1976, Rößler 1980, Dubowsky and DesForges 1979).

In this paper an adaptive concept for the control of an electro-hydraulic industrial robot will be represented. The proposed control scheme has the following properties:
- to overcome the effect of nonlinearities
- to estimate the unknown and/or time-varying parameters from the real input and output data disturbed by system and measurement noises
- to follow a desired reference signal.

The algorithm will be so simple that its implementation in process computers or microprocessors for on line control is possible.

For this purpose a mathematical model of an industrial robot with three-degrees-of-freedom (two translational and one rotational movement) will be reviewed. Based on these nonlinear or linearized equations of the model an analytical solution of the control problem would be possible. However, because of the uncertainties of the system dynamics and the effect of real conditions an adaptive control scheme is considered. In the first step a structure equivalent to a Kalman filter for each of the three degrees of freedom will be formulated. This structure contains two main characteristics:
- the predition of output is obtained separately from the state vector
- the state estimate is a linear function of the estimated parameters.

The application of the proposed structure enables the solution of many standard control problems such as state or output feedback control, optimal regulator and tracking problem. Especially a control system is investigated in detail, which results from the application of this equivalent structure of the adaptive Kalman filter and from minimizing a quadratic single-stage costfunction incorporating both control and tracking costs (one-step-ahead controllers) in attemping to force the industrial robot to follow a desired reference signal. A stochastic approximation method is used to estimate the unknown parameters, which determine the controller parameters in a simple manner.

Theoretically the control of the global system can be examined in the same way. Since the subsystems for vertical motion is uncoupled from the others, the coupled part of the industrial robot is described by a (2x2) model. Again a structure equivalent to the Kalman filter for this coupled subsystems will be given. Apart from the increasing computational requirements in comparison with the SISO case, its application for the tracking problem of the global system is included.

Because of its simplicity and flexibility this type of control structure is suitable for process control. Only the reference signal in the next sampling interval must be known. This adaptive control scheme has been applied to control the subsystems of an industrial robot. In the first study harmonic testsignals as reference functions have been used. As a result, a periodic testsignal with f<5Hz has been followed after 50 sampling intervals (sampling interval: 10 msec). Also for a simple movement the control performances are satisfactory. Similar to the most indirect adaptive control methods any resultant bias in the parameter estimates could unfortunately lead to control inconsistency. Given a "unfavourable" initial estimate of the system parameters the control could become unstable. The circumvention of this effect is one of the aims of further investigations.

2. DYNAMICAL MODEL OF AN INDUSTRIAL ROBOT

As shown in figure 2.1 the considered electro-hydraulic industrial robot (FIBRO) has three degrees-of-freedom. The coordinates $r(t)$ and $z(t)$ describe translational motions; $\varphi(t)$ describes the angle of rotation.

With the notation:

$$x_1(t)=r(t), \quad x_3(t)=\varphi(t), \quad x_5(t)=z(t)$$
$$x_2(t)=\dot{r}(t), \quad x_4(t)=\dot{\varphi}(t), \quad x_6(t)=\dot{z}(t)$$
$$y_1(t)=x_1(t), \quad y_2(t)=x_3(t), \quad y_3(t)=x_5(t)$$
$$u_1(t)=K_r(t), \quad u_2(t)=M_\varphi(t), \quad u_3(t)=K_z(t)-(m+m_L)g \;,$$

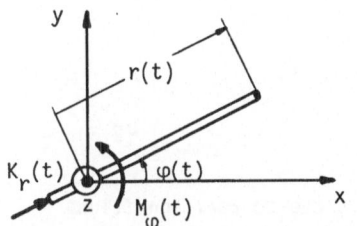

Fig. 2.1 principle configuration of the industrial robot

where m and m_L are the masses of the arm and the load, g is the gravity coefficient , $K_r(t), M_\varphi(t)$ and $K_z(t)$ are the forces respectively the torque corresponding to $r(t), \varphi(t)$ and $z(t)$, the dynamics of the system can be represented by the following state space model (Freund and Syrbe 1976, Rößler and Schwarz 1978):

$$
\begin{bmatrix} \dot{x}_1 \\ \dot{x}_2 \\ \dot{x}_3 \\ \dot{x}_4 \\ \dot{x}_5 \\ \dot{x}_6 \end{bmatrix}
=
\begin{bmatrix} x_2 \\ (x_1-\frac{c}{2a})x_4^2 \\ x_4 \\ h(x_1)[(c-2ax_1)x_2x_4] \\ x_6 \\ 0 \end{bmatrix}
+
\begin{bmatrix} 0 & 0 & 0 \\ \frac{1}{a} & 0 & 0 \\ 0 & 0 & 0 \\ 0 & h(x_1) & 0 \\ 0 & 0 & 0 \\ 0 & 0 & \frac{1}{a} \end{bmatrix}
\begin{bmatrix} u_1 \\ u_2 \\ u_3 \end{bmatrix}
+
\begin{bmatrix} 0 \\ -\frac{K_{hr}}{a}x_2-\frac{R_r}{a}\,\mathrm{sgn}\Delta p_r \\ 0 \\ (-K_{h\varphi}x_4-R_\varphi \mathrm{sgn}\Delta p_\varphi)h(x_1) \\ 0 \\ -\frac{K_{hz}}{a}x_6-\frac{R_z}{a}\,\mathrm{sgn}\Delta p_z \end{bmatrix}
$$

$$\overleftrightarrow{\text{frictions}}$$

$$
\begin{bmatrix} y_1 \\ y_2 \\ y_3 \end{bmatrix} = \begin{bmatrix} 1 & 0 & 0 & 0 & 0 & 0 \\ 0 & 0 & 1 & 0 & 0 & 0 \\ 0 & 0 & 0 & 0 & 1 & 0 \end{bmatrix} \underline{x} \qquad , \qquad (2.1)
$$

where $c=ml$, $a=m+m_L$, $h(x_1) = \dfrac{1}{k-mlx_1+(m+m_L)x_1^2}$, $k= \dfrac{m*r*^2}{2} + \dfrac{ml^2}{3}$, l is the length of

the arm, m* is an equivalent mass according to the radius of inertia r* of the upright
column of the robot.

From these equations it is possible to see the highly nonlinear, complex and coupled
nature of the robot dynamics. For convenience these nonlinear equations of motion may
be linearized with respect to a trajectory characterized by a mean radius R, a mean
translational velocity V_r and a mean angular velocity V_φ. This leads to the following
linearized equations:

$$
\begin{bmatrix} \dot{\bar{x}}_1 \\ \dot{\bar{x}}_2 \\ \dot{\bar{x}}_3 \\ \dot{\bar{x}}_4 \\ \dot{\bar{x}}_5 \\ \dot{\bar{x}}_6 \end{bmatrix} = \begin{bmatrix} 0 & 1 & 0 & 0 & 0 & 0 \\ \bar{a}_{21} & \bar{a}_{22} & 0 & \bar{a}_{24} & 0 & 0 \\ 0 & 0 & 0 & 1 & 0 & 0 \\ a^*_{41} & \bar{a}_{42} & 0 & a^*_{44} & 0 & 0 \\ 0 & 0 & 0 & 0 & 0 & 1 \\ 0 & 0 & 0 & 0 & 0 & \bar{a}_{66} \end{bmatrix} \begin{bmatrix} \bar{x}_1 \\ \bar{x}_2 \\ \bar{x}_3 \\ \bar{x}_4 \\ \bar{x}_5 \\ \bar{x}_6 \end{bmatrix} + \begin{bmatrix} 0 & 0 & 0 \\ \bar{b}_{21} & 0 & 0 \\ 0 & 0 & 0 \\ 0 & \bar{b}_{42} & 0 \\ 0 & 0 & 0 \\ 0 & 0 & b^*_{63} \end{bmatrix} \begin{bmatrix} u_1 \\ u_2 \\ u_3 \end{bmatrix} + \begin{bmatrix} 0 \\ 0 \\ 0 \\ \bar{r}_4 \\ 0 \\ 0 \end{bmatrix}
$$

$$(2.2)$$

$$
y_1 = \bar{x}_1 \ , \ y_2 = \bar{x}_3, \ y_3 = \bar{x}_5
$$

Taking measurement noises into account (2.2) can be denoted as:

$$
\dot{\underline{x}}(t) = \underline{A}(\theta)\bar{\underline{x}}(t) + \underline{B}(\theta)\underline{u}(t) + \underline{r}(t)
$$
$$
\underline{y}(t) = \underline{C}\bar{\underline{x}}(t) + \underline{v}
$$

$$(2.3)$$

where $\underline{A}(\theta)$ and $\underline{B}(\theta)$ contain unknown or variable parameters due to real conditions (
e.g. frictions), which could be obtained only by expensive experiments (Rößler and
Schwarz 1979). The variables $\underline{r}(t)$ and $\underline{v}(t)$ can be considered as system and measurement
disturbances, respectively.

Based on these nonlinear or linearized equations an analytical solution for the control of the industrial robot would be possible (Freund and Syrbe 1976, Rößler 1980).
However, because of the uncertainties of the system dynamics and real conditions an
adaptive control scheme may be specially favoured. In the following proposed control
procedure only a simple model of the system will be considered. To illustrate the configuration of the control scheme one of the subsystems will be first investigated. The
extension to control the global system will be shown in the same way. Since the control system will be realized by a process computer or microprocessor, an equivalent
discrete time model will be choosen as basis for the design of the controller.

3. ADAPTIVE CONTROL SCHEMES

Each of the subsystems of the industrial robot is assumed to be completely controllable and observable. Its dynamics can be approximated as:

$$x(k+1) = \underline{A}\underline{x}(k) + \underline{b}u(k) + \underline{h}r(k)$$

$$y(k) = \underline{c}^T\underline{x}(k) + v(k)$$

(3.1)

$$\underline{x} \in R^n, \ u \in R^1, \ y \in R^1, \ n \geq 2 \qquad ,$$

where $r(k)$, $v(k)$ are white Gaussian random sequences with

$$E\{r\} = 0, \ E\{r(j)r(k)\} = V_r(k)\delta_{jk}$$

$$E\{v\} = 0, \ E\{v(j)v(k)\} = V_v(k)\delta_{jk}$$

(3.2)

$$E\{r(j)v(k)\} = 0 \qquad ,$$

δ_{jk} is the Kronecker Delta, the initial value of $\underline{x}$ is a random variable with

$$E\{\underline{x}(0)\} = \hat{\underline{x}}(0), \ E\{\underline{x}(0)\underline{x}^T(0)\} = \underline{V}_x(0)$$

(3.3)

and $\underline{x}(0)$ is uncorrelated with $r(k)$ and $v(k)$. Further it will be assumed that the system parameters $\underline{A}$, $\underline{b}$, $\underline{c}^T$ are unknown and/or time-varying. Its structure can be given by:

$$\underline{A} = \begin{bmatrix} -a_1 & 1 & & & \\ -a_2 & 0 & 1 & & \\ \cdot & & & & \\ \cdot & & & & 1 \\ -a_n & 0 & \cdots & & 0 \end{bmatrix} , \quad \underline{b} = \begin{bmatrix} b_1 \\ b_2 \\ \cdot \\ \cdot \\ b_n \end{bmatrix} , \quad \underline{c}^T = \begin{bmatrix} 1 & 0 & \cdots & 0 \end{bmatrix} . \quad (3.4)$$

For this system a structure equivalent to a Kalman filter will be introduced at first.

3.1 An equivalent representation of the Kalman filter

The estimate of the state at time k of (3.1) based on observations of the output $y(0)$, $y(1),\ldots,y(k-1)$, which minimizes

$$E\{\tilde{\underline{x}}(k/k-1)\tilde{\underline{x}}^T(k/k-1)\}$$

(3.5)

where $\qquad \tilde{\underline{x}}(k/k-1) = \underline{x}(k) - \hat{\underline{x}}(k/k-1) \qquad ,$

(3.6)

is the conditional mean $\hat{\underline{x}}(k/k-1)$ which satisfies the recursive equations (Astrom 1970):

$$\hat{\underline{x}}(k+1/k) = \underline{A}\hat{\underline{x}}(k/k-1) + \underline{b}u(k) + \underline{g}(k)e(k)$$

(3.7)

$$e(k) = y(k) - \underline{c}^T\hat{\underline{x}}(k/k-1) = y(k) - \hat{y}(k/k-1)$$

(3.8)

$$\hat{\underline{x}}(0/-1) = \hat{\underline{x}}(0) \qquad .$$

If $\lim_{k\to\infty} \underline{g}(k) = \underline{g}^\infty = \underline{g} = \text{const.}$, (3.7) describes a steady state Kalman filter. In this case (3.7) can be rewritten as:

$$\hat{x}(k+1/k) = F\hat{x}(k/k-1) + gy(k) + \underline{b}u(k) \tag{3.9}$$

where
$$F = A - g\underline{c}^T \tag{3.10}$$

Under consideration of the canonical structure of $\underline{A}$ and $\underline{c}^T$ the "systemmatrix" $\underline{F}$ of the steady state Kalman filter can be obtained as:

$$\underline{F} = \begin{bmatrix} -f_1 & 1 & & & \\ -f_2 & 0 & 1 & & \\ \cdot & & & & \\ \cdot & & & & 1 \\ -f_n & 0 & 0 & \cdots & 0 \end{bmatrix} \tag{3.11}$$

with
$$f_i = a_i + g_i \quad , \quad i = 1,2,\ldots,n \quad , \tag{3.12}$$

where g_i are the elements of the vector $g = \begin{bmatrix} g_1, g_2, \ldots, g_n \end{bmatrix}^T$.

As it has been shown in (Cao 1980) an equivalent representation to (3.7) is given under consideration of (3.9) - (3.12):

$$\hat{y}(k/k-1) = \underline{z}^T(k)\underline{p} \tag{3.13}$$

with
$$\underline{z}^T(k) = \begin{bmatrix} \underline{\varsigma}_1^T(k) , \underline{\varsigma}_2^T(k) \end{bmatrix} \tag{3.14}$$

$$\underline{p}^T = \begin{bmatrix} \underline{g}^T , \underline{b}^T \end{bmatrix} \tag{3.15}$$

and
$$\hat{x}(k/k-1) = \begin{bmatrix} \underline{I}_1\varsigma_1(k), \underline{I}_2\varsigma_1(k),\ldots,\underline{I}_n\varsigma_1(k),\underline{I}_1\varsigma_2(k),\underline{I}_2\varsigma_2(k),\ldots,\underline{I}_n\varsigma_2(k) \end{bmatrix} \underline{p} \quad ,$$

where $\varsigma_1(k)$, $\varsigma_2(k)$ satisfy the recursive equations: (3.16)

$$\underline{\varsigma}_1(k+1) = \underline{F}^T\underline{\varsigma}_1(k) + \begin{bmatrix} 1 \\ 0 \\ \vdots \\ 0 \end{bmatrix} y(k) \qquad , \ \underline{\varsigma}_1(0) = \underline{0} \quad , \tag{3.17a}$$

$$\underline{\varsigma}_2(k+1) = \underline{F}^T\underline{\varsigma}_2(k) + \begin{bmatrix} 1 \\ 0 \\ \vdots \\ 0 \end{bmatrix} u(k) \qquad , \ \underline{\varsigma}_2(0) = \underline{0} \quad , \tag{3.17b}$$

and $\underline{I}_i$ are matrices obtainable from $\begin{bmatrix} \underline{I} - \tau^{-1}\underline{F} \end{bmatrix}^{-1}\underline{e}_i$, with τ as the shift operator defined by

$$\tau^{-1}\hat{y}(k/k-1) = \hat{y}(k-1/k-2) \qquad , \tag{3.18}$$

and
$$\underline{e}_i = \begin{bmatrix} 0 \\ 0 \\ \vdots \\ 1 \\ 0 \end{bmatrix} \quad \rightarrow \quad i^{th} \text{ element}$$

Consequently the representation (3.13),(3.16) describes a steady state Kalman filter, whereas it has another structure and contains two main properties:

- the prediction of output $\hat{y}(k/k-1)$ can be obtained by (3.13) independently from the state estimate
- the state estimate $\hat{\underline{x}}(k/k-1)$ is a linear function of the parametervector $\underline{p}$.

Based on these properties many of the standard adaptive control schemes can be solved.

3.2 Adaptive schemes using Kalman filter

3.2.1 Adaptive Kalman filter

Under the assumption that $\underline{A}, \underline{b}$ are uncertain, the estimation of the state $\underline{x}$ in (3.1) is in general no trivial problem. Nevertheless on the basis of the relation (3.16) a simple solution can be suggested. Since the data $\underline{I}_i$, $\underline{\varsigma}_1(k)$ and $\underline{\varsigma}_2(k)$ are computable for any arbitrary choosen stable matrix $\underline{F}$, the estimate $\hat{\underline{x}}(k/k-1)$ can be obtained from the estimate of the parametervector $\hat{\underline{p}}$, which results from any standard methods of recursive parameter estimation (e.g. stochastic approximation, least square). In comparison witn other adaptive Kalman filters (such as Quigley 1973, Saridis and Lobbia 1972) this solution ist easy for the implementation (Cao 1980).

3.2.2 Adaptive stabilization

In principle an adaptive stabilization scheme can be constructed as below:

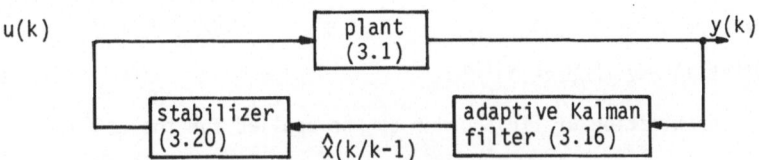

Fig. 3.1   adaptive stabilization scheme

The stabilizer is e.g. a "per interval" optimal controller (Saridis and Lobbia 1972), which minimizes the costfunction:

$$J = E\{\underline{x}^T(k+1)\underline{P}\underline{x}(k+1) + qu^2(k)\} \qquad (3.19)$$

The (sub)optimal input can be then given as:

$$u(k) = -\underline{k}_R^T(k)\hat{\underline{x}}(k/k-1) \qquad (3.20)$$

where $\qquad \underline{k}_R^T(k) = [\hat{\underline{b}}^T(k)\underline{P}\hat{\underline{b}}(k) + q]^{-1}\hat{\underline{b}}(k)q\hat{\underline{A}}(k) \qquad (3.21)$

Because of the relations (3.12), (3.15) the values $\underline{k}_R^T(k)$ can be calculated at every time k from the estimated parametervector $\hat{\underline{p}}(k)$.

3.2.3 Adaptive control

For tracking problems the following configuration is considered:

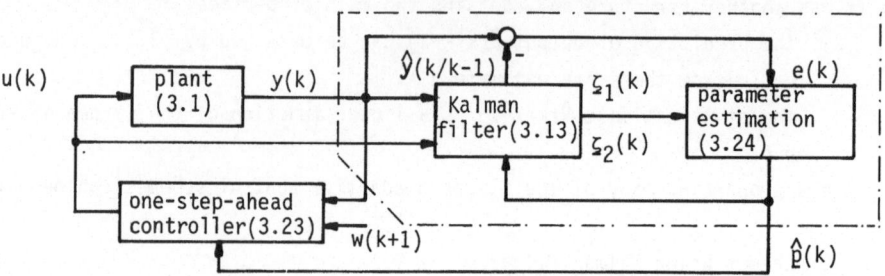

Fig. 3.2 adaptive control

In this case the "one-step-ahead" optimal controller has the costfunction:

$$J = \frac{1}{2}\{ p(w(k+1)-y(k+1))^2 + qu^2(k)\} \rightarrow min. \quad , \tag{3.22}$$

where $w(k+1)$ denotes the reference signal at time $k+1$ and $p>0$, $q>0$.

Then the optimal input $u(k)$ can be given as (Johnson and Tse 1978):

$$u(k) =[q+p\hat{b}_1^2(k)]^{-1}\hat{b}_1(k)p [w(k+1)- \sum_{i=1}^{n} (-\hat{a}_i(k))y(k-i+1)- \sum_{j=2}^{n} \hat{b}_j(k)u(k-j+1)] \tag{3.23}$$

which requires only the a priori informations of the reference w at time $k+1$, the data of output y until time k, input u until time $k-1$, and the estimated values of $\hat{a}_i(k)$, $\hat{b}_j(k)$. For that reason the central problem of this configuration is the adaptation of the controller-parameters via the estimated parametervetor $\hat{\underline{p}}(k)$.

With the linear relation (3.13):

$$\hat{\underline{y}}(k/k-1) = \underline{z}^T(k)\hat{\underline{p}}(k-1) \tag{3.13}$$

a stochastic approximation of first order can be applied for the estimate of $\underline{p}$:

$$\hat{\underline{p}}(k) = \hat{\underline{p}}(k-1) + \alpha(k)\begin{bmatrix} \zeta_1(k) \\ -\frac{1}{\zeta_2(k)} \end{bmatrix} (y(k) - \hat{y}(k/k-1)) \quad , \tag{3.24}$$

where $\alpha(k)$ will be choosen such that the Dvoretzky's condition (Saridis 1974) is fulfiled.

The control system is theoretically BIBO stable, if p and q are appropriately selected such that the characteristic polynom

$$C(z) = q + \hat{b}_1 pz \frac{\sum\limits_{i=1}^{n} \hat{b}_i z^{-i}}{1+ \sum\limits_{i=1}^{n} \hat{a}_i z^{-i}} \tag{3.25}$$

has only root locus with $|z|< 1$ (Johnson 1978, Cao 1980).

Unfortunately the stability will be really assured if the parameters $\hat{a}_i$, $\hat{b}_j$ were a priori exactly known. Nevertheless the simplicity of this adaptive control concept is still one of its major advantages with regard to practical implementation.

3.3 Extensions

The results in 3.1 and 3.2 can be extended to the global system of the considered industrial robot. Since the subsystem for vertical motion is uncoupled with the others (see(2.2)) , the coupled part of the industrial robot is described by a (2x2) model. Under the assumption of completely controllability and observability its dynamics can be approximated as:

$$\underline{x}(k+1) = \underline{A}\underline{x}(k) + \underline{B}\underline{u}(k) + \underline{H}\underline{r}(k)$$
$$\underline{y}(k) = \underline{C}\underline{x}(k) + \underline{v}(k)$$

$$x \in R^4, \quad u \in R^2, \quad y \in R^2 \qquad , \qquad (3.26)$$

where

$$\underline{A} = \begin{bmatrix} -a_{11} & 1 & -a_{13} & 0 \\ -a_{21} & 0 & -a_{23} & 0 \\ \hline -a_{31} & 0 & -a_{33} & 1 \\ -a_{41} & 0 & -a_{43} & 0 \end{bmatrix} \qquad \underline{B} = \begin{bmatrix} b_{11} & b_{13} \\ b_{21} & b_{23} \\ b_{31} & b_{33} \\ b_{41} & b_{43} \end{bmatrix} \qquad C = \begin{bmatrix} 1 & 0 & 0 & 0 \\ 0 & 0 & 1 & 0 \end{bmatrix} \quad (3.27)$$

With the notations ($\underline{F}$ is the systemmatrix of the steady Kalman filter):

$$\underline{G} = \begin{bmatrix} g_{11} & g_{13} \\ g_{21} & g_{23} \\ g_{31} & g_{33} \\ g_{41} & g_{43} \end{bmatrix} \qquad , \qquad \underline{F} = \begin{bmatrix} -f_{11} & 1 & -f_{13} & 0 \\ -f_{21} & 0 & -f_{23} & 0 \\ \hline -f_{31} & 0 & -f_{33} & 1 \\ -f_{41} & 0 & -f_{43} & 0 \end{bmatrix} \qquad (3.28),(3.29)$$

f_{ij} can be given as below:

$$f_{ij} = a_{ij} + g_{ij} \quad , \qquad i=1,2,3,4 \quad , \quad j=1,3 \qquad . \quad (3.30)$$

It can be shown that:

$$\hat{\underline{y}}(k/k-1) = \begin{bmatrix} \underline{\zeta}_1^T(k), \underline{\zeta}_2^T(k) \end{bmatrix} \begin{bmatrix} \underline{g}(k-1) \\ \hline \underline{b}(k-1) \end{bmatrix} \qquad (3.31a)$$

where

$$\underline{g}^T = [\ g_{11}\ g_{13}\ g_{31}\ g_{33}\ g_{21}\ g_{23}\ g_{41}\ g_{43}\ ] \qquad (3.31b)$$
$$\underline{b}^T = [\ b_{11}\ b_{13}\ b_{31}\ b_{33}\ b_{21}\ b_{23}\ b_{41}\ b_{43}\ ] \qquad (3.31c)$$

$$\underline{\zeta}_1^T(k) = [\ \underline{I} + \underline{F}_1\ \tau^{-1} + \underline{F}_2\tau^{-2}\ ]^{-1} \begin{bmatrix} \underline{y}^T(k-1) & 0 & \underline{y}^T(k-2) & 0 \\ 0 & \underline{y}^T(k-1) & 0 & \underline{y}^T(k-2) \end{bmatrix} \quad (3.31d)$$

$$\underline{\zeta}_2^T(k) = [\ \underline{I} + \underline{F}_1\ \tau^{-1} + \underline{F}_2\tau^{-2}\ ]^{-1} \begin{bmatrix} \underline{u}^T(k-1) & 0 & \underline{u}^T(k-2) & 0 \\ 0 & \underline{u}^T(k-1) & 0 & \underline{u}^T(k-2) \end{bmatrix} \quad (3.31e)$$

$$\underline{F}_1 = \begin{bmatrix} f_{11} & f_{13} \\ f_{31} & f_{33} \end{bmatrix} , \quad \underline{F}_2 = \begin{bmatrix} f_{21} & f_{23} \\ f_{41} & f_{43} \end{bmatrix} \qquad (3.31f)$$

and the optimal input for tracking problems is given by:

$$\underline{u}(k) = [Q + \hat{\underline{B}}_1^T P \hat{\underline{B}}_1]^{-1} \hat{\underline{B}}_1^T P [\underline{w}(k+1) - \sum_{i=1}^{2} \hat{\underline{A}}_i \underline{y}(k-i+1) - \hat{\underline{B}}_2 \underline{u}(k-1)] \qquad (3.32)$$

where

$$\hat{\underline{A}}_1 = \begin{bmatrix} -\hat{a}_{11} & -\hat{a}_{13} \\ -\hat{a}_{31} & -\hat{a}_{33} \end{bmatrix} , \quad \hat{\underline{A}}_2 = \begin{bmatrix} -\hat{a}_{21} & -\hat{a}_{23} \\ -\hat{a}_{41} & -\hat{a}_{43} \end{bmatrix} , \quad \hat{\underline{B}}_1 = \begin{bmatrix} \hat{b}_{11} & \hat{b}_{13} \\ \hat{b}_{31} & \hat{b}_{33} \end{bmatrix} , \quad \hat{\underline{B}}_2 = \begin{bmatrix} \hat{b}_{21} & \hat{b}_{23} \\ \hat{b}_{41} & \hat{b}_{43} \end{bmatrix} \qquad (3.33)$$

$$\underline{P} > \underline{0} \quad , \quad \underline{Q} > \underline{0} \qquad . \qquad (3.34)$$

4. ADAPTIVE CONTROL OF THE SUBSYSTEMS OF AN INDUSTRIAL ROBOT

The adaptive concept under 3.2.3 has been implemented for the control of the industrial robot. There are no assumptions about a priori informations of the system. For the
control design of the subsystems a linear model of third order as given by (3.1),(3.4)
has been selected. In the first study the control performance of only one of the subsystems has been considered. The control has been realized by coupling a process computer (Krantz Mulby 3, Sperry Univac) with the industrial robot via A/D- respectively
D/A-converters. Harmonic functions were selected as reference signals. A simple typical movement has also been tested. Representative experimental results are shown as
below (——— :w(k), ⋯⋯⋯ :y(k)):

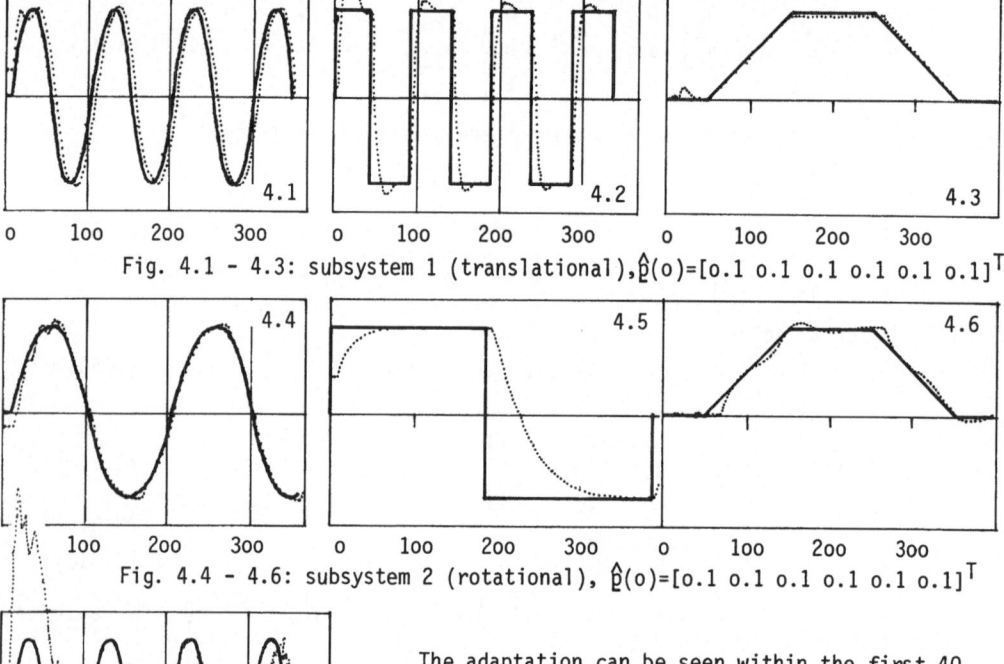

Fig. 4.1 - 4.3: subsystem 1 (translational),$\hat{\underline{\beta}}(o)=[o.1 \ o.1 \ o.1 \ o.1 \ o.1 \ o.1]^T$

Fig. 4.4 - 4.6: subsystem 2 (rotational), $\hat{\underline{\beta}}(o)=[o.1 \ o.1 \ o.1 \ o.1 \ o.1 \ o.1]^T$

Fig. 4.7: subsytem 1
$\hat{\underline{\beta}}(o)=[o.1 \ o.2 \ o.3 \ o.4 \ o.5 \ o.6]^T$

The adaptation can be seen within the first 40
sampling intervals (sampling interval:10msec). As
reported in detail in(Cao 1980) a periodic testsignal with f<5Hz for subsystem 1 has been followed after 50 sampling intervals. For the simple
movement also the control performence is satisfactory. Similar to the most indirect control methods any resultant bias in the parameter estimates
could unfortunately lead to control inconsistency.This can be shown in fig.4.7, where
$\hat{\underline{\beta}}(0)$ is replaced by $[0.1 \ 0.2 \ 0.3 \ 0.4 \ 0.5 \ 0.6]^T$ in comparison with fig.4.1. This disadvantage may be the price for the simplicity of the adaptive concept.

5. CONCLUSIONS AND FURTHER INVESTIGATIONS

An adaptive concept incorporating an adaptive Kalman filter and an "one-step-ahead-controller" has been represented. Because of its simplicity and flexibility this type of control structure is especially suitable for process control. It has been implemented for controlling the arm of an industrial robot. The global results are satisfactory. Nevertheless improvements must be used to circumvent the limitation with regard to the stability of the control scheme. The author is studying the control performance of the global system and the influence of manipulated objects to the system behaviour. Further results will be reported later.

ACKNOWLEDGEMENT

The author would like to express his gratitude to Prof.Dr.H.Schwarz and J.Rößler for the useful discussions and to the Deutsche Forschungsgemeinschaft which supported the research to this paper (DFG Schw 120/24).

REFERENCES

Astrom,K.J.(1970): Introduction to stochastic control theory. Academic Press, New York, 1970.

Cao,C.T.(1980): Eine äquivalente Kalman-Filter-Struktur und ihre Anwendung zur adaptiven Regelung eines elektro-hydraulischen Handhabungssystems. Interner Bericht Nr. 1/80 des Fachgebiets Meß-, Steuer- und Regelungstechnik der Universität Duisburg.

Dubowsky,S.;Des Forges,D.T.(1979): The Application of Model-Referenced Adaptive Control to Robotic Manipulators. Journal of Dynamic Systems, Measurments and Control,101,pp.193-200.

Freund,E.;Syrbe,M.(1976): Control of Industrial Robots by Means of Microprocessors. In Lecture Notes in Control and Information Sciences, 2, Springer-Verlag, Berlin-Heidelberg-New York, pp.167-185.

Johnson,C.R.(1978): On Single-Stage Optimal Control. Proceedings 1978 IEEE Southeastcon.,Atlanta, GA, April.

Johnson,C.R.;Tse,E.(1978): Adaptive Implementation of One-Step-Ahead Optimal Control via Input Matching. IEEE Trans. on Automatic Control, 23,5,pp.865-872.

Quigley,A.L.C.(1973): An Approach to the Control of Divergence in Kalman Filter Algorithms, Int. J. Contr.,17, pp.741-746.

Rößler,J.(1980): A decentralized hierarchical control concept for large-scale systems. In IFAC Symposium for Large-Scale Systems Theory and Applications, Toulouse, France, June 1980.

Rößler,J.;Schwarz,H.(1978): Industrie-Roboter-Regelung. Arbeitsbericht zum Forschungsvorhaben DFG Schw 120/17.

Rößler,J.;Schwarz,H.(1979): Industrie-Roboter-Regelung. Arbeitsbericht zum Forschungsvorhaben DFG Schw 120/22.

Saridis,G.N.(1974): Stochastic Approximation Methods for Identification and Control — a Survey. IEEE Trans. on Automatic Control, 19, pp.798-809.

Saridis,G.N.;Lobbia,R.N.(1972): Parameter Identification and Control of Linear Discrete -Time Systems. Trans. on Automatic Control, 17, pp.52-56.

USING THE SELF-OPTIMISING CONTROL OF AN ELECTRO-HYDRAULIC SERVO SYSTEM

TO MINIMISE THE POWER LOSS:

W L Green and D J Sanger
Dept. Aeronautical & Mechanical Engineering
University of Salford

SALFORD, United Kingdom

B N Suresh
Vikram Sarabhai Space Centre
Trivandrum

INDIA

There is an increasing use of electro-hydraulic servo systems to perform mechanical operations within complete machine concepts at ever higher power levels. The most commonly used system comprises a set of electro-hydraulic servo valves controlling the position, or speed, of hydraulic actuators within a given specification. The valves draw their supply of hydraulic fluid from a constant pressure source and a typical system incorporating velocity or position control is shown in Fig.1. The output signals from the actuators are returned to the input of their respective valve units to form a closed loop control. In the past the use of a fixed delivery pump meant that the level of power dissipation was fairly high due to the necessary pressure used to drive the fluid through the valve units.

A particular application is that for aircraft flying controls. In these the actuators are selected to withstand the maximum 'stall' load on the aerodynamic surfaces and also to move at a specified speed under 'no-load' conditions. The valves are selected using the 'load speed' locus specified. As the authors were examining the feasibility of the proposed system and there were no typical load locus curves available, it was decided to specify the valves on the basis of maximum power application when one third of the supply pressure forces the fluid through the valves at the specified load which then absorbs two thirds of the pressure. The selection of the valves to perform in this way means that the maximum efficiency of the valve actuator units is 67.7%. If they are used with a fixed delivery pump the total system efficiency is then only 38% (Ref.1). The advent of the variable delivery pump has improved this overall efficiency, as indicated in Fig.2., but when it is realised that for most of an aircraft's flight time when cruising, the loads on the flying controls are quite small then the power dissipation remains at a high level and can prove to be an embarrassment as this lost power always raises the temperature of the transmitting fluid. In earlier days oil-fuel heat exchangers were used to dissipate this waste heat but with the development of very high performance small aircraft it is proving impossible to do this and designers have turned their thoughts to possible ways of alleviating the problem.

Two possible solutions have been suggested and have been discussed in a previous

publication by the authors (Ref.2). One of these entails the use of a variable
pressure relief valve so that the system pressure is always set at 1.5 times the
maximum differential pressure which corresponds to the maximum operating load. The
required flow demands to the servos are set by the respective valves and then the
pump flow is adjusted accordingly. The system has not yet been analysed in detail
and certainly warrants further investigation. A second method involves using a
'search' strategy and self-optimising control built into the aircraft so that
signals from the output loads and speeds are continually monitored and fed back to
the valves which then position themselves at the optimum position for minimum power
dissipation in conjunction with the setting of a variable delivery pump. The analysis
and implementation of such a system has been discussed and demonstrated as feasible
within the context of the aforementioned publication. (Ref.2).

The purpose of this present paper is to extend the results of the previous report
and to present some experimental results obtained on a facility which was unique in
showing the application of a self-optimising control to a machine producing a reasonable level of power.

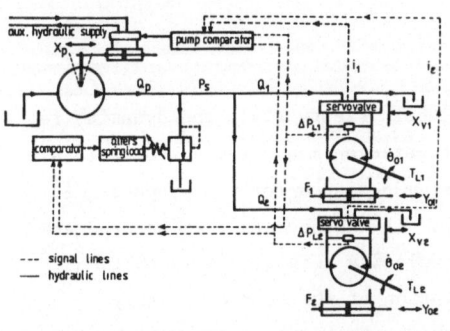

Fig.1.

A Schematic Circuit Diagram of an
Electro Hydraulic System

--- signal lines
— hydraulic lines

NOTATION

PI – Performance index
w_1, w_2 weighting factors Eqn.(1)
H_o – power output
H_i – total input power
n – total number of servo actuators
Q – flow rate
X – displacement
X* – non-dimensionalised displacement
p – a) suffix – pump control or b) pressure.
S – suffix – supply pressure
v – suffix – valves
$\dot{\theta}_o$ – actual output speed
$\dot{\theta}_i$ – desired speed
F – output load
Y – output displacement
T – output torque
L – suffix refers to load

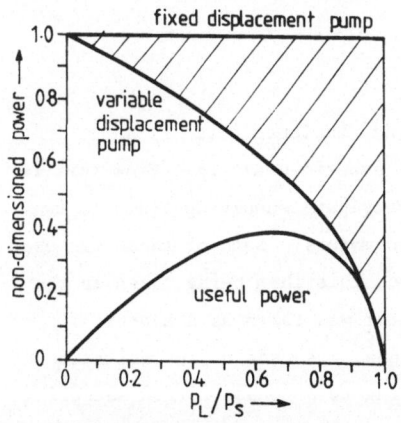

Fig.2. Use of a variable pump to reduce power losses

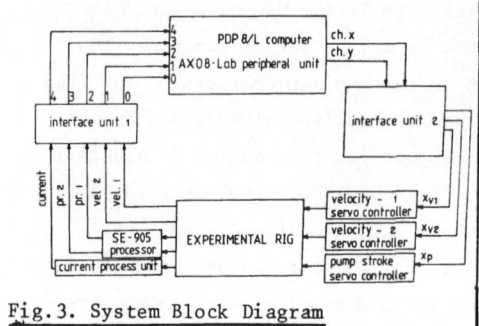

Fig.3. System Block Diagram

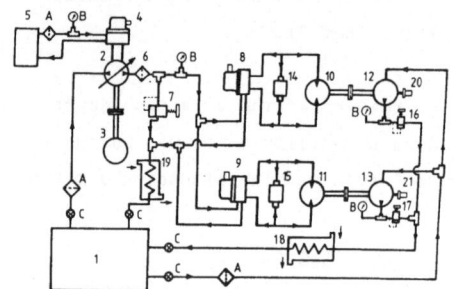

Fig.4. Details of Experimental Rig

Key to Fig.4.

1	Reservoir
2	Lucas PM-125 variable pump
3	Crompton FDK-5 Induction motor
4	Moog 76-101 servovalve
5	External hydraulic power pack
6	High pressure filter
7	Vickers 210 Bar relief valve
8,9	Dowty 2551 series servovalves
10,11	Lucas Pml25 & PM60 fixed motors
12	Plessy A-54Y gear pump
13	Keelavite CPD-2000 gear pump
14,15	SE 165-D pressure transducers
16,17	Keelavite load relief valves
18,19	Heat exchangers
20,21	Evershed D.C tachogenerators
A	Low pressure line filters
B	Pressure gauges
C	Hand operated needle valves

The analysis of electro-hydraulic servo systems, both steady state and dynamic, is well documented and has been analysed and investigated on many previous occasions. The particular methods used to stabilise the system designed for the purposes of this particular rig are indicated in (Ref. 3, 4). It was necessary to incorporate electronic shaping networks within the selected velocity servo loops which then provided the response required for the implementation of the search strategy eventually selected. As rotary motion was easier to transduce rotary motors were used.

THE SELF-OPTIMISING SYSTEM

The block diagram of the system used for the 'on-line' implementation is shown in Figure 3 with Figure 4 providing the details of the experimental rig. From this it can be seen that this consisted of a variable delivery pump supplying fluid to two electro-hydraulic servo valves controlling hydraulic motors. Each of these was used to drive a pump forcing fluid through a relief valve, this then being taken as the load on the motor. The pressure drop across the motor was taken as a measure of this load and a differential pressure transducer signal was used as an indication of its magnitude. This obviated difficulties with having to purchase torque measuring instrumentation. Tachometers measured the output speed of the motors and these four signals together with that obtained from the current used by the electric motor driving the variable delivery pump were fed to an interface unit then into the

laboratory PDP8/L computer. The output signals from the computer are fed into the
servo controllers corresponding to the pump stroke control and velocity servos,
through the second interface unit. It was therefore possible for the operator to
set the loads by means of the pressure relief valves and to select the speeds required.
It would be possible to modify the rig so that both the loads and speed could be
varied in a random manner. The computer using a search strategy which has been
selected as the result of an extensive analytical comparison used calculated values
to minimise the power losses and to set the servo valves and the pump stroke to a
position for minimum power dissipation. This did not occur instantaneously and the
selection of a suitable optimising strategy was based on the minimum number of
adjustments which had to be made in order to find this position.

For this particular application the computer was fairly old and slow which inhibited
the method of control. In addition it only provided two variable controlling outputs
and the rig required three. It is achieved by sending the signal corresponding to X_p^*
through the X-channel whereas signals corresponding to X_{v1}^* and X_{v2}^* were sent in
sequence with a time delay of 9ms along the Y channel. The corresponding relays 1
and 2 were then pulsed with a delay of 3ms after each signal was sent in order to
drive the relays in the interface unit. They were held for 3ms and then switched
off. This again slowed down the control and resulted in a much slower response than
would normally be obtained from a specifically designed controller. The interactive
language was 'Focal' and a special programme had to be developed to send the signals
in sequence. The flow chart for the 'send routine' (FSEND) is shown in Figure 5,
together with the sampling routine (FSAM). At the end of (FSEND) routine a delay
of 120ms corresponding to the sampling time required between the two iterations was
introduced. The timing sequence is indicated in Figure 6, details of the interface programmes being given in Ref.5. The analogue to digital conversion is stored
in the core as an integer corresponding to the input voltage read and calibrated on
the channel.

POWER MINIMISING

Finding the point or zone of minimum power dissipation is a mathematical problem
which has been examined in detail in many previous publications, a reasonably concise
description of the various methods available for digital computation being given in
(Ref.6).

In order to assess the performance of each selected search strategy on the response
surface describing the behaviour of the system, a digital computer simulation was
carried out. A suitable performance index had already been determined in the form

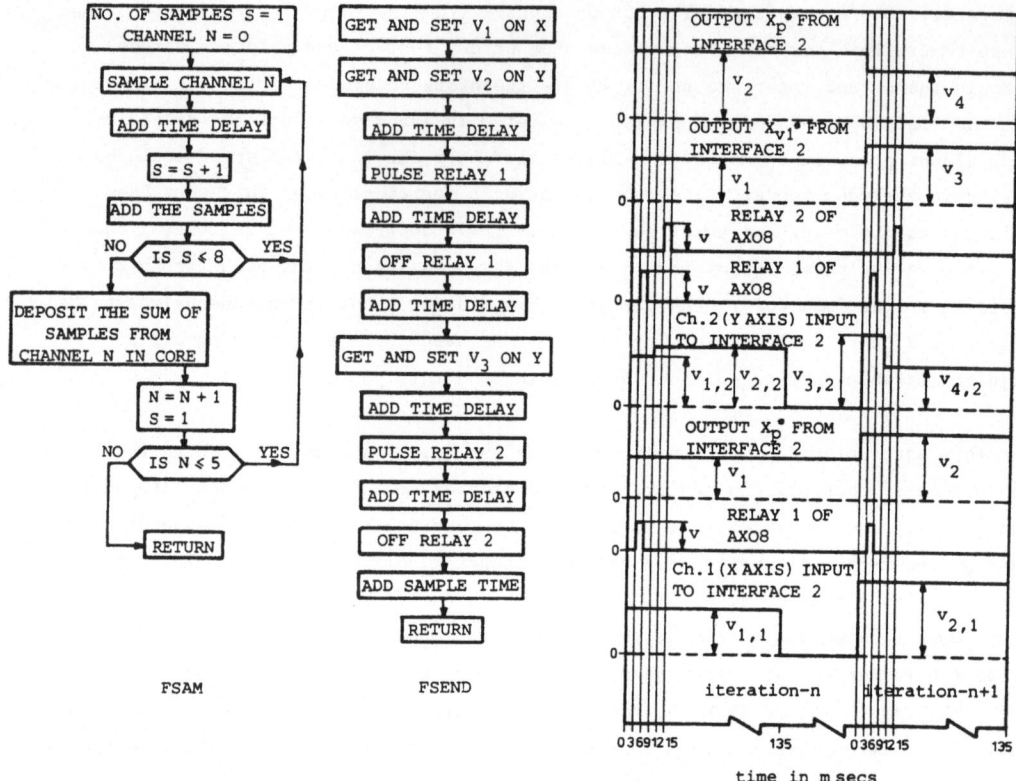

Fig. 5 - Flow Charts   Fig. 6 - Timing Sequence

of the following equation, as indicated in (Ref.5).

$$PI = w_1 \left[\sum_1^n (1 - \frac{\dot\theta_o}{\dot\theta_i})^2 \right] + w_2 \left[(1 - \frac{\sum_1^n (H_o)}{H_i}) \right]^{0.125} \qquad (1)$$

(In this case n = 2 and w_1 = 0.4 and w_2 = 0.6 were chosen).

Each iteration in any strategy corresponds to one set of servo adjustments by the
controlled variables and thus the total number of iterations indicates the time
required by that strategy to achieve the minimum in a practical system. This time
also depends on the complexity of the logical operations involved in a particular
strategy and apart from the speed of convergence the effectiveness of strategy lies
in its ability to explore the true minimum both in 'noise' and 'noise free' conditions.
These properties form the vital factors in assessing the suitability of a particular
strategy for 'on-line' implementation.

Five strategies were examined, using the digital computer, to study their ability to

minimise the power losses inherent in the electro-hydraulic systems within the specified speed errors of the hydraulic motors. A three-dimensional model of the system
with three different load and speed combinations was examined, these being a) 0.9 and
0.9 b) 0.5 and 0.5 c) 0.9 and 0.5 for the servos 1 and 2 respectively, the values
referring to the non-dimensionalised torques on the two hydraulic motors, the maximum
ratio being 1. The corresponding response surfaces for these load conditions are
different in shape, even though the servo speed was kept constant at 1500 revs per
minute for all the tests. The various constraints imposed upon the models during
the simulation were:

    i)   0 < pump stroke ratio < 1

   ii)   0 < valve displacement 1 ratio < 1

  iii)   0 < valve displacement 2 ratio < 1

   iv) Motor speed > 0.

These constraints were dealt with by setting the servo adjustments well within the
boundary limits and although this method could confuse gradient strategies it is
believed that they are quite acceptable for direct search methods.

To study the performance of each search strategy two suitable starting points of
1) 0.8, 0.6, 0.6 and 2) 0.7, 0.5, 0.4 were chosen for X_p*, $X_{v1}*$ and $X_{v2}*$ respectively
by utilising the earlier results obtained from an examination of trial and error
methods.

It was clear when the results were examined that only the 'Simplex' (Ref.7) and
'Pattern' (Ref.8) methods were successful in reaching the minimum both with 10%
noise and in the 'zero noise' case. Although the performance of the other methods
was good when there was no noise they failed to converge to a minimum when it was
introduced. If only the final value obtained for the performance index is considered
when examining the performance of these search strategies there is no doubt that the
'Pattern' method demonstrated consistency in obtaining the lowest value in almost
all cases, closely followed by the 'Simplex'. If on the other hand the efficiency
with which the minimum is achieved is the consideration which differentiates between
the strategies it is evident that the 'Simplex' method not only exhibited the
fastest convergence speed towards a minimum value but was more or less consistent in
all the cases studied when considering the total number of servo adjustments required
by that strategy. However, there was no significant difference in the speed with
which the strategies approached the minimum region. The only difference was that
the 'Pattern' method took many more servo adjustments than the 'Simplex' in the
region of the minimum before reaching the point. Comparing the fluctuation of the
servo speeds during these adjustments it is apparent that they are pronounced in both

the 'Pattern' and 'Simplex' methods.

Both the 'Pattern' and 'Simplex' methods were reasonably successful in the simulation studies and both justify their consideration for on-line implementation. However, if a balance is to be drawn between these two successful strategies considering the final value which must be obtained and the efficiency with which it is to be achieved the most successful strategy is the 'Simplex'. Thus it is recommended as the best strategy for this 'on-line' implementation. This method is the modified 'Simplex' as recommended by Nelder and Mead (Ref.7).

THE ON-LINE IMPLEMENTATION OF THE SELF-OPTIMISING CONTROL FOR THE EXPERIMENTAL RIG AND RESULTS THEREFROM

To implement the on-line strategy a laboratory model PDP8 digital computer with a 4K memory was used.

In all the search strategies discussed previously, the controlled variables X_p*, $X_{v1}*$, and $X_{v2}*$ were adjusted between 0 and 1. To carry out these adjustments it was essential to measure the voltage levels corresponding to the full stroke of the pump and full displacements of the valves. In the case of the servos these values do change with the variation of the system pressure. From the previous simulation results it was clear that the supply pressure variation ranges from 140 to 210 bars in the majority of cases and thus the input voltage to these servos was measured at the mean value of the pressure, i.e. 175 bar in order to minimise the error within the operating region. The resolution of all the three controlled variables was checked and was observed to be less than 0.5% of the respective signals corresponding to unity.

The implementation of the 'Simplex' strategy was carried out successfully on the rig and a series of experiments were conducted at several combinations of servo speeds and loads. The effectiveness of the strategy in seeking the optimum point and the effect of noise on the search performance were studied and a typical set of experimental results are now discussed.

DISCUSSION OF RESULTS

Some of these experimental results are summarised in Table 1 and were obtained by starting the search from the same initial point of $X_p* = 0.6$, $X_{v1}* = 0.8$ and $X_{v2}* = 0.6$. The reduction in the power loss given in this table was obtained by comparing the power input to the present system with one operating at a constant pressure and fixed displacement. The table shows the performance of the strategy when the servo loads were varied keeping the speeds constant. In almost all cases the system attained the optimum efficiency within 36 iterations, the minimum being 20. It was able to maintain the servo system speed errors within 5% in all cases except one. The reduction of the power loss was quite significant and became substantial when the servo was operating at small loads. These results are in good agreement with those of the theoretical simulation in spite of small variations in the system load experienced during the experiments which were caused by pump flow variations with the change in servo speeds, compared with the constant load assumed in the simulation.

Table 2 indicates the performance when the servo speeds are varied at constant load. The reduction in power loss was maximum when the servo speeds were very low. The effectiveness of the strategy in seeking an optimum was then tested by starting the search from different initial locations keeping the speeds and loads constant. In almost all these cases the search was able to reach the optimum region successfully, and the results are given in Table 3.

The effect of the 'noise' on the search performance was studied by repeating the search several times starting from the same initial point and keeping all the load and speed parameters unchanged. The results of these tests are shown in Table 4. The maximum variations in the speed errors for servos 1 and 2 were 4.43 and 4.29% respectively and the power saving was around 4%. Due to the noise the servo valve operating at the maximum load was unable to reach the fully open position although it tended to reach 0.9 of this point and beyond in the majority of cases as can be observed.

GENERAL CONCLUSIONS

The analysis presented in (Ref.5) had already indicated that the mode of operation of this system was for one of the valves to move to the 'fully open' position and then for the pressure to set itself to provide the necessary flow through this valve to provide the required speed at the hydraulic motor. The second valve then had as its input this pressure and moves to a position where the pressure drop across it, determined by the load, allowed the required flow through it to provide the selected speed. All the previous analyses had indicated that this was the case and the experimental results confirmed this.

The investigation, therefore, has helped to answer many questions concerning the design and practical application of self-optimising control to an electro-hydraulic system, wherein the power losses are minimised in even larger power installations.

TABLE 2: PERFORMANCE OF SIMPLEX FOR DIFFERENT SPEEDS

SL. NO	SPEED 1 RPM	SPEED 2 RPM	FINAL POSITIONS X_p^*	X_{v1}^*	X_{v2}^*	SPEED ERROR 1 %	SPEED ERROR 2 %	POWER LOSS REDUCTION %	SUPPLY PR. IN BARS	NO.OF SERVO ADJ.MENTS
1	800	800	0.488	0.862	0.330	3.57	7.14	44.71	137	40
2	900	900	0.567	0.786	0.425	2.38	4.13	33.60	158	36
3	1000	1000	0.634	0.951	0.491	0.71	0.71	32.10	155	36
4	1200	1200	0.81	0.961	0.591	0.71	2.95	21.20	158	29
5	1300	1300	0.918	0.991	0.59	1.54	2.20	17.00	155	20
6	700	1400	0.508	0.871	0.705	1.40	0.31	42.90	140	20
7	1400	700	0.973	0.998	0.310	3.06	1.22	13.40	176	28
8	800	1200	0.600	0.800	0.600	2.69	2.38	38.10	140	20
9	1000	1400	0.744	0.988	0.694	5.00	4.09	23.10	167	24
10	1200	800	0.715	0.969	0.394	1.19	5.36	28.30	158	36

LOADS: SERVO 1 - 1.0 ; SERVO 2 - 0.25

TABLE 1: PERFORMANCE OF 'SIMPLEX' - DIFFERENT LOADS

SL. NO	LOAD ON SERVO 1	LOAD ON SERVO 2	FINAL POSITIONS X_p^*	X_{v1}^*	X_{v2}^*	SPEED ERROR 1	SPEED ERROR 2	POWER LOSS REDUCTION	SUPPLY PR. IN BARS	NO.OF SERVO ADJ.MENTS
1	0.25	0.25	0.61	1.00	0.70	6.43	2.57	51.40	95	28
2	0.50	0.50	0.66	0.65	0.889	5.71	2.14	42.70	116	36
3	0.75	0.75	0.62	0.883	0.627	0.86	5.71	36.70	130	32
4	1	0.25	0.634	0.951	0.491	0.71	0.71	32.40	155	36
5	1	0.50	0.662	0.944	0.620	1.50	6.40	33.80	149	20
6	1	0.60	0.717	0.757	0.866	5.00	0.30	25.00	173	36
7	0.80	0.40	0.636	0.953	0.509	2.70	3.20	39.80	140	32
8	0.25	0.70	0.643	0.505	0.682	4.30	1.55	28.60	155	36
9	0.50	0.70	0.634	0.477	0.722	2.57	3.70	27.70	165	36
10	0.70	0.75	0.603	0.725	0.693	5.71	5.71	30.30	158	28

SET SPEEDS: SERVO 1 = 1000 rpm; SERVO 2 = 1000 rpm

TABLE 4: EFFECT OF NOISE ON 'SIMPLEX'

SL. NO	FINAL POSITIONS X_p^*	X_{v1}^*	X_{v2}^*	SPEED ERROR 1 %	SPEED ERROR 2 %	POWER LOSS REDUCTION %	FINAL SYS.PR. BARS	NO.OF SERVO ADJ.MENTS
1	0.717	0.904	0.578	0.57	2.17	24.3	172	32
2	0.722	0.942	0.560	1.43	1.19	23.6	172	32
3	0.710	0.969	0.562	1.43	0.95	24.9	172	28
4	0.733	0.938	0.581	2.14	1.79	23.1	172	36
5	0.684	0.890	0.605	0.71	4.55	24.0	172	28
6	0.693	0.982	0.602	0.43	4.29	26.4	170	32
7	0.716	0.954	0.574	4.57	0.95	24.6	172	28
8	0.692	0.960	0.593	0.71	2.63	26.4	172	32
9	0.682	0.892	0.607	3.57	4.55	27.1	172	24
10	0.720	0.916	0.596	5.00	5.24	23.6	172	24

SET SPEEDS (rpm) - 1000 1200
LOADS - 1.0 0.25

TABLE 3: EFFECTIVENESS OF 'SIMPLEX' IN SEEKING THE OPTIMUM

SL. NO	STARTING POINT X_p^*	X_{v1}^*	X_{v2}^*	FINAL VALUES X_p^*	X_{v1}^*	X_{v2}^*	SPEED ERROR 1	SPEED ERROR 2	POWER LOSS REDUCTION	SYSTEM PR. IN BARS	NO.OF SERVO ADJ.MENTS
1	0.6	0.8	0.6	0.626	0.894	0.610	3.86	5.88	23.6	172	24
2	0.7	0.9	0.6	0.697	0.933	0.598	0.29	3.81	27.4	173	28
3	0.8	0.9	0.7	0.667	0.732	0.566	1.14	2.12	24.6	180	32
4	0.5	0.5	0.5	0.681	0.705	0.579	1.14	4.17	27.7	180	32
5	0.7	0.7	0.7	0.695	0.990	0.580	1.71	0.71	27.1	169	32
6	0.8	0.8	0.8	0.716	0.917	0.587	1.86	7.14	25.0	176	36
7	0.7	0.6	0.5	0.707	0.887	0.620	1.86	4.00	24.3	176	36
8	0.6	0.6	0.6	0.710	0.930	0.601	2.29	2.98	24.9	172	40
9	0.7	0.5	0.5	0.713	0.935	0.583	2.14	1.79	25.0	172	44
10	0.4	0.6	0.5	0.704	0.955	0.633	0.57	7.86	27.1	173	48

SET SPEEDS (rpm) - 1000 1200
LOADS - 1.0 0.50

It establishes the feasibility of using such a system in practice and focuses attention on the problems associated with the design of an on-line controller.

The choice of the hydraulic elements and associated transducers is a most important factor when designing this type of system. Certain hydraulic components used on this particular rig had a low efficiency, particularly the hydraulic motors, and hence inherent power losses were encountered of the order 37%. It is, therefore, absolutely necessary to select components with a high efficiency. Similarly, certain problems were experienced with the poor dynamic behaviour of the pressure transducers chosen and it is essential to base the choice of the transducers on a good dynamic performance.

The formation of the mathematical model incorporating true velocity servos and the theoretical investigation using this model resulted in a general description of the nature of the response surfaces associated with the defined performance index. The preliminary simulation study on this model using trial and error methods showed that the considerable power loss can be reduced by using this type of control and the servo speed errors can be maintained within 5 to 6% by assigning a suitable weighting to the speed error and power loss factors.

The digital computer simulation studies using different search strategies indicated that the 'Simplex' method was the most successful and efficient optimising strategy for this particular on-line power minimisation. The method also possesses the added advantage that a new 'Simplex' can easily be formed by the addition of new points to any of the phases within the mathematical model. The strategy was therefore used for implementing the 'on-line' control of the rig using the small PDP8 digital computer and the experimental results proved that significant reductions in power losses can be achieved. The strategy was successful in seeking a minimum in nearly all the cases selected and has shown its ability to attain this minimum in a practical system characterised by noise.

Unfortunately, the time taken by the computer between two successive servo adjustments during the practical implementation was approximately 800ms. This long delay is mainly attributable to the slow speed of the computer and the interactive language 'Focal' together with the signal delay of 3ms giving a total time loss of 120ms when it was necessary to sample the 5 signals 8 times. This could be considerably reduced by using a fast computer and suitable machine coded language, and substantially reducing the signal to noise ratio by a careful selection of the transducers and associated electronics.

This particular application was mainly concerned with rotary elements used in velocity servos but the same techniques can be extended to minimise the power losses in linear servo systems. Although the 'Simplex' strategy has proved successful in this present investigation doubts have been raised about its efficiency when more than 5 or 6 multivariables are encountered. Secondly, the mathematical definition used for the performance index which sums speed errors 1 and 2 before multiplying by the weighting factor leads to a discrepancy.

It is hoped to pursue this investigation further to determine whether the system is feasible for a large scale application such as an aircraft where it is possible to incorporate the computational process within the auto pilot for example. Or is it possible that a much simpler system can be used whereby the relief valve which sets the inlet pressure to the servos can be adjusted so that this considerable saving in power loss can be achieved.

ACKNOWLEDGEMENT

Acknowledgement is made to Professor J L Livesey for his encouragement, to Mr W Reynolds for his help in the laboratory at the Department of Aeronautical and Mechanical Engineering at the University of Salford and to Commonwealth Scholarship Commission for the financial support to Dr B N Suresh during the tenure of this work.

REFERENCES

1. WALTERS, R. - Hydraulic and Electro-Hydraulic Servo Systems - Iliffe, 1967.

2. GREEN, W. L., SANGER, J. D., AND SURESH, B. N. - The reduction of fluid power losses in electro-hydraulic servo systems. - 21st Israel Annual Conference on Aviation and Astronautics, Tel-Aviv - Haifa 1979, pp. 111-119.

3. BELL, R., De PENNINGTON, A., - 'Active compensation of lightly damped electro-hydraulic cylinder drives using derivative Signals', Automation Control gp., Proc. Inst. Mech. Engrs. Vol.184, Pt 1, pp. 83-98, 1970.

4. WELCH, T. R., - 'The use of derivative pressure feedback in high performance hydraulic servo mechanism', J. Engng. Indstr., Trans. ASME 84, Series 8, 1962.

5. SURESH, B. N. - The Minimisation of Power Losses in Electro-hydraulic Servo Systems - Ph.D. Thesis, Department of Aeronautical and Mechanical Engineering, University of Salford, 1978.

6. ADBY, P. R. AND DEMPSTER, M. A. H. - Introduction to Optimisation Methods - Chapman & Hall 1974.

7. NELDER, J. A. AND MEAD, R. - A Simplex method for function minimisation. Comput. J. 7 pp. 308-313, 1965.

8. HOOKE, R., & JEEVES, T. A., - 'Direct Search Solution of numerical and statistical problems', J. Assoc. Comp. Machinery, 8 pp. 212-219, 1961.

ADAPTIVE CONTROL BY A SENSITIVITY METHOD
WITHOUT NEED FOR ON-LINE IDENTIFICATION

E.G. Kunze
Fraunhofer-Institut für Informations-
und Datenverarbeitung
7500 Karlsruhe, Germany

Summary

The continuous pseudo gradient method is used for the design of self-optimizing control systems. For plant parameter variations within certain
limits on-line identification is not needed. An important improvement
of performance is achieved by a special normalization technique of the
gradient and by an automatic control of the parameters of the gradient
algorithm itself. Simulation and application results show the performance of the method.

1. Introduction

There are many industrial processes with variable parameters and signal
properties which can be controlled in a suitable way by use of self-optimizing systems. For a problem description consider the system given
by fig. 1.

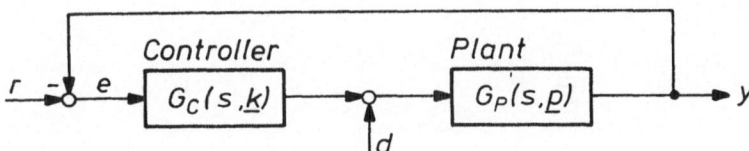

Fig. 1   System

The blocks $G_C(s,\underline{k})$ and $G_P(s,\underline{p})$ contain the transfer-functions of controller and plant with the parameter vectors $\underline{k}$ and $\underline{p}$. The vector $\underline{p}$ - or
at least some of its elements - are only known so far as to lie or to
vary slowly within a limited range of the parameter space. The inputs r
and d are the reference and the disturbance signal. Conditions encountered in practice often show r to be constant while d is unknown.

Self-optimiźation means that the parameter vector $\underline{k}$ is automatically adjusted in order to minimize some performance criterion of the error signal e. This criterion is usually a time integral which may be slowly time varying if $\underline{p}$ and the properties of d change with time, e. g.

$$I(\underline{k},\underline{p},t) = \int_{t-T}^{t} f[e(\underline{k},\underline{p},\lambda)]d\lambda \qquad (1)$$

where f is a positive definite function. Solutions to the problem of self-optimization have been developed in /1/ and /2/ making use of a continuous gradient method requiring on-line identification of $\underline{p}$.

In the following a continuous pseudo gradient method will be proposed which operates with fixed nominal values $\underline{p}_o$ and therefore avoids on-line identification. Additionally an important improvement of performance is achieved by a special normalization technique of the gradient and by an automatic control of the parameters of the gradient algorithm itself. The performance of the method is demonstrated by simulation results and by results obtained with the adaptive control of an industrial cement mill.

2. Description of the Pseudo Gradient Method

In /1/ and /2/ it has been shown that the integral of equ.(1) can automatically be minimized with respect to $\underline{k}$ by use of the continuous gradient equation

$$\frac{dk}{dt} = - \underline{H} \int_{t-T}^{t} \frac{d}{de} f \cdot \frac{\partial e}{\partial \underline{k}} d\lambda \qquad (2)$$

where $\underline{H}$ is a matrix of design parameters of the same order as $\underline{k}$ with all off-diagonal elements equal to zero, and $\partial e/\partial \underline{k}$ is a vector of sensitivity functions. Equ.(2) determines the time derivative of $\underline{k}$ in dependence upon the error signal e and therewith specifies the design of the adaptive loop. Equ.(2) will also work - even with improved performance /1/ - if the integration is omitted.

Assuming quasi stationary parameters $\underline{k}$ and $\underline{p}$ the sensitivity functions can be generated with a sensitivity model as shown in fig. 2 /2/, /5/. The realization of the model requires the knowledge of G_p and $\underline{p}$. Thus on-line identification is needed. This can be avoided, however, by using a pseudo sensitivity function (PSF) instead of the normal sensitivity function /3/. The PSF $\hat{\partial}$ is defined according to /4/ as an approximate

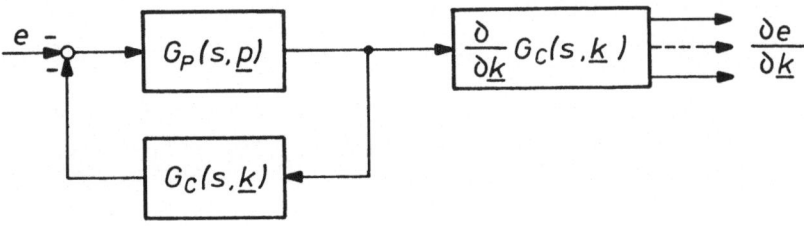

Fig. 2   Sensitivity Model

sensitivity function obtained by replacing the exact sensitivity model by some suitable approximation. (A different definition is given in /5/).

The approximation suggested here is to substitute the parameter vector $\underline{p}$ in fig. 2 by a fixed nominal vector $\underline{p}_o$ lying within the range of possible parameter variations and to optimize the model with respect to $\underline{k}$ in the sense of the integral criterion (1). This leads to a fixed pseudo sensitivity model with fixed parameter vectors $\underline{p}_o$ and $\underline{k}_o$. The outputs of the model then supply the vector $\hat{\underline{\sigma}}$ of PSFs.

It should be noted that the pseudo sensitivity model obtained in this way is always stable which contributes to the stability of the overall adaptive system. This is expecially important if a simplified model of the plant has to be used.

The continuous pseudo gradient equation is now obtained from equ.(2) by substituting $\partial e/\partial \underline{k}$ by $\hat{\underline{\sigma}}$ and omitting the integration, i. e.

$$\frac{d\underline{k}}{dt} = - \underline{H} \frac{d}{de} f \cdot \hat{\underline{\sigma}} \tag{3}$$

3. Scaling of the Gradient

The gradients of equ.(2) and (3) are dependent on the size of the input signals. It is of great practical importance to avoid this dependence by scaling the gradient function as follows:

Let g_i be the gradient of the i-th parameter, i. e.

$$g_i = \frac{d}{de} f \cdot \hat{\sigma}_i \tag{4}$$

Consider a scaling function s_i that has the property to confine the quotient $g_{is} = g_i/s_i$ to the interval $-1 \le g_{is} \le 1$ independent of the size of the input signals r and d. This function s_i has to be always somewhat larger as or at least to be as large as the absolute value of g_i.

Good results have been obtained by using a piecewise exponentially time decaying function described by fig. 3 and equ.(5) with n = 1,2,... .

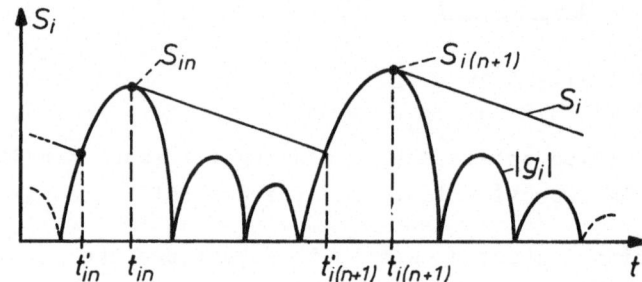

Fig. 3   Example of a Scaling Function

$$s_i = s_{in} e^{(t_{in}-t)/T} \qquad \text{for } t_{in} \leq t < t'_{i(n+1)} \tag{5a}$$

$$s_i = |g_i| \qquad \text{for } t'_{i(n+1)} \leq t < t_{i(n+1)} \tag{5b}$$

From a practical point of view the time constant T should have about the magnitude of the settling time of the system.

4. Control of Parameter Variation

The parameters of the vector $\underline{k}$ oscillate about their mean value. These variations have to be kept within certain limits for performance and stability reasons by a proper choice of the design parameters of the diagonal matrix $\underline{H}$. The variations are dependent on the frequency spectrum of the input signals and on the properties of the system so that a fixed choice of $\underline{H}$ cannot satisfy different signal and system conditions equally well.

An increase in performance can therefore be achieved if $\underline{H}$ is adjusted continuously in order to keep the relative parameter variations constant. As measure for the momentary relative variation the term

$$u_i = \frac{|h_{ii} g_{is}|}{k_i} \tag{6}$$

can be used. h_{ii} is the diagonal element of $\underline{H}$ belonging to k_i and $k_i > 0$

is assumed. A desired value u_{di} of u_i can in practice be maintained on the average by the relation

$$\frac{dh_{ii}}{dt} = q_i \{u_{di} - u_i\} \qquad , \quad h_{ii}(t_o) = h_{iio} \qquad (7)$$

where q_i is a gain factor to be properly chosen.

5. Simulation Results

Consider the adaptive system of fig. 4 which has been designed according to equ.(3). The blocks A1 and A2 perform the scaling of the gradients and the control of variations. The results of a system simulation with a symmetrical rectangular wave (period 96 min.) as disturbance signal d shows fig. 5. In 5a the results are shown without control of the parameter variations ($q_i=0$) and in 5b with this control ($q_i=1$). Fig. 5c shows the time functions of h_{11} and h_{22}. For the same speed of adaption the control of variations results in a better control performance.

6. Adaptive Control of a Cement Mill

The method was applied to the adaptive control of a cement mill /6/ using a laboratory computer. Fig. 6 shows the structure of the cascade control scheme with five adaptive parameters and two adaptation algorithms minimizing independently the performance integrals $I_1(|e_1|)$ and $I_2(|e_2|)$. The control of parameter variations has only been realized in algorithm 2.

The results of an experiment during the production of the plant are collected in fig. 7. Fig. 7a shows the measuring signals of the plant and 7b the time functions of the five adaptive parameters. The controlled variables are the power of the elevator system (EP) and the sound level of the mill. At 23.45 h. production is shifted to a knew sort of cement. At this point the parameters were adjusted close to their optimal values. The setpoint of the system was adjusted in small steps until 1.00 a.m. A few minutes later the parameters K_4 and K_5 were purposely offset to observe the adaptation process. K_5 immediately turns back towards its prior position. At about 1.20 a.m. an operator control action took the cooling water off from the main feed conveyer belt since the temperature had dropped too far. This impairs the material flow through the mill, which, however, is brought back to normal by the controllers. During this process K_4 has increased considerably. At about 2.10 a.m.

the internal control loop becomes unstable indecated by a small oscillation of the new material flow. During this time K_4 gets drastically reduced so that a normal system operation is achieved again.

It should be emphasized that the reference signal was kept constant and
that adaptation took place completely due to the normal operating signals
of the plant. A simplified model of the plant had been used according
to fig. 6b.

7. Conclusion

It has been shown by simulation and practical results that the continuous
pseudo gradient method can be used in self-optimizing systems. The actual
values of the plant parameters need not be known - and therefore on-line
identification is not required - if the parameters are restricted to a
limited range of the parameter space. Specifications for this range are
not yet available. However, it seems to be large enough (about 200 % parameter variation) to make the method interesting for applications and
further investigation. Beyond this the method is well suited for the use
of simplified plant models (see example of the cement mill) and can adapt
in the influence of unknown disturbance signals occuring during the normal operation of the plant.

Literature
/1/   Narendra, K.S., McBride, L.E.: Multiparameter Self-Optimizing
      Systems Using Correlation Techniques. IEEE Trans. AC-10, Jan.1964,
      pp. 31-38

/2/   Rake, H.: Selbsteinstellende Systeme nach dem Gradientenverfahren.
      Regelungstechnik 15 (1967), pp. 211-217

/3/   Kunze, E.G.: Entwurf und Eigenschaften direkt adaptierender und
      modelladaptiver Regelungssysteme. Dissertation TU Karlsruhe, 1976

/4/   Vuskoviĉ, M.I., Bingulac, S.P., Djoroviĉ, M.: Application of the
      pseudosensitivity functions in linear dynamic systems identifica-
      tion. 2nd IFAC Symposium on Identification and Process Parameter
      Estimation, Prague, 1970

/5/   Frank, P.M.: Introduction to System Sensitivity Theory. Academic
      Press, 1977

/6/   Kunze, E.G., Salaba, M.: Praktische Erprobung eines adaptiven Re-
      gelungsverfahrens an einer Zementmahlanlage. PDV-Bericht KfK-PDV
      158, Jan. 1979

The author owes thanks to Prof. P.M. Frank for the discussion of the
manuscript.

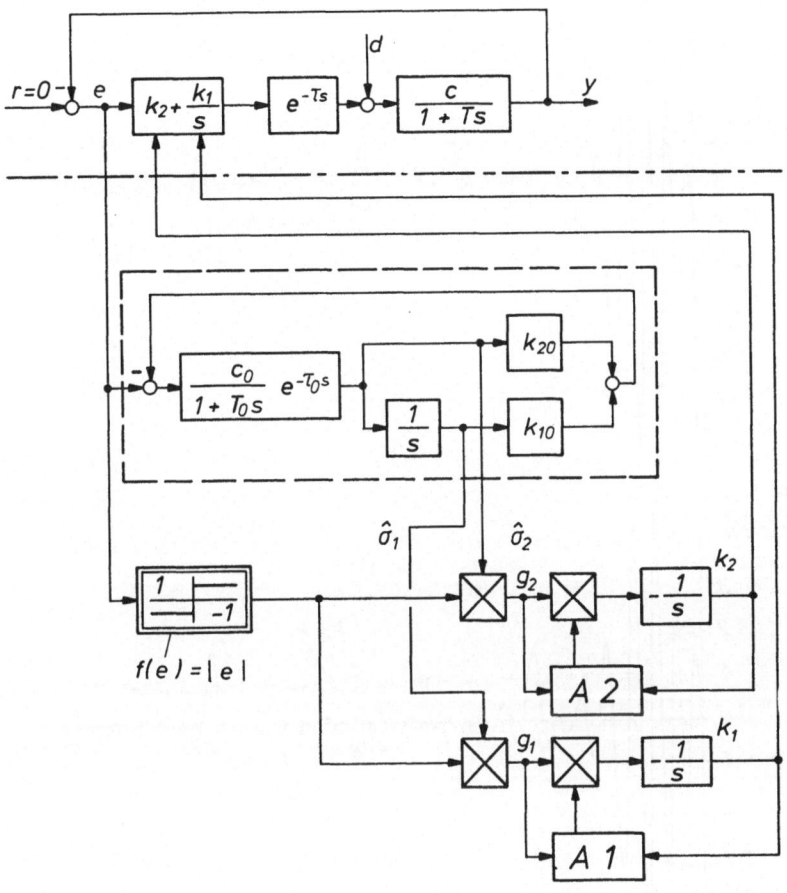

$$C = C_o = 3,35$$
$$\tau = \tau_o = 3 \text{ min.}$$
$$T = 10 \text{ min.} \qquad K_{1o} = 0,173 \text{ min}^{-1}$$
$$T_o = 14 \text{ min.} \qquad K_{2o} = 1,342$$

Fig. 4: Self-Optimizing Control System

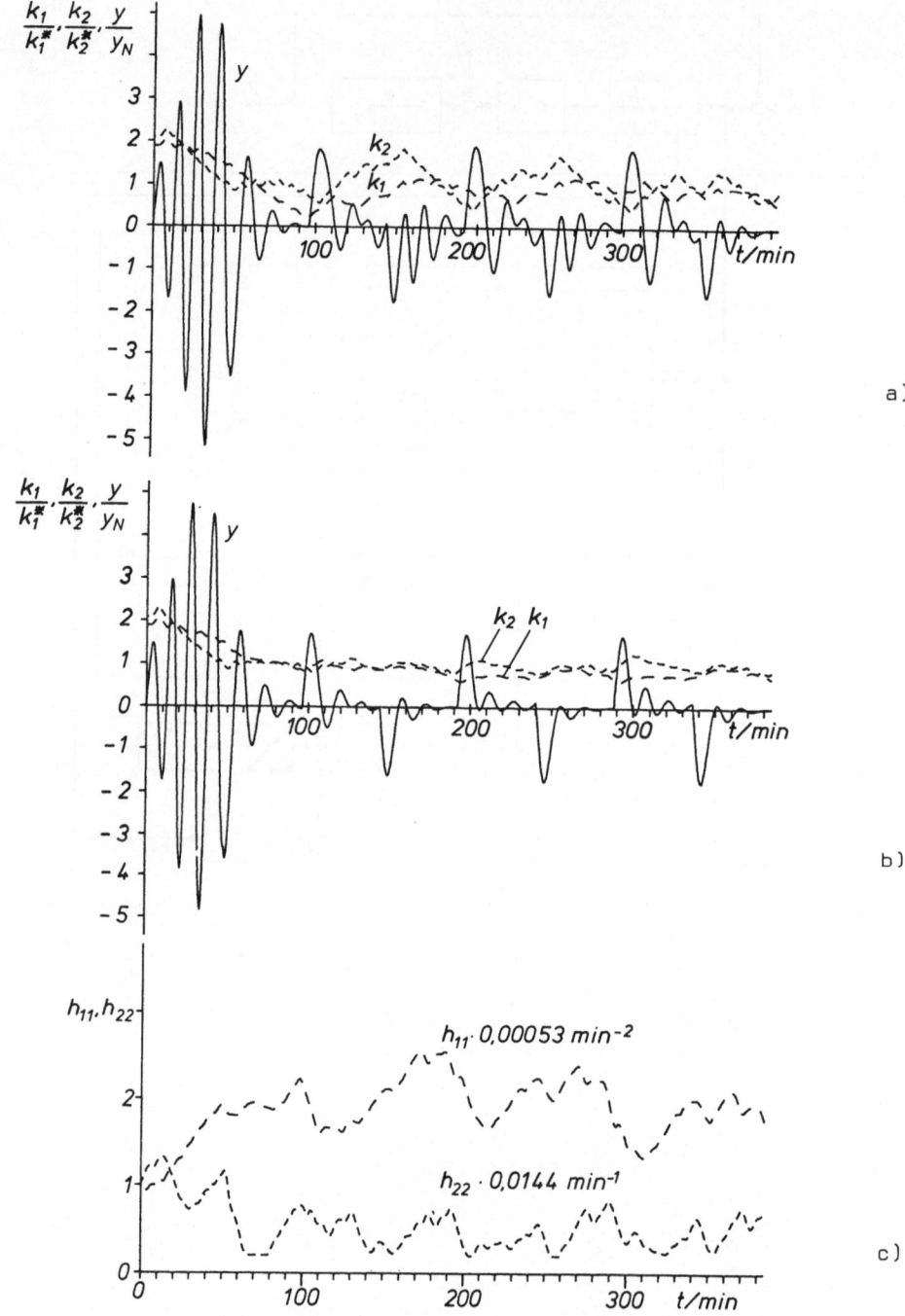

Fig. 5: Simulation Results to System of Fig. 4:
a) without, b) with control of parameter variations,
c) time functions of h_{11} and h_{22} to case b.
(* indicates optimal values)

299

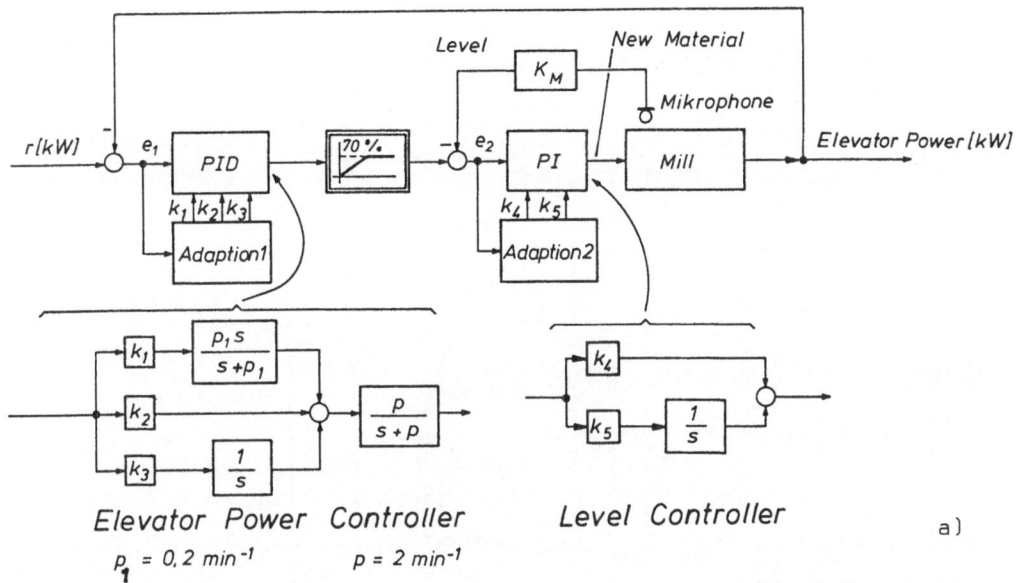

Elevator Power Controller

$p_1 = 0.2\ min^{-1}$

Level Controller

$p = 2\ min^{-1}$

a)

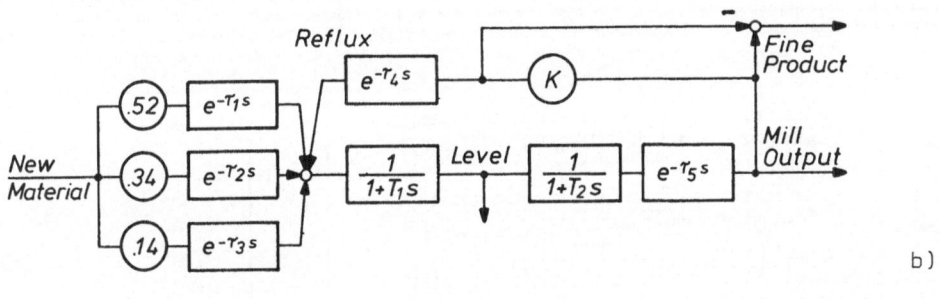

b)

$\tau_1 = 1,15\ min.$ $T_1 = 3,3\ min.$

$\tau_2 = 0,6\ min.$ $T_2 = 1\ min.$

$\tau_3 = 0,75\ min.$ $K = 0,4$

$\tau_4 = 0,6\ min.$

$\tau_5 = 0,45\ min.$

Fig. 6: a) Structure of Adaptive Ball Mill Control System
        b) Simplified Model of the Mill used in Adaptation
           Algorithms

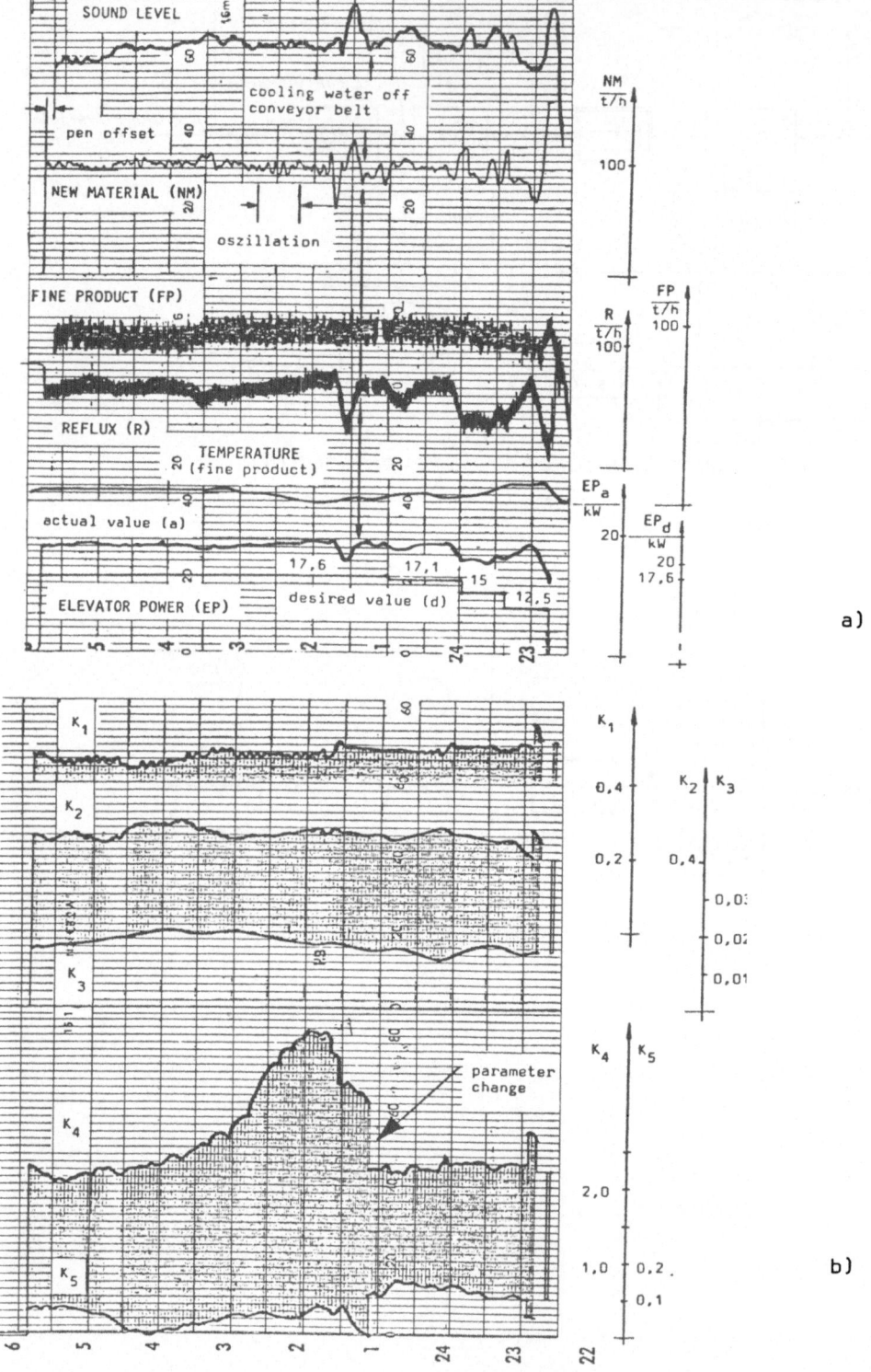

Fig. 7: Operating Signals of the Cement Mill with Adaptive Control
a) process signals   b) adaptive parameters

EXPERIMENTAL EVALUATION OF SELF-TUNING CONTROLLERS
APPLIED TO PILOT PLANT UNITS

H. Lieuson, A.J. Morris*, Y. Nazer* and R.K. Wood

Department of Chemical Engineering
University of Alberta
Edmonton, Alberta
Canada T6G 2E1

ABSTRACT

This paper describes the results of experimental tests performed to evaluate the performance of self-tuning controllers for temperature control of a continuous stirred tank heater and composition control of a pilot plant distillation column. Both units have been controlled using single variant self-tuning control and multivariant self-tuning control used for control of the terminal compositions of the distillation column. The control performance of each of the units operating under self-tuning control is compared to that obtained using very well tuned proportional plus integral or proportional plus integral plus derivative conventional controllers.

INTRODUCTION

The satisfactory control of process systems that exhibit nonlinear dynamic behavior and also frequently time delays continues to present a challenge to the control engineer. One approach to the control of such systems that shows promise is the adaptive control scheme known as the self-tuning controller developed by Clarke and Gawthrop[1] as a generalization of the self-tuning regulator proposed by Astrom and Wittenmark[2]. Since Gawthrop[3] and more recently Clarke and Gawthrop[4] have considered the relationship between these and other adaptive control strategies, only a brief outline of the fundamental theory of the self-tuning controller will be presented herein.

THEORY
Single Variant Self-Tuning Control

The application of the self-tuning controller is based upon representing the system by means of a general discrete time model of the form

$$Ay_t = z^{-k}Bu_t + C\xi_t + z^{-\ell}Lv_t \tag{1}$$

where A, B, C and L are polynomials in z^{-1}, with $a_o = 1$, $b_o \neq 0$ and $c_o = 1$. C^{-1} is assumed to be a stable transfer function with a zero mean uncorrelated random input. The system time delay is denoted by k sample intervals (k>1). The term v_t allows the inclusion of measurable loop disturbances in a feedforward manner, the associated feedforward time delay being denoted by ℓ sample intervals. Controller design is

---

*Permanent address: Department of Chemical Engineering
University of Newcastle Upon Tyne
NEWCASTLE UPON TYNE, England
NE1 7RU

based upon the minimization of the general performance function

$$J = E \{ [Py_{t+k} - Rw_t]^2 + [Q'u_t]^2 \}$$ (2)

where P, R and Q' are weighting functions on system output (y), set point (w) and control effort (u) respectively. It can be shown[1,3,7,8] using the definition

$$P \triangleq \frac{P_n}{P_d}$$ (3)

and

$$P \frac{C}{A} \triangleq E + z^{-k} \frac{F}{AP_d}$$ (4)

that the best k-step ahead prediction of the weighted system output is given by

$$(Py_{t+k/t})^* = \frac{F}{CP_d} y_t + \frac{EB}{C} u_t + \frac{EL}{C} z^{k-\ell} v_t$$ (5)

The control law is derived[8] by minimizing J with respect to u_t to give

$$u_t = -\frac{1}{G} [\frac{F}{P_d} y_t - (Qu_t - Rw_t) H + Dz^{k-\ell} v_t$$ (6)

where the controller polynomials, in z^{-1}, G H and D are defined as

$$G = EB, \quad H = -C \text{ and } D = EL$$

This control law can be written, in terms of the feedback of weighted predicted output, as

$$u_t = \frac{1}{2} [Rw_t - (Py_{t + k/t})^*]$$ (7)

It has been found useful at this point to define the inverse of the control effort weighting, Q, as being a dynamic compensating function G_c. The substitution $G_c = \frac{1}{Q}$ is advantageous in that it allows the system closed loop response to be modified by suitable design of G_c using conventional compensator design procedures.

The system closed loop characteristic equation is then given by $G_c P B + A = 0$ (8)

with the closed loop response as

$$y_t = \frac{(G_c EB + C)\xi_t + z^{-k} G_c BRw_t + z^{-\ell} Lv_t}{(G_c P B + A)}$$ (9)

The control law given by Equations (6) and (7) for a known fixed parameter plant can be made to self-tune or adapt to slowly changing plant dynamics by allowing the controller polynomials F, G, H and D to be estimated by a suitable on-line parameter estimator.

The control policy is then calculated assuming that the estimated parameters F, G, H and D are correct. Convergence of the estimated parameters has been demonstrated and proved elsewhere[1,8]. In this work a recursive least squares based estimator is used with covariance matrix factorization to ensure satisfactory numerical behavior of the algorithm[8,9].

Figure 1 shows a block diagram repre-

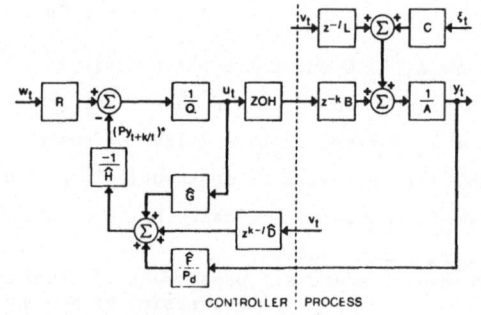

Fig.1 - Block Diagram of the Self-Tuning Controller

sentation of the controller given by Equation (6) while Figure 2 shows the controller configured so as to include the k-step ahead output predictor in the feedback loop. Such a structure eliminates the time delay from the system characteristic equation and in so doing the self-tuning controller provides better time delay compensation for systems with time varying dynamics than say the Smith predictor since the controller parameters F, G, H and D are continuously being adapted to chan-ging process conditions.

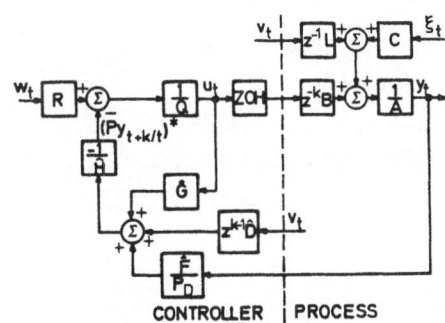

Fig. 2 - Block Diagram of the Self-Tuning Controller with Output Predictive Feedback

Multivariant Self-Tuning Control

The single variant control strategy can be extended to include the control of a class of interactive multivariable systems described by the matrix equation

$$[A][y_t] = [B][u_{t-k}] + [C][\xi_t] + [L][v_{t-\ell}] \qquad (10)$$

where [A], [B], [C] and [L] are polynomial matrices with $[A_o] = [I]$, $[C_o] = [I]$ and $[B_o]$ nonsingular. Vectors $[y_t]$, $[u_{t-k}]$, $[\xi_t]$ and $[v_{t-\ell}]$ represent the system outputs, control inputs, noise sequences and measurable disturbances respectively. The objective function to be minimized is given by

$$J = E\{[P_{y_{t+k}} - Rw_t]^T[P_{y_{t+k}} - Rw_t] + [Q'u_t]^T[Q'u_t]\} \qquad (11)$$

where P, R and Q' are weighting functions on the individual loop outputs, set points and control inputs respectively. The control vector $[u_t]$ chosen to minimize Equation (11) becomes (7,8)

$$[G-HQ][u_t] =- [F]\begin{bmatrix} y_t \\ P_d \end{bmatrix} + [H][Rw_t] + [D][v_{t+k-1}] \qquad (12)$$

or

$$[Q][u_t] = [R][w_t] - [(Py_{t+k/t})^*] \qquad (13)$$

The multivariant control policy outlined above is currently under intensive evaluation for distillation column terminal composition control. Some of the preliminary results obtained by simulation and experimental testing are presented in this work.

EQUIPMENT

The continuous stirred tank heater used in this study, which was interfaced with a Z-80 microcomputer, is shown schematically in Figure 3. Constant holdup of 6.3 kg of water was maintained in the 15 cm diameter by 55 cm height tank by averaging liquid level control. The temperature of the outlet water stream could be measured by either thermocouple A, at the discharge of the tank, or at point B, thus introducing a deliberate time delay to the system. Water flow rate, adjusted by a hand valve and measured by

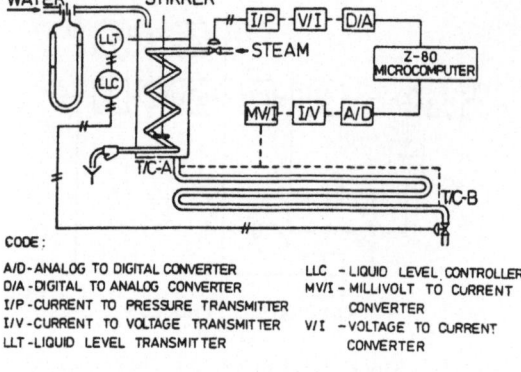

CODE:

A/D-ANALOG TO DIGITAL CONVERTER   LLC - LIQUID LEVEL CONTROLLER
D/A -DIGITAL TO ANALOG CONVERTER   MV/I - MILLIVOLT TO CURRENT
I/P -CURRENT TO PRESSURE TRANSMITTER   CONVERTER
I/V -CURRENT TO VOLTAGE TRANSMITTER   V/I -VOLTAGE TO CURRENT
LLT -LIQUID LEVEL TRANSMITTER   CONVERTER

Fig. 3 - Schematic Diagram of the Continuous
Stirred Tank Heater

an orifice/manometer, was maintained at 125 g/s giving a dominant system time constant of about 50 s at steady state conditions. The control valve on the steam line was of the equal percentage type giving rise to the inherent nonlinear input (pressure) to steam flow rate relationship.

A simplified schematic diagram of the pilot scale distillation column used in this study is shown in Figure 4. The column, which is completely instrumented and interfaced with the Department's HP-1000 distributed computer network, was operated with a 50 weight percent methanol-water feed at a rate of 18.22 g/s. The 22.5 cm diameter bubble cap column which has been used in previous control investigations[5,6] contains eight trays at a tray spacing of 30.5 cm. Each tray contains four bubble caps. Top composition is measured continuously by means of an in-line capacitance probe. An HP-5702A gas chromatograph provides bottom composition analysis on a 180 s cycle time.

RESULTS

Evaluation of the single variant self-tuning controller for temperature control of the stirred tank heater was carried out with the algorithm programmed into the Z-80 microprocessor. The control performance was studied for 20% step changes in water flow rate and for 25% step changes in set point. The mathematical model required for the simulation studies was obtained by experimental dynamic testing of the system. At the reference operating point corresponding to a water flow rate of 125 g/s, and a steam rate resulting in an outlet water temperature of 42°C (chart reading of 60%) the transfer function model was found to be

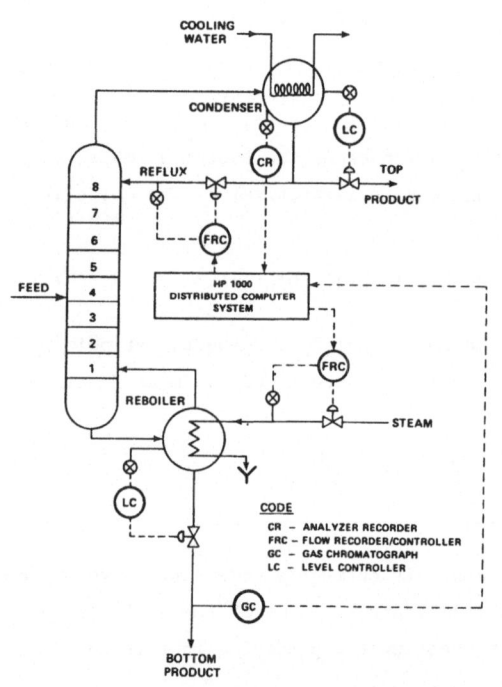

CODE
CR - ANALYZER RECORDER
FRC - FLOW RECORDER/CONTROLLER
GC - GAS CHROMATOGRAPH
LC - LEVEL CONTROLLER

Fig. 4 - Schematic Diagram of the Pilot
Scale Distillation Column

$$Y(s) = \frac{1.75e^{-5.7s}}{45s+1} U(s) - \frac{2.29e^{-6.1s}}{50s+1} V(s) \qquad (14)$$

where Y(s), U(s) and V(s) correspond to outlet water temperature, steam flow rate and water flow rate respectively. Sample time for temperature measurement and applied control action was 5 s for all tests. Figure 5 shows the simulated and experimental responses of tank outlet temperature (1°C per 4 chart %) for a 25 chart percent step increase in set point under proportional plus integral (PI) and self-tuning (ST) control. In this figure and all subsequent figures, the time at which a change in set point or disturbance was introduced is denoted by the vertical arrow on the abscissa of the plot, in this case at 5 s. Also shown in this figure is the response of the predicted k-step ahead output, $(Py_{t+k/t})^*$, of the ST control, which shows very close agreement to the actual system output while being in advance of it by the system time delay.

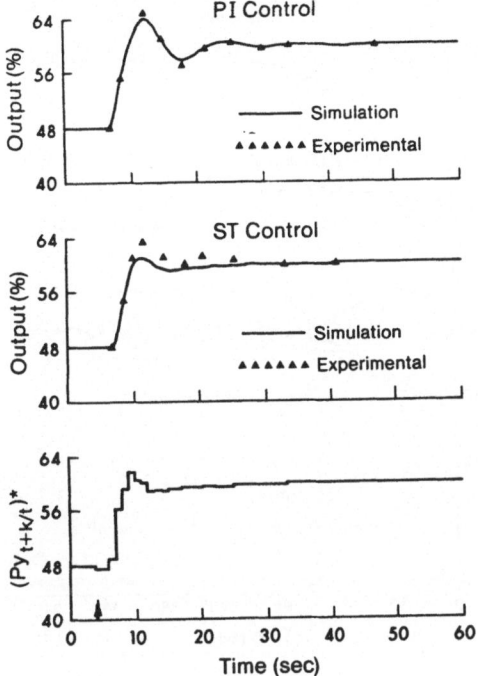

Fig. 5 - Comparison of PI and ST Temperature Control for a 25 Chart % Increase in Set Point

The responses of outlet temperature and predicted output for a 20% increase in water flow rate are shown in Figure 6. It is interesting to note in this figure and Figure 5 that ST control shows little improvement over that of conventional well tuned PI control. This is to be expected since in any practical situation where plant dynamics are relatively well known, it should always be possible to design a perfectly acceptable PI regulator providing the system time delay is small compared to the dominant system time constant.

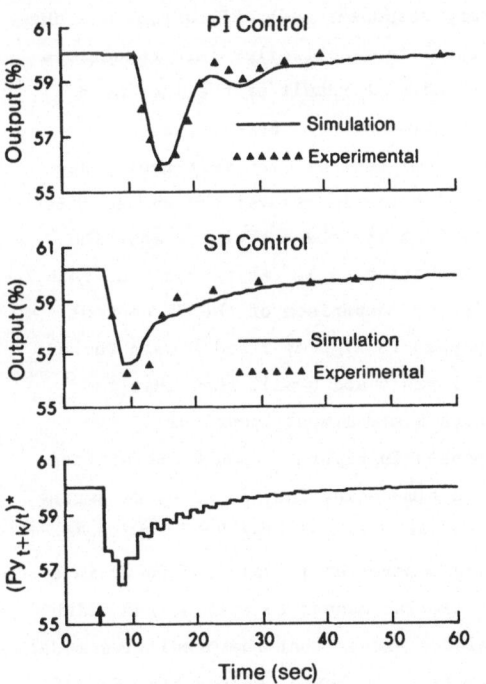

Fig. 6 - Comparison of PI and ST Temperature Control for a 20% Increase in Water Flow Rate

The difficulty in controlling processes with long time delays can readily be demonstrated with the heater system by measuring the water outlet temperature by thermocouple

"B" which is located 3.1 m downstream from the actual tank outlet resulting in a time delay of 13 s at steady state. This time delay changes with water flow rate disturbances, as do the dynamics of the process. Figure 7 shows the experimental responses for PI and ST control for 20 chart percent step changes in set point. It can be seen that a more rapid and damped response has been obtained with ST control than with PI control due to the anticipative nature of the self-tuning control action despite the effect of changing system dynamics evident from the different responses. The experimental response of the outlet water temperature to 20% step changes in water flow rate are shown in Figure 8. The slow oscillatory responses under PI control are due to the larger controller gain reductions required as a result of the increased system time delay. However, in the case of ST control, the self-tuning action effectively removes the system time delay from the characteristic equation (cf. Equation (8)). As a result of this action, a comparison of the ST control responses in Figures 7 and 8 with those in Figures 5 and 6 will show that the responses are almost identical if the responses in Figures 7 and 8 are shifted by the time delay introduced by measuring the temperature at point "B". The self-tuning controller is seen to behave in a very similar manner to that of the Smith predictor except that the predictive model is continually being updated and adapted to changing plant conditions by an on-line least squares estimator.

The single variant self-tuning controller was further evaluated by studying the control of a single product compo-

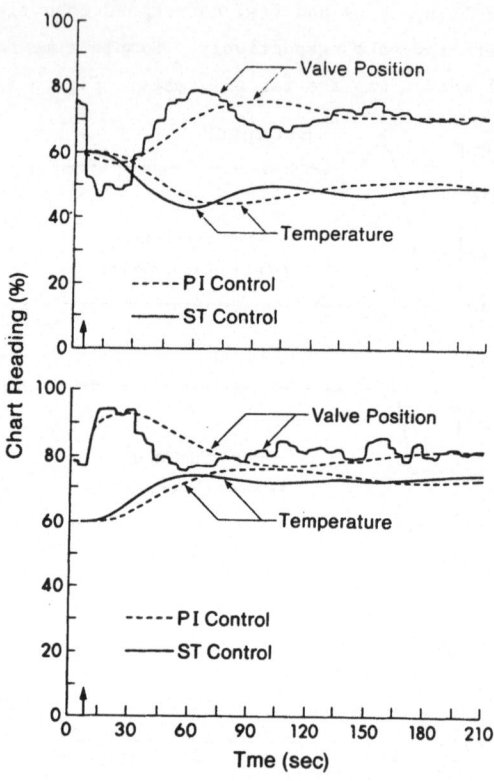

Fig. 7 – Effect of Time Delay on the PI and ST Temperature Control for a 20 Chart % Decrease and Increase in Set Point

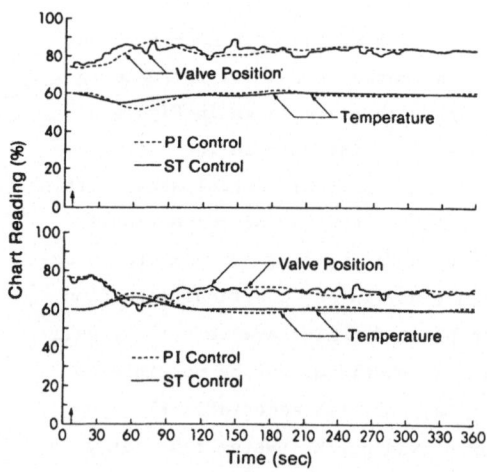

Fig. 8 – Effect of Time Delay on the PI and ST Temperature Control for a 20% Increase and Decrease in Water Flow Rate

sition of the binary distillation column.
Figure 9 shows simulated and experimental
top composition responses with the column
operated under PI and ST control for a 25%
step decrease in feed rate. In comparing
the experimental responses it can be seen
(note different ordinate scales) that the
maximum deviation of the composition from
its set point is smaller and the composition returns to its set point more rapidly under ST control than PI control.
The same conclusion applies to the simulated responses, which show poor agreement
with those obtained experimentally. Further simulations are presently underway
to investigate this discrepancy, using
a variable heat loss model representation of the column. These results do
however substantiate the general conclusion that can be drawn from the previous

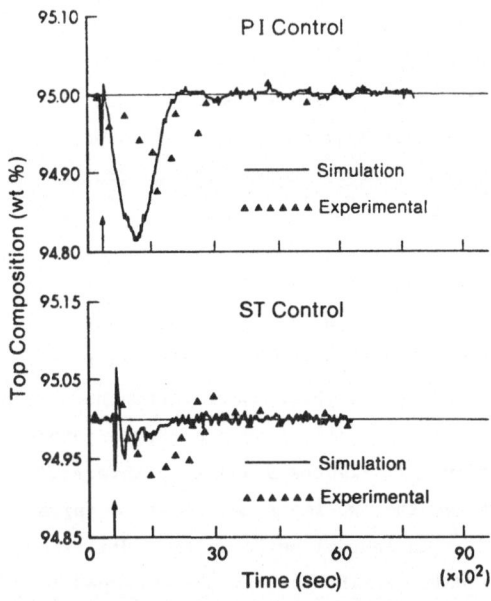

Fig. 9 – Comparison of PI and ST Control
of Top Composition for a 25% Decrease in Feed Flow Rate

stirred tank heater results. Unless the ratio of dominant system time constant to time
delay is small (lower than 10:1) there will be little difference in control performance
using ST control compared to well tuned PI control. Experimental dynamic testing of

the column for disturbances in feed flow
rate show the dominant time constant to be
of the order of 250 s with a time delay of
29 s for top composition while the corresponding values for bottom composition are
900 s and 400 s respectively. As would be
expected ST control of bottom composition
is clearly superior to PI control as can be
seen from the responses given in Figure 10
for a 25% step decrease in feed flow rate.
These responses are typical of numerous
bottom composition control experiments.
The self-tuning control action is able to
follow the severe nonlinearities of column
dynamic behavior and can handle loops
that contain large time delays so better
control performance can be achieved than
is possible using conventional PI control.

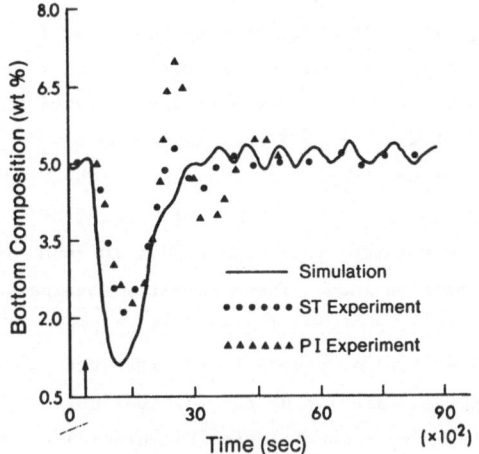

Fig. 10 – Comparison of PI and ST Control of
Bottom Composition for a 25% Decrease in Feed Flow Rate

    Evaluation of the extension of the single variant self-tuning control to the class of interactive multivariant systems has been performed by studying the simultaneous composition control of both product streams of the distillation column. Incorporating a single variant self-tuning controller into each of the composition loops (multi-loop ST control) results in the control performance shown in Figure 11 for a 25% step decrease in feed flow rate. As can be seen by comparing (note different ordinate scales) these control responses with those for control of only a single composition (cf. Figures 9 and 10), interaction between the two loops has led to a degradation in control performance. Utilizing the multivariable algorithm developed to handle such a multivariant interactive system, when the feed flow rate was abruptly decreased by 25%, resulted in the composition responses shown in Figure 12. A comparison of this control behavior (note different ordinate scales) with that in Figure 11 shows that much better regulation has been achieved. Furthermore, improved control performance would likely result if the limit on the maximum allowable rate of valve movement were relaxed. Also shown in this figure is the best control performance that could be achieved using a proportional plus integral plus derivative (PID) controller in each composition loop. These results, obtained only after an extensive program of controller tuning, reveal (note different ordinate scales) that the multivariable self-tuning control action has provided the best control of top composition but the control behavior for bottom composition shows only a marginal improvement.

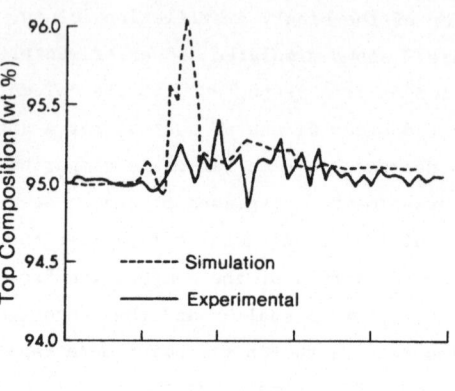

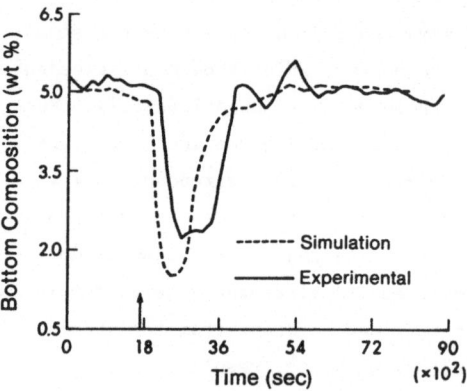

Fig. 11 – Multiloop Self-Tuning Control of Top and Bottom Composition for a 25% Decrease in Feed Flow Rate

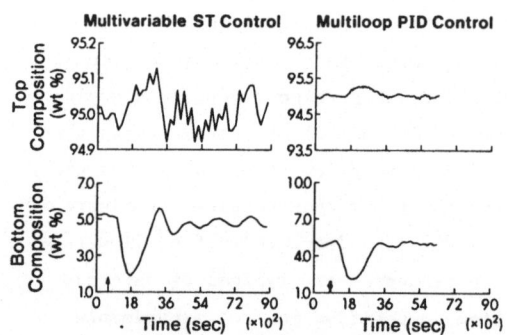

Fig. 12 – Comparison of Multivariable ST Control with Multiloop PID Control of Product Compositions for a 25% Decrease in Feed Flow Rate

CONCLUSIONS

Experimental and simulation results have shown that for temperature control of the continuous stirred tank heater, without the additional measurement time delay, or top composition control of the distillation column, there is little difference in the control performance achieved using ST or PI control. However in the case of temperature control of the heater with the additional time delay and bottom composition control of the distillation column, that is the magnitude of the time delay is significant compared to the dominant system time constant, ST control clearly provides superior control behavior. The results from testing of the multivariable extension of the self-tuning controller for composition control of both products of the distillation column are promising. It is anticipated that experimental and simulation studies presently in progress will lead to refinements to the algorithm that will provide even better terminal composition control performance.

ACKNOWLEDGEMENT

The authors gratefully acknowledge the financial support of the Natural Sciences and Engineering Research Council under grant A-1944 and use of the facilities provided by the Departments of Chemical Engineering at the Universities of Alberta and Newcastle Upon Tyne.

REFERENCES

(1) Clarke, D.W. and Gawthrop, P.J. "Self-tuning Controller", Proc. IEE, 122, p. 929-934, 1975.

(2) Åström K.J. and Wittenmark, B. "On Self-tuning Regulators", Automatica, 9, pg. 185-199, 1973.

(3) Gawthrop, P.J. "Some Interpretations of the Self-tuning Controller", Proc. IEE, 124, p. 889-894, Oct. 1977.

(4) Clarke, D.W. and Gawthrop, B.A. "Self-tuning Control", Proc. IEE, 126, p. 633-640, 1979.

(5) Meyer, C., Seborg, D.E. and Wood, R.K. "An Experimental Application of Time Delay Compensation Techniques to Distillation Column Control", Digital Computer Applications to Process Control, p. 439-446, 1977.

(6) Sastry, V.A., Seborg, D.E. and Wood, R.K. "Self-tuning Regulator Applied to a Binary Distillation Column", Automatica, 13, p. 417-424, 1977.

(7) Morris, A.J. and Nazer, Y. "Self-tuning Control of a Binary Distillation Column", Int. Report 1977, Dept. Chem. Eng., Univ. of Newcastle Upon Tyne, England.

(8) Morris, A.J. and Nazer, Y. "Self-tuning Process Controllers for Single Variable and Multivariable Systems", Submitted to Automatica 1979.

(9) Morris, A.J., Nazer, Y., Chisholm, K. "A comparison of identification techniques for robust self-tuning control", 5th IFAC Symp. on Identification and System Parameter Estimation, Darmstadt, Germany, Sept. 1979.

Lecture Notes in Control and Information Sciences

Edited by A. V. Balakrishnan and M. Thoma